Innovations in Plant Science for Better Health: From Soil to Fork

HEALTH BENEFITS OF SECONDARY PHYTOCOMPOUNDS FROM PLANT AND MARINE SOURCES

Edited by

Hafiz Ansar Rasul Suleria, PhD

Megh R. Goyal, PhD, PE

APPLE
ACADEMIC
PRESS

First edition published 2021

Apple Academic Press Inc.
1265 Goldenrod Circle, NE,
Palm Bay, FL 32905 USA
4164 Lakeshore Road, Burlington,
ON, L7L 1A4 Canada

CRC Press
6000 Broken Sound Parkway NW,
Suite 300, Boca Raton, FL 33487-2742 USA
4 Park Square, Milton Park,
Abingdon, Oxon OX14 4RN

First issued in paperback 2023

© 2021 Apple Academic Press, Inc.

Apple Academic Press exclusively co-publishes with CRC Press, an imprint of Taylor & Francis Group, LLC

Publisher's Note
The publisher has gone to great lengths to ensure the quality of this reprint but points out that some imperfections in the original copies may be apparent.

Library and Archives Canada Cataloguing in Publication

Title: Health benefits of secondary phytocompounds from plant and marine sources / edited by Hafiz Ansar Rasul Suleria, PhD, Megh R. Goyal, PhD, PE.

Names: Suleria, Hafiz, editor. | Goyal, Megh R., editor.

Series: Innovations in plant science for better health.

Description: First edition. | Series statement: Innovations in plant science for better health : from soil to fork | Includes bibliographical references and index.

Identifiers: Canadiana (print) 20200365479 | Canadiana (ebook) 20200365517 | ISBN 9781771888981 (hardcover) | ISBN 9781003019602 (ebook)

Subjects: LCSH: Phytochemicals. | LCSH: Bioactive compounds. | LCSH: Medicinal plants.

Classification: LCC RS160 .H43 2021 | DDC 615.3/21—dc23

Library of Congress Cataloging-in-Publication Data

Names: Suleria, Hafiz, editor. | Goyal, Megh Raj, editor.

Title: Health benefits of secondary phytocompounds from plant and marine sources / edited by Hafiz Ansar Rasul Suleria, Megh R. Goyal.

Other titles: Innovations in plant science for better health.

Description: First edition. | Palm Bay : Apple Academic Press, 2021. | Series: Innovations in plant science for better health: from soil to fork | Includes bibliographical references and index. | Summary: "This new volume, Health Benefits of Secondary Phytocompounds from Plant and Marine Sources, looks at a selection of important issues and research topics on phytochemicals in plant-based therapeutics, covering bioactive compounds from both plant and marine sources. Natural products and their bioactive compounds are increasingly utilized in preventive and therapeutic medication, as pharmaceutical supplements, as well as in functional foods and nutraceuticals, all of which have potentially positive effects on health and have preventive and curative properties for various diseases and health conditions. The first section of the book, on Bioactive Compounds from Plant Sources, describes the concept of extraction of bioactive molecules from plant sources, both conventional and modern extraction techniques, available sources, biochemistry, structural composition, and potential biological activities. Advanced extraction techniques, such as enzyme-assisted, microwave-assisted, ultrasound-assisted, pressurized liquid extraction, and super critical extraction techniques, are described in detail. Part 2, on Bioactive Compounds from Marine Sources, discusses the isolation of potential bioactive molecules from marine sources, their importance, and health perspectives. This section explains the marine bioactivity, physical characteristics, uniqueness, uses, distribution, importance, traditional importance, nutritional importance, bioactivities, and future trends of different functional foods. This book volume sheds light on the potential of secondary phytocompounds from plant and marine sources for human health and will be a valuable reference for faculty and students, researchers and scientists, industry professionals, and others in this important and growing field"-- Provided by publisher.

Identifiers: LCCN 2020046629 (print) | LCCN 2020046630 (ebook) | ISBN 9781771888981 (hbk) | ISBN 9781003019602 (ebk)

Subjects: MESH: Phytochemicals--pharmacology | Plants, Medicinal--chemistry

Classification: LCC RS160 (print) | LCC RS160 (ebook) | NLM QV 766 | DDC 615.3/21--dc23

LC record available at https://lccn.loc.gov/2020046629
LC ebook record available at https://lccn.loc.gov/2020046630

ISBN: 978-1-77188-898-1 (hbk)
ISBN: 978-1-77463-766-1 (pbk)
ISBN: 978-1-00301-960-2 (ebk)

DOI: 10.1201/9781003019602

OTHER BOOKS ON PLANT SCIENCE FOR BETTER HEALTH BY APPLE ACADEMIC PRESS, INC.

Book Series: *Innovations in Plant Science for Better Health: From Soil to Fork*
Editor-in-Chief: Hafiz Ansar Rasul Suleria, PhD

Assessment of Medicinal Plants for Human Health: Phytochemistry, Disease Management, and Novel Applications
Editors: Megh R. Goyal, PhD, and Durgesh Nandini Chauhan, MPharm

Bioactive Compounds of Medicinal Plants: Properties and Potential for Human Health
Editors: Megh R. Goyal, PhD, and Ademola O. Ayeleso

Bioactive Compounds from Plant Origin: Extraction, Applications, and Potential Health Claims
Editors: Hafiz Ansar Rasul Suleria, PhD, and Colin Barrow, PhD

Cereals and Cereal-Based Foods: Functional Benefits and Technological Advances for Nutrition and Healthcare
Editors: Megh Goyal, PhD, Kamaljit Kaur, PhD, and Jaspreet Kaur, PhD

Health Benefits of Secondary Phytocompounds from Plant and Marine Sources
Editors: Hafiz Ansar Rasul Suleria, PhD, and Megh Goyal, PhD

Human Health Benefits of Plant Bioactive Compounds: Potentials and Prospects
Editors: Megh R. Goyal, PhD, and Hafiz Ansar Rasul Suleria, PhD

Plant- and Marine-Based Phytochemicals for Human Health: Attributes, Potential, and Use
Editors: Megh R. Goyal, PhD, and Durgesh Nandini Chauhan, MPharm

Plant-Based Functional Foods and Phytochemicals: From Traditional Knowledge to Present Innovation
Editors: Megh R. Goyal, PhD, Arijit Nath, PhD, and Hafiz Ansar Rasul Suleria, PhD

Plant Secondary Metabolites for Human Health: Extraction of Bioactive Compounds
Editors: Megh R. Goyal, PhD, P. P. Joy, PhD, and Hafiz Ansar Rasul Suleria, PhD

Phytochemicals from Medicinal Plants: Scope, Applications, and Potential Health Claims
Editors: Hafiz Ansar Rasul Suleria, PhD, Megh R. Goyal, PhD, and
Masood Sadiq Butt, PhD

The Role of Phytoconstitutents in Health Care: Biocompounds in Medicinal Plants
Editors: Megh R. Goyal, PhD, Hafiz Ansar Rasul Suleria, PhD,
and Ramasamy Harikrishnan, PhD

The Therapeutic Properties of Medicinal Plants: Health-Rejuvenating Bioactive Compounds of Native Flora
Editors: Megh R. Goyal, PhD, PE, Hafiz Ansar Rasul Suleria, PhD, Ademola Olabode Ayeleso, PhD, T. Jesse Joel, and Sujogya Kumar Panda

ABOUT THE LEAD EDITOR

Hafiz Ansar Rasul Suleria, PhD
McKenzie Fellow, School of Agriculture and Food,
Faculty of Veterinary and Agricultural Science,
University of Melbourne, Australia

Hafiz Suleria, PhD, is working in the School of Agriculture and Food as a McKenzie Fellow. He has completed his postdoctoral research at the Department of Food, Nutrition, Dietetics and Health, Kansas State University, USA. Previously, he was awarded the Alfred Deakin Postdoctoral Research Fellowship at the Deakin University, Australia. He earned his PhD from the University of Queensland in collaboration with Translational Research Institute, UQ School of Medicine and Commonwealth and Scientific and Industrial Research Organization, Australia. Before joining the UQ, he worked as a Lecturer in the Department of Food Sciences, Government College University Faisalabad, Pakistan. He also worked as a Research Associate in "PAK-US Joint Project" at the National Institute of Food Science and Technology, University of Agriculture Faisalabad, Pakistan, and the Department of Food Science, University of Massachusetts, USA, funded by the Higher Education Commission, Pakistan, and the Department of State, USA.

He has a major research focus on food science and nutrition, particularly in the screening of phytochemicals/bioactive molecules from different plant, marine, and animal sources. His research interest includes isolation, purification, and characterization of bioactive compounds using various cutting-edge techniques followed by their in vitro bioactivity, in vivo, cell culture, and animal studies.

Dr. Suleria has published more than 100 peer-reviewed scientific papers in various high reputed journals. He is also in collaboration with more than five universities where he is working as a co-supervisor/special member for PhD and postgraduate students. He is also involved in joint publications, projects, and grants. Readers may contact him at: hafiz.suleria@uqconnect.edu.au.

ABOUT THE SENIOR EDITOR-IN-CHIEF

Megh R. Goyal, PhD

Retired Professor in Agricultural and Biomedical Engineering, University of Puerto Rico, Mayaguez Campus; Senior Acquisitions Editor, Biomedical Engineering and Agricultural Science, Apple Academic Press, Inc.

Megh R. Goyal, PhD, PE, is a Retired Professor in Agricultural and Biomedical Engineering from the General Engineering Department in the College of Engineering at the University of Puerto Rico–Mayaguez Campus; and Senior Acquisitions Editor and Senior Technical Editor-in-Chief in Agriculture and Biomedical Engineering for Apple Academic Press, Inc. He has worked as a Soil Conservation Inspector and as a Research Assistant at Haryana Agricultural University and Ohio State University.

During his professional career of 49 years, Dr. Goyal has received many prestigious awards and honors. He was the first agricultural engineer to receive the professional license in Agricultural Engineering in 1986 from the College of Engineers and Surveyors of Puerto Rico. In 2005, he was proclaimed as "Father of Irrigation Engineering in Puerto Rico for the Twentieth Century" by the American Society of Agricultural and Biological Engineers (ASABE), Puerto Rico Section, for his pioneering work on micro irrigation, evapotranspiration, agroclimatology, and soil and water engineering. The Water Technology Centre of Tamil Nadu Agricultural University in Coimbatore, India, recognized Dr. Goyal as one of the experts "who rendered meritorious service for the development of micro irrigation sector in India" by bestowing the Award of Outstanding Contribution in Micro Irrigation. This award was presented to Dr. Goyal during the inaugural session of the National Congress on "New Challenges and Advances in Sustainable Micro Irrigation" held at Tamil Nadu Agricultural University. Dr. Goyal received the Netafim Award for Advancements in Microirrigation: 2018 from the American Society of Agricultural Engineers at the ASABE International Meeting in August 2018.

A prolific author and editor, he has written more than 200 journal articles and textbooks and has edited over 80 books. He is the editor of three book series published by Apple Academic Press: Innovations in Agricultural & Biological Engineering, Innovations and Challenges in Micro Irrigation, and Research Advances in Sustainable Micro Irrigation. He is also instrumental in the development of the new book series Innovations in Plant Science for Better Health: From Soil to Fork.

Dr. Goyal received his BSc degree in engineering from Punjab Agricultural University, Ludhiana, India; his MSc and PhD degrees from Ohio State University, Columbus; and his Master of Divinity degree from Puerto Rico Evangelical Seminary, Hato Rey, Puerto Rico, USA.

CONTENTS

CONTRIBUTORS

Huma Bader Ul Ain
Research Associate, Institute of Home and Food Sciences, Government College University, Faisalabad, Pakistan, E-mail: humahums@yahoo.com

Tawheed Amin
Division of Food Science and Technology, Sher-e-Kashmir University of Agricultural Sciences and Technology-Kashmir, Shalimar Campus, Srinagar – 190 025, Jammu & Kashmir, India, Tel.: +91-7006448091; +91-9419043640, E-mail: tawheed.amin@gmail.com

Saud Bawazeer
Assistant Professor, Department of Pharmaceutical Chemistry, Faculty of Pharmacy, Umm Al-Qura University, Makkah, P.O. Box 42, Saudi Arabia, Tel.: +966500052002, E-mail: saud.bawazeer@hotmail.co.uk

Megh R. Goyal
PE, Retired Professor in Agricultural and Biomedical Engineering from College of Engineering at University of Puerto Rico-Mayaguez Campus; and Senior Technical Editor-in-Chief in Agricultural and Biomedical Engineering for Apple Academic Press Inc.; PO Box 86, Rincon-PR – 006770086, USA, E-mail: goyalmegh@gmail.com

Sadia Hassan
Institute of Home and Food Sciences, Government College University, Faisalabad, Pakistan, E-mail: sadiahassan88@gmail.com

Syed Shams Ul Hassan
PhD student (Medicinal Chemistry), School of Pharmacy, Shanghai Jiao Tong University, Shanghai, China, Tel.:+ 0086-15695712713, E-mail: shams1327@yahoo.com

Zohaib Hassan
Research Associate, Institute of Home and Food Sciences, Government College University, Faisalabad, Pakistan, E-mail: zohaib.hassan@gcuf.edu.pk

R. L. Helen
PhD Research Scholar, School of Biosciences, Mahatma Gandhi University, Kottayam – 686560, Kerala, India, Tel.: +91-9497377149, E-mail: lalhelenraisa@gmail.com

Syed Zameer Hussain
Division of Food Science and Technology, Sher-e-Kashmir University of Agricultural Sciences and Technology-Kashmir, Shalimar Campus, Srinagar – 190 025, Jammu & Kashmir, India, Tel.: +91-7780827474, E-mail: zameerskuastj@rediffmail.com

Muhammad Imran
Associate Professor, University Institute of Diet and Nutritional Sciences, Faculty of Allied Health Sciences, The University of Lahore, Pakistan, Tel.: +92-332-4746613, E-mail: mic_1661@yahoo.com

K. Jayesh
PhD Research Scholar, School of Biosciences, Mahatma Gandhi University, Kottayam – 686560, Kerala, India, Tel.: +91-9447120534, E-mail: jkbiotechno@gmail.com

Hui-Zi Jin
Professor, School of Pharmacy, Shanghai Jiao Tong University, Shanghai, China,
E-mail: kimhz@stu.edu.cn

Ninan Jisha
Research Scholar, Pharmacognosy Laboratory, School of Biosciences, Mahatma Gandhi University,
Kottayam – 686560, Kerala, India, Tel.: 8589075151, E-mail: jishaninan@gmail.com

Anees Ahmed Khalil
Assistant Professor, University Institute of Diet and Nutritional Sciences, Faculty of Allied
Health Sciences, The University of Lahore, Pakistan, Tel.: 192-333-7000757,
E-mail: aneesahmedkhalil@gmail.com

Muhammad Kamran Khan
Assistant Professor, Institute of Home and Food Sciences, Government College University, Faisalabad,
Pakistan, E-mail: mk.khan@gcuf.edu.pk

M. S. Latha
Professor, School of Biosciences, Mahatma Gandhi University, Kottayam – 686560, Kerala, India,
Tel.: +91-9446190331, E-mail: mslathasbs@yahoo.com

H. R. Naik
Division of Food Science and Technology, Sher-e-Kashmir University of Agricultural Sciences and
Technology-Kashmir, Shalimar Campus, Srinagar – 190 025, Jammu & Kashmir, India,
Tel.: +91-7006277094, E-mail: haroonnaik@gmail.com

Bazila Naseer
Division of Food Science and Technology, Sher-e-Kashmir University of Agricultural Sciences and
Technology-Kashmir, Shalimar Campus, Srinagar – 190 025, Jammu & Kashmir, India,
Tel.: +91-9419381715, E-mail: sheikhbazila@gmail.com

Seema Patel
PhD Biotech (IITG-India); MS Bioinformatics (SDSU-USA), Bioinformatics and Medical
Informatics Research Center, San Diego State University, San Diego – 92182, USA,
E-mail: seemabiotech83@gmail.com

R. N. Raji
Research Scholar, Pharmacognosy Laboratory, School of Biosciences, Mahatma Gandhi University,
Kottayam – 686560, Kerala, India, Tel.: 9633839935, E-mail: mitra.rrn@gmail.com

Abdur Rauf
Institute of Chemical Sciences University of Peshawar, Pakistan, Assistant Professor,
Department of Chemistry, University of Swabi, Anbar – 23561, Khyber Pakhtunkhwa, Pakistan,
Tel.: +923469488944, E-mails: mashaljcs@yahoo.com; kimhz@sjtu.edu.cn

Farhan Saeed
Assistant Professor, Institute of Home and Food Sciences, Government College University, Faisalabad,
Pakistan, E-mail: f.saeed@gcuf.edu.pk

Zafar Ali Shah
HEJ, University of Karachi, Pakistan, Assistant Professor, Department of Chemistry,
University of Swabi, Anbar – 23561, Khyber Pakhtunkhwa, Pakistan, Tel.: +923349134454,
E-mail: zafarwazir30@yahoo.com

K. L. Sreejamole
Assistant Professor, P.G and Research Department of Zoology, S.N. College, Cherthala – 688582,
Alappuzha District, Kerala, India, Tel.: +91-9447782231; +91-484-2340231,
E-mail: sreejakl@gmail.com

Hafiz Ansar Rasul Suleria
McKenzie Fellow, Department of Agriculture and Food Systems, The University of Melbourne, Level 3, 780 Elizabeth Street, Parkville, Victoria – 3010, Australia, Tel.: +61 470-439-670,
E-mails: h.suleria@hotmail.com; hafiz.suleria@uqconnect.edu.au

D. Suma
Research Scholar, Pharmacognosy Laboratory, School of Biosciences, Mahatma Gandhi University, Kottayam – 686560, Kerala, India, Tel.: 9048860137, E-mail: sumadwarai@yahoo.com

S. Syama
PhD Research Scholar, School of Biosciences, Mahatma Gandhi University, Kottayam – 686560, Kerala, India, Tel.: +91-9605635134, E-mail: syama@macfast.org

Tabussam Tufail
Visiting Lecturer, Research Associate, Institute of Home and Food Sciences, Government College University, Faisalabad, Pakistan, E-mail: tabussamtufail@gcuf.edu.pk

A. Vysakh
Research Scholar, Pharmacognosy Laboratory, School of Biosciences, Mahatma Gandhi University, Kottayam – 686560, Kerala, India, Tel.: 947358718, E-mail: vysakh15@gmail.com

Hafiza Sidra Yaseen
MPhil Pharmacology, Department of Pharmacology, Government College University, Faisalabad, Pakistan, E-mail: hsidra2181@gmail.com

ABBREVIATIONS

μM	micro moles
3-Ph-3-SG	3-phenyl-3-shogaol
AA	arachidonic acid
ABTS	2,2′-azino-bis(3-cthylbcnzothiazoline-6-sulfonate)
ACC1	acetyl-CoΛ carboxylasc 1
ACE	angiotensin converting enzyme
AChE	acetylcholinesterase
AHR	ameliorated airway hyper-responsiveness
AI	anti-inflammatory
AIDS	acquired immunodeficiency syndrome
AKI	acute kidney injury
ALA	alpha linolenic acid
ALI	acute lung injury
ALT	alanine aminotransferase
AME	*abrusmollis* extracts
AMK	AMP-activated protein kinase
AMP	activated protein kinase
AMPK	adenosine monophosphate-protein kinase
AOAC	Association of Official Analytical Chemists
AOE	antioxidative enzymes
AP	acute pancreatitis
ASP	amnesic shellfish poisoning
AST	aspartate aminotransferase
ATP	adenosine tri-phosphate
AuNPs	gold nanoparticles
BA	betulinic acid
BALF	bronchoalveolar lavage fluid
BDNF	brain-derived neurotropic factor
BFC	bioactive food component
BHA	butylated hydroxyl anisole
BHT	butylated hydroxyl toluene
BMD	bone mineral density
BMDMs	bone marrow-derived macrophages
BMI	body mass index

CA	coussaric acid
CAM	complementary and alternative medicines
cAMP	cyclic adenosine monophosphate
CAR	carvacrol
CAT	catalase
CBC	complete blood count
CBD	cannabinoids
CCI	chronic constriction injured
CCl_4	carbon tetrachloride
CD4	cluster of differentiation-4
CDKs	cyclin-dependent kinases
CFIA	Canadian Food Inspection Agency
CFP	ciguatera fish poisoning
CHD	coronary heart disease
CHOP	C/EBP homologous protein
CLA	conjugated linoleic acid
CM	cerebral malaria
CMP	comprehensive metabolic panel
CNS	central nervous system
Con A	concanavalin A
COSY	correlated spectroscopy
COX	cyclooxygenase
COX-2	cyclooxygenase-2
CP	cyclophosphamide
CPE	cytopathic effect
CR E-binding	cAMP response element-binding
CT	computed tomography
CVD	cardiovascular disease
DA	domoic acid
DAD	diode array detector
DADS	diallyl disulfide
DAI	disease activity index
DAP	domoic acid poisoning
DATS	diallyl trisulfide
DBH	debromohymenialdisine
DEP	diesel exhaust particles
DHA	docosahexaenoic acid
DMAPP	dimethylallyl pyrophosphate
DMF	digital microscopy facility

DMSO	dimethyl sulfoxide
DNA	deoxyribonucleic acid
DNP	dictionary of natural products
DPA	docosapentaenoic acid
DPPH	1,1-diphenyl-2-picrylhydrazyl
DSP	diarrheic shellfish poisoning
DSS	dextran sodium sulfate
DTX	dinophysistoxins
DW	dry weight
EC	European Commission
EFSA	European Food Safety Association
ELISA	enzyme-linked immunosorbent assay
ELSD	evaporative light-scattering detector
eNOS	endothelial nitric oxide synthase
EPA	eicosapentaenoic acid
EPTT	end point titration technique
ERK	extracellular signal-regulated kinases
ESI	electrospray ionization
EU	European Union
EURL	European Reference Laboratories
FAO	Food and Agricultural Organization
FAS	fatty acid synthase
FAs	fatty acids
FDA	Food and Drug Administration
FFA	free fatty acid
FH	*Ficus hispida*
FID	flame ionization detector
FN	fibronectin
FPD	flame photometric detector
FRAP	ferric reducing antioxidant power
GA	Ginkgolide-A
GABA	gamma-aminobutyric acid
GB	Ginkgolide-B
GBE	*G. biloba* extract
GC-MS	gas chromatography-mass spectrometry
GI	gastrointestinal
GIT	gastro intestinal tract
GJC	gap junction communication
GLUT	glucose transporter

GLUT4	glucose transporter type-4
GM	Ginkgolides mixture
GMCs	glomerular mesangial cells
GPx	glutathione peroxidase
GR	glutathione reductase
GRAS	generally recognized as safe
GRP	glucose-regulated protein
GRx	glutathion reductase
GSH	glutathione
GSH-Px	glutathione peroxidase
GSTs	glutathione S-transferase
H_2O_2	hydrogen peroxide
HCV	hepatitis C virus
HIV	human immunodeficiency virus
HMBC	heteronuclear multiple bond correlation
HMGR	3-hydroxy-3-methylglutaryl-CoA reductase
HMQC	hetero-nuclear multiple quantum correlation
HO-1	heme oxygenasse-1
HPLC	high performance liquid chromatography
HPP	high pressure processing
HRS	hydroxyl radical scavenging
HSC	hepatic stellate cell
HSV-1	herpes simplex virus-1
HTS	high throughput screening
IBD	inflammatory bowel disease
IC50	inhibitory concentration 50
ICAMPs	intercellular adhesion molecule-1
ICH	intracerebral hemorrhage
IL	interleukin
IL	intromission latency
IL-1β	interleukin-1 beta
IL-6	interleukin-6
IPP	isopentenyl pyrophosphate
ISSFAL	International Society for the Study of Fatty Acids and Lipids
IVD	intervertebral disc
Iκβ	inhibitor of κβ
JNK	c-Jun N-terminal kinases
KCl	potassium chloride
LAM-D	Lamellarin D

LC-MS	liquid chromatography coupled with MS
LDH	lactate dehydrogenase
LDL	low density lipoproteins
LG	*Lippia graveolens*
LLE	liquid-liquid extraction
LLL	combination of linoleic, linoleic and linoleic
LO	lipooxygenase activity
LOX	lipoxygenase
LP	*Lippia palmeri*
LPS	lipopolysaccharides
M.E.	methanolic extract
MAE	microwave assisted extraction
MAP	microwave-assisted processing
MAPK	mitogen-activated protein kinase
MC	Myrtucommulone
MCE	*M. citrifolia* extract
MCF-7	human mammary cancer cell
MDA	malondialdehyde
MDM-LDL	malondialdehyde-modified-LDL
MDR-1	multidrug resistant protein-1
MEP	methylerythritol phosphate
MEP	mevalonic acid pathway
MF	mounting frame
MFRM	mango fruit reject meal
MHG	microwave hydrodiffusion and gravity
MIC	minimum inhibitory concentration
MMC	mitomycin-C
mMECs	mouse mammary epithelial cell
MMP 9	matrix metalloproteinase 9
MMP	matrix-metalloproteinase
MMP	mitochondrial membrane potential
MNP	marine natural products
MP	methylprednisolone
MPO	myeloperoxidase
MPP	microparticulated proteins
MRI	magnetic resonance imaging
mRNA	messenger ribonucleic acid
MRS	magnetic resonance spectroscopy
MS	mass spectrometry

MSF   methanolic soluble fraction
MSU   mono-sodium urate
MVA   mevalonic acid
NaCl   sodium chloride
NAPRALERT   natural product alert
NCI   National Cancer Institute
NF-κB   nuclear factor-kappa B
NGF   nerve growth factor
NIDDM   non-insulin dependent diabetes mellitus
NM   neuromuscular
NMDA   N-methyl-D-aspartate
NMR   nuclear magnetic resonance
NO   nitric oxide
NOESY   nuclear over Hauser enhancement spectroscopy
NOS   nitric oxide synthase
NP   natural product
NPD   nitrogen phosphorus detector
NPQ   non-photosynthetic quenching
NPs   natural products
NQO-1   NADPH: quinone oxidoreductase 1
NRL   National Reference Laboratories
NSP   neurotoxic shellfish poisoning
NSS   neurological severity score
OA   okadaic acid
ODC   ornithine decarboxylase
OH   ohmic heating
OLL   combination of oleic, linoleic and linoleic
ORAC   oxygen radical absorbance capacity
ORCs   olfactory receptor cells
OSI   oxidative stress index
OVA   ovalbumin
OVX   ovariectomized
PABC   pro-oxidant-antioxidant balance
PAL   phenylalanine ammonialyase
PBMCs   peripheral blood mononuclear cells
PC   phosphatidylcholine
PCA   passive cutaneous anaphylaxis
PEF   pulsed electric field
PEI   Prince Edward Island

PET	positron emission tomography
PG	propylene glycol
PGE2	prostaglandin E$_2$
PGs	propylene glycols
Phospho-AKT	protein kinase B (PKB) or serine/threonine-specific protein kinase
PLA2	phospholipase A2
PLE	pressurized liquid extraction
PLL	combination of palmitic, linoleic and linoleic
PLs	phospholipids
PMA	phorbol myristate acetate
POL	combination of palmitic, oleic and linoleic
POP	persistent organic pollutants
PP	phenylpropanoid
PPARy	peroxisome proliferator-activated receptors
PSP	paralytic shellfish poisoning
PTP1B	protein tyrosine phosphatase 1B
PTX2	pectenotoxin-2
PTX2sa	pectenotoxin-2-seco acid
PTXs	pectenotoxins
PTZ	pentylenetetrazole
PUFAs	polyunsaturated fatty acids
QSFR	quantitative structure function relationships
RAAS	renin angiotensin aldosterone system
RE	retinol equivalents
ROCCs	receptor operated Ca2þ channels
ROS	reactive oxygen species
RPAE	rosemary polyphenols alcoholic extract
R-PE	r-phycoerythrin
RSV	respiratory syncytial virus
SAC	S-allyl cysteine
SACN	scientific advisory committee for nutrition
SAMC	s-allylmercaptocysteine
SCE	spinal cord edema
SCFs	supercritical fluids
SCI	spinal cord injury
SF	supercritical fluids
SFA	saturated fatty acids
SFE	supercritical fluid extraction

SGLT1	sodium glucose linked transporter 1
SI	stimulation index
SIRT	sirtuin (silent mating type information regulation 2 homolog)
S-MC	semimyrtucommulone
SNPs	silver nanoparticles
SnRK1	SNF1-related protein kinase-1
SOD	superoxide dismutase
SPs	sulfated polysaccharides
ST	*Salmonella typhimurium*
STX	saxitoxin
STZ	streptozotocin
T2D	type-2 diabetes
TAGs	triacylglycerols
TBARS	thiobarbituraic acid reactive species
TBHQ	tert-butylhydroquinone
TBI	iraumatic brain injury
TBIL	total bilirubin
TC	total cholesterol
TCA	trans-cinnamaldehyde
TG	triglycerides
TGF	tumor growth factor
THC	tetrahydrocannabidiol
THC	tetrahydrocurcumin
THCV	tetrahydrocannabivarin
TLC	thin layer chromatography
TLR	toll-like receptor
TNBS	trinitrobenzene sulfonic acid
TNF	tumor necrosis factor
TNF-α	tumor necrosis factor α
TPA	12-*O*-tetradecanoylphorbol-13-acetate
TPC	total phenolic content
TRAP	radical-trapping antioxidant parameter
TrKB	tropomyosin receptor kinase B
TRP	transient receptor potential
TSG	total saponins ginseng
TTX	tetrodotoxin
TXB2	thromboxane
TXs	thromboxanes
UAE	ultrasound assisted extraction

UCP	uncoupling protein
UCP1	uncoupling protein-1
UFA	unsaturated fatty acids
UPE	ultrahigh pressure extraction
USFDA	U.S Food and Drug Administration
UV	ultraviolet
UVB	ultraviolet B
VLDL	very low-density lipoproteins
WAT	white adipose tissue
WB	white button
WHO	World Health Organization
WHR	waist to hip ratio
WOE's	welsh onion green leaves
XO	xanthine oxidase
YTXs	yesso toxins
ZJRB	*Ziziphus jujube* root bark

PREFACE 1

We introduce this book volume under the book series *Innovations in Plant Science for Better Health: From Soil to Fork*. This book mainly covers the current scenario of the research and case studies and the importance of phytochemicals from plant-based therapeutics, divided into two main parts: Part I: Bioactive Compounds from Plant Sources and Part II: Bioactive Compounds from Marine Sources.

Part I: Bioactive Compounds from Plant Sources describe the concept of extraction of bioactive molecules from plant sources, both conventional and modernizations extraction techniques, available sources, biochemistry, structural composition, and potential biological activities. Advanced extraction techniques such as enzyme-assisted, microwave-assisted, ultrasound-assisted, pressurized liquid extraction, and supercritical extraction techniques are described in detail. Natural products and their bioactive compounds are increasingly utilized in preventive and therapeutic medication. Bioactive compounds have been utilized for the production of pharmaceutical supplements and more recently as food additives to increase the functionality of foods.

Part II: Bioactive Compounds from Marine Sources covers the isolation of potentially bioactive molecules from marine sources along with their importance and health perspectives. The incorporation of any functional foods, nutraceuticals, and bioactives in the daily diet is a beneficial endeavor to prevent the progression of chronic disorders. This section explains the marine bioactivity, physical characteristics, uniqueness, uses, distribution, importance, traditional importance, nutritional importance, bioactivities, and future trends of different functional foods. Functional foods, beyond providing basic nutrition, may offer a potentially positive effect on health and may cure various disease conditions such as metabolic disorders, cancer, and chronic inflammatory reactions.

This book volume sheds light on the potential of both plant and marine sources for human health for different technological aspects, and it contributes to the ocean of knowledge on food science and nutrition. We hope that this compendium will be useful for the students and academic researchers as well as the persons working with the food, nutraceuticals, and herbal industries.

The contributions of the cooperating authors to this book volume have been most valuable in the compilation. Their names are mentioned in each chapter and in the list of contributors. We appreciate you all for having patience with our editorial skills. This book would not have been written without the valuable cooperation of these investigators, many of whom are renowned scientists who have worked in the field of food science, biochemistry, and nutrition throughout their professional career.

The goal of this book volume is to guide the world science community on how bioactive compounds can alleviate us from various conditions and diseases.

I thank Dr. Megh R. Goyal for his leadership qualities and for inviting me to join his team. He is a world-known scientist and engineer with expertise in agricultural and biological engineering. Truly, he is a giver and a model for budding scientists. I am on the board to learn.

—*Hafiz Suleria, PhD*
Lead Editor

PREFACE 2

To be healthy is our moral responsibility towards Almighty God,
ourselves, our family, and our society; and no life without nature.
Eating fruits and vegetables makes us healthy, believe, and have a faith.
Reduction of food waste can reduce the world hunger and
can make our planet eco-friendly.

—**Megh R. Goyal, PhD**
Lover of Mother Nature and Healthy Foods
Senior Editor-in-Chief

The use of plant and marine-based natural products has largely increased because they are locally accessible are economical and are vital in promoting health. However, scientific data and information regarding the safety and efficacy of these medicinal plants are inadequate. This book mainly covers the current scenario of the research and case studies and contains scientific evidence on the health benefits that can be derived from natural medicines and how their efficacies can be improved. The literature reviews in this book can be useful in health policy decisions and will also motivate the development of health care products from plant and marine ecosystems for health benefits. The book will further encourage the preservation of traditional medical knowledge. Therefore, natural products in this book volume are drawing the attention of researchers and policymakers because of their demonstrated beneficial effects against diseases that are high global burdens, such as diabetes, hypertension, cancer, and neurodegenerative diseases, etc.

This book volume is a treasure house of information and excellent reference material for researchers, scientists, students, growers, traders, processors, industries, dieticians, medical practitioners, users, and others. We hope that this compendium will be useful for students and researchers as well as those working in the food, nutraceutical, and herbal industries.

The goal of this book volume is to guide the world science community on how plant-based secondary metabolites can alleviate us from various conditions and diseases.

We will like to thank editorial and production staff, Sandy Jones Sickles, Vice President, and Ashish Kumar, Publisher and President at Apple

Academic Press, Inc., for making every effort to publish this book when all are concerned with health issues.

We express our admiration to our families and colleagues for their understanding and collaboration during the preparation of this book volume.

As an educator, there is a piece of advice to one and all in the world: *"Permit that our almighty God, our Creator, provider of all and excellent Teacher, feed our life with Healthy Food Products and His Grace...; and Get married to your profession..."*

—Megh R. Goyal, PhD, PE
Senior Editor-in-Chief

Part I:
Plant-Based Secondary Phytocompounds

CHAPTER 1

FUNCTIONAL FOODS: BIOAVAILABILITY, STRUCTURE, AND NUTRITIONAL PROPERTIES

TAWHEED AMIN, H. R. NAIK, SYED ZAMEER HUSSAIN, and BAZILA NASEER

ABSTRACT

Consumers demand food products that are wholesome, nutritious, good tasting, and processed in such a way so that the functionality and bioavailability of their bioactive components are retained to the maximum. The food industry is responding to this demand from time to time by developing functional food products. However, the bioactive components of such functional foods are losing their functionality and bioavailability of ingredients, thereby depriving consumers from the added health benefits. This loss in functionality and bioavailability is attributed to food structure and the interaction among different food components on bioactive components. It is, therefore, important to understand and control food structure and the interactions among different food components within a food matrix. Thus, it is important to assess the effects of food structure on functionality and bioavailability of bioactive components, so as to develop a personalized diet to get more added health benefits.

1.1 INTRODUCTION

These days, consumer's awareness about the relationship between health and diet is increasing [90]. An increasing demand exists for a balanced diet [90], functional food products [90] and high nutritive value food products [60], which address both specific human health needs [60, 90] and sustainable produced foods according to high ethical standards [60]. Several attributes characterize health-beneficial food products, such as lower caloric value, lower or moderate trans-fat, low sodium and sugar contents [90]. There are numerous food products, known as functional foods with enhanced quantities of bioactive food components (BFCs) [25]. Functional foods include whole food, fortified food, enriched foods, or enhanced food that have potential health benefits [25, 26].

Keeping in mind the emphasis of the nutritionists on food behaviors and potential health benefits of BFCs, functional foods with added BFCs are continuously being designed and developed with such matrices to improve the stability, bioactivity, and bioavailability. It is very important to optimize the delivery of BFCs and the original food matrix has a role to play [25]. Structurally, a food matrix involves a diverse network of nutrients and non-nutrients and interacts chemically/ physically in a given food. Within each matrix, the physicochemical properties affect the release, mass transfer, accessibility, digestibility, and stability of different food components (such as phenolic compounds) present in an ample amount in almost all plant-based foods [2].

Besides, the nutrient-nutrient and nutrient-nonnutrient interactions are affected by their physicochemical properties. Since each food matrix involves a myriad of interactions (such as synergistic, additive, antagonizing, or neutralizing effect) therefore, it is important to fully understand the dynamics of the original matrix and its influence on bioactive components. This may serve as a guide in the design and development of functional food matrices, specifically to protect the active BFC from the harsh environment, food processing, storage, and digestion. The tailor-made and structurally designed matrices ensure that the BFCs reach the intended site of action inside the human body [25].

Since processed foods are usually multi-component and multi-phase systems, their microstructure is affected by the interplay among the ingredients, which in turn, affects the physical properties of the food product. During processing and storage, food microstructure could be oriented through the control of various interactions (inter-molecular and inter-particle) among various food components [30, 31, 64]. These interactions may be utilized lucidly to design desired structures to develop gel-networks and emulsions with tailor-made surface-active properties [24, 33, 36, 82, 87, 104, 109], which would potentially result in the efficacious delivery of encapsulated components with enhanced stability [74, 106]. Tailor-made structuring requires the knowledge of the long- and short-term molecular assemblies of different components and other physical characteristics, e.g., molecular size, hydrophobicity, density, charge, and conformation under different environmental conditions that in turn affects the functionality during designing [104].

This chapter focuses on the functional foods and their food structure, bioavailability, functionality, and within the food matrix.

1.2   FOOD STRUCTURE, BIOAVAILABILITY, AND FUNCTIONALITY

In addition to food processing operations and environmental conditions, the physicochemical properties of bioactive components strongly affect and determine their functionality and bioavailability [64]. Heat and mass transfer processes affect the food microstructure. The texture of food and the release of BFCs are determined by the complexity of the food product. The relationship between the food structure and functionality, in terms of the quantitative aspect, commonly known as quantitative structure-function relationships (QSFR) that helps in the efficient design and development of such functional food systems [63].

1.2.1   FOOD STRUCTURE AND BIOAVAILABILITY

During the past few years, the bioavailability of nutrients has gained considerable attention in the food industry. One of the biggest and ongoing challenges and concerns to the food industry is to make available safe and wholesome food to the consumer. Further, the nutritional and caloric composition is equally important. Since the last decade, considerable research is being carried out to design food matrices with enhanced physicochemical properties and a positive effect on food digestion. The effect of physical properties (such as particle size, microstructure, and physical state of foods on nutrient absorption) has also been enunciated. Therefore, food technologists can induce different modifications to fabricate the food products possessing controlled bioavailability of BFCs.

The food matrix influences both the ingredient stability and their bioavailability. Research studies confirm that only a fraction of carotenoids present in vegetables and fruits is absorbed in the intestine, due to the fact that mostly plant carotenoids are bound to proteins or are present in the form of crystals. However, the carotenoids dispersed in oils especially vegetable oils, have been reported to show improved bioavailability [91]. The crystallinity and solubility, and thus the absorption of carotenoids are influenced by their incorporation into nano- and micro-structures in food matrices. The formulation of carotenoids into the particulate system makes them more soluble and easily releasable during digestion for easy and smooth delivery of them into the components of the cell.

Faulks et al. [39] have reviewed the effects of food processing methods on the bioavailability of individual BFCs. For the design and development of novel functional food products, the following considerations are important:

- Original BFCs are present in a form that is not available directly within the GI system;
- Food matrix has a profound impact on the bioavailability and release of BFCs;
- An extra carrier is required to improve the solubility of BFC;
- Released components might be absorbed partially;

Functional responses to BFCs vary throughout the population depending upon the genetic make-up.

These considerations can help to predict the rates of absorption, metabolism, and bioavailability of BFCs, when ingested by humans. It is believed that the food interactions inside the human digestive tract are hugely intricated due to various physicochemical processes along with their impact on the metabolism characteristics of an individual BFC and food structure. However, an understanding of the digestion process of a single food component is lacking [39]. Through proper selection and development of the protection and delivery methods for BFCs, an increase in their bioavailability has been suggested by recent advancements of functional/medical foods. Therefore, the food matrix has a role to play in deciding the bioavailability of BFCs. For example, the absorption of plant sterols varies in different food matrices. Milk is believed to be an efficient carrier with efficiency about three times than bread or any other cereal product [56]. It is of utmost importance to fully understand the normal gastrointestinal (GI) system to design and fabricate the functional foods with improved nutritional value, the bioavailability of BFCs, and specific beneficial functionality [102].

1.2.2 FOOD STRUCTURE AND FUNCTIONALITY

While developing a food product, a new challenge and an emerging opportunity are to efficiently, control the functionality of a BFC in a regular food product throughout the use of appropriate vehicles for delivery [104]. It is important to maintain or improve the functionality of bioactive components from its source until it is digested (Figure 1.1).

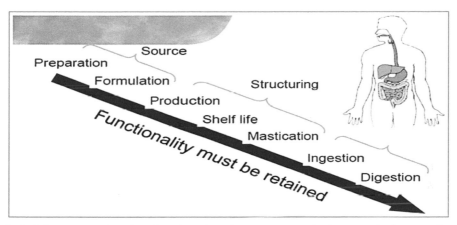

FIGURE 1.1 The functionality of a bioactive component should maintain functionality from source until it is digested.

The functional and physical properties of food products (e.g., the sensorial attributes, structural breakdown during digestion, and the storage stability) are predominantly determined by the microstructure of complex food matrix. The development of delivery systems for the delivery and protection of BFC from several environmental stresses (such as hydrolysis, oxidation, polymerization) or to moderate some processes (e.g., digestion of lipids or controlled release for agricultural, pharmaceutical, and food applications) is the option to combat such problems [19, 37, 58, 92, 104].

1.3 CLASSIFICATION OF HEALTH FOODS

1. **Excipient Foods:** These are usually beverage products possessing such ingredients or structures that increase the bioavailability of BFCs or nutraceuticals that it is consumed with [81].
2. **Functional Foods:** These are foods or beverage products containing a nutraceutical with health benefits over and above normal nutritional function [81].
3. **Medical Foods:** These are foods or beverage products containing a drug to prevent or treat a specific disease [81].

Differences among functional foods, excipient foods, and medical foods are described in Figure 1.2. Development of medical or functional food usually involves the encapsulation of lipophilic BFCs into the food matrix; however, in excipient foods, it is ingested along with different food matrix.

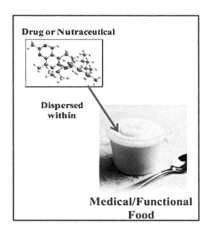
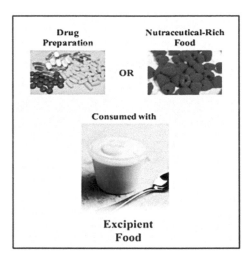

FIGURE 1.2 Difference among functional, excipient, and medical foods.

Source: Reprinted with permission from Ref. [81]. © 2014 Royal Society of Chemistry.

1.3.1 DESIGNING EXCIPIENT FOODS

Excipient is any bioactive component, which does not possess bioactivity in itself and is mostly used in pharmaceutical preparations for improving the efficacy of drugs [42, 48, 81]. Excipients facilitate administration, modulate the release of an active component, and prevent its degradation from the environment by improving its stability [9].

The most commonly used excipients in pharmaceutical preparations are carbohydrates, proteins, lipids, surfactants, synthetic polymers, salts, and co-solvents [81]. As an analogous term, excipient food is the food possessing no bioavailability in-itself but may increase the efficacy of any nutraceutical or functional components [81]. Excipient foods have structures and compositions specially fabricated to improve the bioaccessibility of BFCs present in other foods co-ingested with them [119]. Therefore, excipient foods may be taken along with dietary supplements (such as nuts, seeds, fruits, vegetables, fish, meat, grains) and some processed foods. It is obvious that the development of different kinds of matrices is needed for BFCs.

Table 1.1 shows examples of excipient foods. The bioavailability of carotenoids present in a salad can be improved by taking it along with special dressings containing such food components that increase its bioavailability. Intestinal solubility of carotenoids is increased by lipids, whereas their chemical transformation is impeded by antioxidants, metabolism is delayed by enzyme inhibitors, and absorption is increased by permeation enhancers [81]. Some studies have suggested the improvement in the bioavailability of fat-soluble carotenoids and vitamins when consumed with fat or oil containing dressings [13].

TABLE 1.1 Excipient Food Products to Increase the Bioavailability of Various Bioactive Food Components (BFCs) from Different Food Sources

Nutraceuticals	Source	Excipient Food
Carotenoids	Salad (kale, lettuce, peppers, carrot, tomato etc.).	Oil containing salad dressings
	Cooked vegetables (kale, spinach, carrot, etc.).	Sauces
Conjugated linoleic acid (CLA)	Dairy (cheese); meat products (beef)	Sauces
Omega-3 fatty acids/oils	Fish	Sauces
Stanols/phytosterols	Nuts	Edible coatings; Sauces
Vitamins	Raspberries, strawberries, blueberries, apple, pear, etc.	Yogurt; Ice-cream
Vitamins, Stanols, phytosterols	*Arachis hypogaea* (peanuts), *Helianthus* (sunflower) seeds, *Prunus delcis* (almonds), etc.	Edible coatings

1.3.2 DESIGNING FUNCTIONAL/MEDICAL FOODS

Functional food may either be natural or processed containing BFCs conferring several health benefits and essential daily nutritional requirements. Vegetables and fruits are considered natural functional foods and their nutraceutical components, however, can be extracted and purified using several novel technologies. These nutraceutical components can be used as dietary supplements to be taken in concentrated form, or with food products. The addition of this beneficial ingredient into some other food product may, however, result in impairment in sensory properties and the structure of the developed food product. These issues may be averted by employing micro-encapsulation [90].

Microencapsulation is the process of an entrapment of a BFC inside a dispersed material to ensure its immobilization, protection, controlled release, structuration, and functionalization [4]. Fortification has enormous applications in food industries. Examples of some of the fortified food products are fruit juices with added ω-3 fatty acids (FAs) or breakfast cereals to which vitamins such as folic acid have been added. Some food products have added components within its natural matrix; however, some food products need further process modification, e.g., encapsulation of BFCs before it is added to any food product. This includes the delivery of protected BFCs to reach their site of action and consequent liberation when triggered by certain factors, such as enzymes, salts, pH, etc. [22].

1.3.3   CONSIDERATIONS FOR DEVELOPMENT OF FUNCTIONAL FOODS

The behavior of a food product within the gastrointestinal tract is primarily determined by its structure. As discussed earlier in this chapter, a well-defined relationship exists between the food structure and related food product characteristics and has a bearing on the design and development of functional food [27]. It is difficult to maintain the desired and traditional sensory qualities (e.g., texture, and flavor and maintain the functionality of the added component), for the design and development of functional food product.

Another objective of functional food product development is to have a positive response in the market, which could be achieved by considering the consumption habits, the targeted consumer group, and fulfilling the consumer's requirements. If the addition of BFCs is carried out successfully with satisfying results on the sensory quality of a food product, the next step is to ensure its functionality. Apart from *in vitro* and *in vivo* studies on the bioavailability of BFCs, researchers, and scientists have used mathematical models [89]. In addition, current understanding of the activity of an ingredient, release mechanism, and human metabolism of a BFC can be employed to predict the functionality of ingredients to justify the addition of such a component into the food product.

Currently, various studies on the design of the microstructure of a product for process modeling in mouth behavior and several other sensory properties include the development of taste or physical sensation, such as mouth feeling and buildup the supporting data to such functionality studies [70]. A novel domain considers the effect of the nutritional supplements on an individual genome (known as nutrigenomics). Nutrigenomics helps to understand the impact of nutrition on metabolic pathways and homeostatic control. It also serves as a new tool for nutritional research and helps in mitigating the health-related issues [3].

1.4 DESIGNING FOODS WITH DIETARY BIOACTIVE LIPIDS

Functional foods (such as margarines, mayonnaise, yogurt, salad dressings, soy milk, orange juice, etc.) containing dietary bioactive components are becoming increasingly available to consumers [114]. For the successful incorporation of bioactive lipids into certain classes of food products, it is important to overcome major challenges. Proposed approaches to develop functional foods with designed bioactive lipids are indicated in Figure 1.3. The major sources of these dietary bioactive lipids include microbial sources, fish, animal, and plant sources (Stage 1 of Figure 1.3).

Lipids being hydrophobic in nature may either have a positive or negative effect on the health of an individual. Within an individual's diet, triacylglycerols are major energy sources, and several diseases such as coronary heart diseases (CHDs), diabetes, and obesity are associated with its overconsumption. However, the consumption of several other bioactive lipids such as polyunsaturated FAs exerts beneficial effects on human health [19]. Therefore, these beneficial bioactive lipids with health-improving ability can be added into different food products in different forms (Table 1.2).

The beneficial bioactive lipids may be present either naturally inside the multiplex matrix of food (e.g., milk, oil, egg, etc.) or maybe incorporated as functional ingredients (e.g., vitamins or phytosterols). The key element affecting the bioavailability of a BFC is its ability to confer health benefits, in addition to providing the conventional nutritional components. The food industry has responded to this by fortifying various food products with bioactive lipids. Designing and fabricating food products incorporated with bioactive lipids is a challenge to food scientists. The major challenges are [19]:

- To design food products with maximum health benefits from the dietary bioactive lipids by improving their bioavailability;
- To incorporate adequate levels of bioactive dietary lipids into food products;
- To prevent chemical deterioration of bioactive lipids during processing and storage of food products;

- To provide an accurate and clear awareness to the consumer of health beneficial effects of these dietary lipids;
- To understand and quantify the composite relationship between the dietary lipids and associated impact on health benefits.

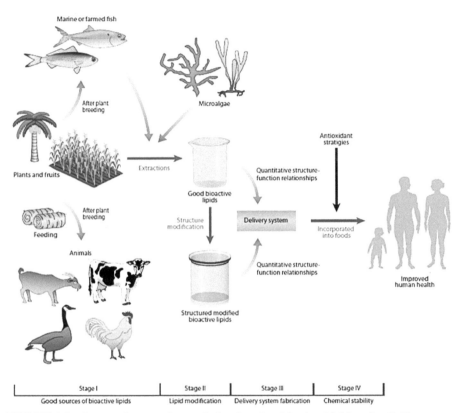

FIGURE 1.3   Proposed approaches to design functional foods with bioactive lipids.

Source: Reprinted with permission from Ref. [81]. © 2014 Royal Society of Chemistry.

To confer several health benefits as shown in Table 1.2, the bioactivity of dietary lipids must be maintained during processing, storage, transportation, and utilization of a food product. However, the desirable sensory and physicochemical attributes, appearance, flavor, and texture stability of food products should not be adversely affected. Some bioactive dietary lipids (e.g., PUFAs, and carotenoids) are chemically liable, thus their bioavailability is influenced by oxygen, light, or pro-oxidants. While carotenoids are not stable physically, crystallize, and thus lose their bioavailability when incorporated into other food matrices. Few following issues emerge while designing and developing the food product containing bioactive lipids:

TABLE 1.2   Major Dietary Bioactive Lipids, Their Source, and Potential Health Benefits

Name	Examples	Sources	Health Benefits
Carotenoids	β-Carotene, lutein, lycopene, and zeaxanthin	Pumpkin, collards, carrot, tomatoes, tangerines, sweet potato, watermelon, etc.	Prevents cataract, cancer, coronary heart disease (CHD), macular degeneration, etc.
Fat-soluble vitamins; phenolic lipids	Vitamins A, D, E, and K; flavonoids	Carrots, spinach, avocados, vegetable oils, fruits, etc.	Prevents cancer, CHD, and urinary tract diseases.
Fatty acids	Docosapentaenoic acid (DPA), Eicosapentaenoic acid (EPA), docosahexaenoic acid (DHA), stearidonic acid, α-linolenic acid (ALA), conjugated linoleic acid (CLA), arachidonic acid (AA)	Fish, certain algae, flaxseeds, walnut oil, collard greens, soybeans, borage, evening primrose (*Oenothera*), etc.	Prevents atherosclerosis, arrhythmias, weight gain, stroke, cancer, and immune response disorders, reduces blood pressure, improves bone health, mental health, visual acuity.
Phytosterols	Stigmasterol, β-sitosterol, Campesterol	Algae, plants, vegetable oils	Coronary heart disease (CHD)

1. **Oxidation:** It is accelerated by heat, light, enzymes, metals, metallo-proteins, and microorganisms [28, 105]. The products after oxidation of the bioactive dietary lipids not only reduce their bioavailability but also result in off-flavor development, color change, nutrient loss, and the formation of toxic compounds. PUFAs are most susceptible to oxidative reactions leading to off-flavor. The development of food products incorporated with bioactive dietary lipids, therefore, face a lot of challenges [19].

Carotenoids are also degraded chemically, either spontaneously or through free-radical initiated oxidations, during storage and processing of food products. Therefore, their biological properties are altered [12]. Several ketones and aldehydes are formed during thermal oxidation of β-carotene

[71]. Peroxyl radicals disrupt the conjugated double bonds thereby resulting in rapid bleaching of β-carotene [117]. These reactions cause undesirable color changes in food products during processing and storage. However, research studies have uncovered the cancer preventive effect of oxidized products from carotenoids by enhancing gap junction communication (GJC) [6]. Free radical scavenging ability is the fundamental function of carotenoids. Therefore, it is essential to prevent their oxidation in food matrices before consumption, in order to maintain their desirable functions [19].

And dietary bioactive lipid, phytosterol, contains an unsaturated ring structure hence making them more vulnerable to oxidation reactions amid processing and storage in mass oil and oil-in-water (O/W) emulsions [15]. Esterified phytosterols have less oxidative stability than free phytosterols [108]. Few investigations have demonstrated that the oxidation products of phytosterols can cross the intestinal barrier at a low level, which might be connected to atherogenesis, cytotoxicity, carcinogenesis, and mutagenesis [111].

Fat-soluble vitamins (vitamin E) operate in the food and in our bodies. Several factors such as temperature during storage, oxygen, transition metals, light, and degree of unsaturation of co-existing lipids may influence the sapping of tocopherols in food matrices. The α-tocopherols (vitamin E) are present in biological membranes at a concentration of one part per 1000 lipid molecules thereby preventing PUFAs from getting preferentially oxidized prior to unsaturated FAs [14]. Restoration of vitamin E (α-tocopherols) is primarily achieved through dietary BFCs. It is thus critical to decrease the degradation of α-tocopherol (vitamin E) in food products to maintain adequate and efficacious concentrations in food products.

2.   **Bioavailability:** It is imperative to know different components influencing the bioavailability of lipophilic bioactive agents since it helps in the efficient designing of food structure [81]. The oral bioavailability of any ingested BFC relies on the part that really achieves the objective site-of-activity in a biologically active form [97]. Assume the general bioavailability of an ingested lipophilic BFC is indicated by "F," which relies on factors [5, 77, 81].

$$F = F_L \times F_A \times F_D \times F_M \times F_E \qquad (1.1)$$

where, F_L represents that fraction of a BFC released from a food matrix into GIT (gastrointestinal tract) with the goal that it ends up bioaccessible (F_L); F_A represents that fraction of the released BFC that is absorbed by the membranous tissue of GIT; F_D represents that fraction of absorbed BFC that reaches the intended site-of-action when it is distributed amongst the various tissues of the body, e.g., liver, blood, heart, kidney, adipose tissue, muscles, etc.; F_M represents that fraction of a BFC that reaches the site of action in a metabolically active form; and F_E represents that fraction of metabolically active BFC that has not been excreted.

All these parameters (F_L, F_A, F_D, F_M, and F_E) vary with time. When BFC is consumed, a graph showing the bioavailability versus time at an indicated site-of-action is obtained. The overall bioavailability (F) of BFC increases sometime after consumption followed by a decrease once it is metabolized, utilized, distributed, stored, or excreted. Thus, the bioavailability of consumed BFC can be obtained from the graph of bioavailability versus time [81]. Many physicochemical and physiological factors (liberation, absorption, metabolism, distribution, excretion, and improving bioavailability) affecting the bioavailability of lipophilic BFCs have been substantiated [7, 40].

3. **Water Solubility:** Most of the bioactive lipids are insoluble, non-polar, and, therefore, cannot be directly dispersed into water-based foods. The solubility of nutrients is one of the most important prerequisites for bioavailability [19].

4. **Molecular Structure and Physical State of Bioactive Lipids:** The physical state and molecular structure of dietary bioactive lipids affect their bioavailability. The bioavailability, absorption, and distribution of carotenoids inside the tissue are being adversely influenced by their molecular weight, polarity, hydrophobicity, and morphology [23]. Lycopene gets converted to *cis*-form during processing possessing varied bioavailability from *trans*-isomer, due to differences in their solubility and crystallization in biological fluids. *Trans*form does not result in tightly packed crystal structures due to their bent configuration [51]. Bioactive lipids in amorphous forms are typically more bioavailable compared to those in crystalline forms [19].

1.4.1 METHODS OF DESIGNING DIETARY BIOACTIVE LIPIDS

1. **Plant Breeding:** It can be employed to minimize levels of undesirable FAs and enhance the chemical stability of bioactive lipids. Classic plant breeding has been employed to reduce the level of α-linolenic acid (ALA) of canola oil and soybean as this fatty acid increases susceptibility to oxidation [35]. The modern plant modifications use

genetic manipulation techniques to produce fats with specific altera-
tions in fatty acid profiles.

2.　**Structuring Bioactive Lipids:** of desirable structural modification is done with the help of enzymatic or chemical reactions. The
structures of TAGs and phospholipids are modified to change the
positional distribution of FAs. The chemical stability and bioavailability of structured bioactive dietary lipids are more compared with
conventional TAGs (Stage 3 of Figure 1.3).

The ω-3 FAs have been successfully incorporated into palm oil at
stereospecific number-2 by a *trans*-esterification process with an average of
21% eicosapentaenoic acid (EPA) and 16% docosahexaenoic acid (DHA)
[38]. TAG containing conjugated linoleic acid (CLA) at the stereospecific
number-2 position has been produced through enzymatically catalyzed
reactions between sn-1,3 diacylglycerols and CLA isomers derived from
sunflower oil [73, 113]. Other bioactive components can also be integrated
into TAG. Structured phenolic lipids with improved health benefits and
oxidative stability could be proved through *trans*-esterification of flaxseed
oil with dihydrocaffeic acid [100].

Structural lipids have been developed by preferentially locating the
bioactive FAs at a stereospecific position in the phospholipid molecule [19].
Lipases are used to catalyze the *trans*-esterification of phosphatidylcholine
(PC) and ω-3 FAs with oil containing 55% EPA and 30% DHA. After the
reaction, the phospholipids produced containing 32% EPA and 16% DHA
[49]. Kim et al. [61] carried out the structural modification of PC by phospholipase A_1-catalyzed trans-esterification with ω-3 FAs from fish oil, which
contains the maximum level of 28 mol% ω-3 polyunsaturated fatty acids
(PUFAs) after reacting for 24 hours. Synthesis of phospholipids with CLA
using an enzymatic process has been carried out to achieve maximum incorporation of 16% CLA [52].

The structured phospholipids have an added benefit of improved
chemical stability since FAs in the form of phospholipids (PLs) are less
likely to get oxidized than when present as triacylglycerols (TAGs) [45]. It
has been reported that the distribution of plasma lipoproteins is being influenced by the chemical form of bioactive fatty acid obtained from different
dietary sources [116]. Phospholipids possess enhanced bioavailability than
TAGs, due to an amphiphilic nature of phospholipids (PLs), thus increasing
dispersibility in water and more reactivity towards phospholipases [50]. The

structural modification of PLs to incorporate into bioactive FAs is a viable option to enhance chemical stability and bioavailability of FAs [19].

1.4.2 DELIVERY SYSTEMS FOR BIOACTIVE LIPIDS

Ingestion of structurally designed desirable dietary bioactive lipids does not guarantee their target delivery to the tissues inside the human body, due to incompatibility of bioactive dietary lipids with water-based food products; and might be ineffectively absorbed. Prior to consumption and in GIT after consumption, the bioactivity of dietary lipids should remain intact inside the food matrix. They must show resentment to changes to environmental conditions such as moisture content, temperature, oxygen levels, enzymatic activities, and ionic strength which otherwise may later affect their structure [78].

Therefore, the intelligent design of the delivery system is a viable option to ensure the effective delivery of dietary lipids (Stage 3 of Figure 1.3). Bioactive lipids are incorporated directly into various oil-based food products, e.g., butter, shortening, spread, and margarine. However, for aqueous-based food products (desserts, beverages, dressings, etc.), proper delivery systems are needed. For example, oil-in-water emulsion could be used for the effective delivery of these bioactive dietary lipids into the aqueous-based food products. Bioactive lipids are used either as such or maybe mixed with carrier oils to be encapsulated into an emulsion with the help of the emulsion technology. The ω-3 FAs, arachidonic acid (AA), CLA, lutein, lycopene, and astaxanthin could be added to various foods, e.g., milk, yogurt, salad dressings, ice cream, etc. [12, 15, 16, 65]. The use of emulsion technology offers the following advantages [18, 55, 93]:

- It imparts oxidative stability to dietary bioactive lipids inside the lipid droplets.
- It increases the absorption and bioavailability of dietary bioactive lipids.
- It enhances the stability of dietary bioactive lipids (e.g., carotenoids) towards a crystallization.
- Structures using emulsion technology are designed in such a way so that digestion could be improved and bioactive components reach the target site of action.

Dietary bioactive lipids (e.g., phytosterols, and antioxidants) could be incorporated into functional food beverages using micro- and nano-emulsions for their effective delivery. Since the wavelength of light is much higher than the dimensions of nanostructures, therefore, these nanostructures cause relatively weak scattering of light, thus, the beverage remains clear [20, 74].

For proper design and selection of delivery systems, the relationship between the properties of particles and the functionality of delivery systems (also called quantitative structure-function relationships, (QSFR), could be used as a guide. QSFR is being used to quantitatively relate the particle properties (such as the composition, thickness charge, permeability) to the functional attributes (e.g., sensory, physicochemical, and biological impact) [64]. The changes being encountered while delivering bioactive dietary lipids into various food products could be conquered by establishing a well-defined link or relationship between the functionality and structure of the delivery system of dietary bioactive lipids.

1.5   FORMULATION OF ANTIOXIDANTS

Using the QSFR principle, delivery systems with physical stability could be designed. However, the long-term chemical stability of unsaturated (double bond containing) lipids (such as PUFAs, carotenoids, and CLA) is still a critical issue. The oxidation of dietary bioactive lipids results not only in the changes in sensory properties, but also may lead to the loss of its bioavailability [59]. There are various ways to combat such oxidation processes and one of the methods is the use of antioxidants [17]. The use of antioxidants in food industries is decreasing since most of the food industries are disinclined to use synthetic antioxidants due to their possible deleterious effects on human health and stern control over their level of incorporation. The aptness to control or prevent an oxidation of bioactive-lipid incorporated food products by synthetic antioxidants is often limited. In contrast, several natural antioxidants (rosemary extract, ascorbic acid, ascorbyl palmitate, and tocopherols) are currently being employed to protect the bioactive lipids from oxidation (stage 4 of Figure 1.3).

Therefore, there is an increasing demand to harness the existing antioxidants smartly within the food products. Regeneration of antioxidants could improve the effectiveness of antioxidants in finished products.

Regeneration of antioxidants is being affected by several factors and is often complicated in food products with multi-component matrices. The activity of an antioxidant molecule is strongly affected by its intrinsic properties (such as stoichiometry of electron transfer, molecular weight, polarity, and free radical scavenging ability). Besides, the partitioning and distribution of antioxidants, and physical location within food matrices or delivery systems also influence the antioxidant regeneration. As a result, the prediction of an antioxidant activity usually varies from the real food system. Free radical scavengers must be localized in such microenvironment, where there is a generation of lipid radicals, to maximize the efficiency. It is thus crucial to develop the simple techniques to easily locate the pairs of antioxidants during their regeneration.

1.6 DESIGNING FOODS WITH ADDED MICRONUTRIENTS

Micronutrients are required in minute quantities, but their absence results in severe consequences. Nowadays, the trend is increasing to incorporate micronutrients into various food products and beverages [76, 101, 112]. The food industry has been practicing addition of micronutrients into various food products since it is a cost-effective measure to alleviate micronutrient malnutrition. The biological effects, molecular characteristics, and physicochemical properties of micronutrients (vitamins and minerals) vary greatly. Iron, zinc, vitamins A and D, and vitamins-B are some of the most common micronutrients added to food products [10].

Due to a variety of physicochemical and biological limitations, some micronutrients cannot be truly integrated into the food products in their pure forms. The water and/or lipid solubility ought to be low, therefore, must be introduced in a unique form. These micronutrients may also be liable to degradation through preparation, processing, storage, or transport and thus are to be protected. Some of the micronutrients possess characteristics and distinguishable off-flavor. Their addition, therefore, can also limit the acceptability of the food products and flavor masking is, therefore, needed [72]. The bioactivity and thus stability of some micronutrients are adversely affected due to their interaction with some other food components. While the oral bioavailability of some micronutrients is inherently low or variable, and therefore their bioavailability must be enhanced by proper designing [53].

The molecular and the physicochemical characteristics and the nature of the food matrix influence the micronutrient degradation. The chemical instability of a micronutrient involves the change in its molecular form and may consequently result in drastic adjustments in the nutritional and physicochemical properties. The chemical degradation of micronutrients is catalyzed by way of enzymes or other activators existing within the food matrix and encompasses reduction, oxidation, isomerization, and hydrolysis [12]. The physical instability includes phase changes, separation, and aggregation. It is important to fully understand the degradation mechanism for a specific micronutrient and to establish the fundamental elements (ionic strength, pH, temperature, oxygen, water activity, and light) accountable for such degradation. As a result, an effective delivery system for micronutrients could be designed to prevent or minimize the degradation [47, 68].

To have the beneficial effects, human body must absorb the active form of a micronutrient after ingestion. Therefore, the delivery systems are to be designed in such a way that it increases the micronutrient fraction that survives in the food. Nanoparticles are being used to encapsulate micronutrients possessing enhanced stability, functionality, and bioavailability [57]. Major advancements have taken place in the design of nanoparticles, which could be utilized to develop coherent and effective delivery system for micronutrients [74, 79, 101, 112].

Extrusion is also one of the technologies for micronutrient fortification and entails the injection of a solution of biopolymer along with encapsulated BFC. Gelation could be carried by way of cross-linking agents, variations in temperature, and/or the extrusion of one of the biopolymers into an oppositely charged biopolymer solution [62, 72]. For example, calcium alginate beads could be formed by administering an alginate solution into a solution of calcium [66]. These biopolymer-based particles have shown excellent protection ability to vitamin D in addition to increasing its bioavailability [47]. Nanoencapsulated of vitamin D in casein is better preserved during long-term cold storage and their level does not change even after thermal processing [47].

Before incorporating them into the food products, selection of a proper encapsulation process is a must. The stability of vitamins and thus their retention in food products depends on light, heat, oxygen, potential interactions, presence of transition metals and packaging [43]. These factors set the criteria for the design and development of microencapsulation systems to ensure the effective delivery of specific micronutrients. Apart from giving due consideration to the cost, scalability of technique, selection of an encapsulant material and regulatory compliance and safety, several other factors (such as stability, bioavailability), sensory characteristics of microencapsulated ingredients should be considered [43, 57].

1.7 STRUCTURING PROTEINS FOR IMPROVING NUTRITIONAL PROFILE OF FOOD

Currently, the food industry is responding to the demands of consumers by reformulating food products, particularly in terms of the salt, sugar, and fat contents. It is also known that the postprandial behavior of any food product not only depends on the relative calorific (energy) content of foods, but also depends on the food structure, breakdown, rheology, and will decide the subsequent digestion process and metabolic response [41, 83, 98, 110].

Among all macronutrients, proteins are most satiating; however, not all proteins are alike. There are few proteins, which are absorbed and digested rapidly, but others may affect metabolism [60]. Protein digestion is a complex method and involves their breakdown into individual amino acids, which might be then absorbed in GIT. The proteolysis begins in the stomach, due to the combined action of low pH and pepsin, chymotrypsin, and carboxy-peptidase. In several cases, the protein system in the solution decides the effectiveness of the action of peptidase thereby exerting the profound influence on the kinetics of the absorption of amino acids [67].

An interest into studying the impact of food structure on behavior of digestion and its association to human nutrition is expanding. The kinetics of lipolysis and proteolysis, two important digestive processes, are controlled by interactions between individual micronutrients. The material properties of food and thus postprandial response are also being influenced by these interactions [67]. Gua et al. [46] reported greater susceptibility of casein to proteolysis than β-lactoglobulin, a whey protein. The open structure of caseins attributes to this susceptibility as enzymes possess greater access to target residues. On the other hand, β-lactoglobulin, a whey protein, has a folded structure and inhibits the access of enzymes to potential cleavage sites. It, therefore, offers considerably greater resistance to proteolysis by either pepsin or trypsin [67].

The native structure of globular protein is resistant to hydrolysis by proteolytic enzymes. Studies have reported that unfolding and denaturation of proteins due to thermal treatment increases the susceptibility of β-lactoglobulin to proteolysis by trypsin [84]. In addition, some food products are treated by high-pressure processing (HPP), which also leads to unfolding of proteins and exposure of active sites to proteolytic enzymes, thereby increasing the rate of proteolysis [54].

Use of proteins (milk or vegetable proteins) is increasing especially for food products meant for weight control. However, the incorporation of proteins into food products results in an increase in viscosity. To limit

this negative impact, distinctive shapes, for example, spherical, microparticles, microgels, and factual gel aggregates, could be assumed by proteins,
contingent on the presence of ions, temperature, and pH [103]. For instance,
blends of soluble aggregates and whey-protein microgels might manipulate
the structure and rheology of food products [32].

1.8   DESIGNING LOW ENERGY DENSITY FOODS

The reduction of the volumetric energy density of foods to reduce obesity can
be acquired either by increasing their water and/or air contents or by decreasing
the fat and/or sugar contents, or replacing the carbohydrates and triglycerides
by water, air, and proteins [90]. Air bubbles or few different gases may scatter
as air pockets in several solid-, semi-liquid, or liquid foods. For example,
gelatin gels are aerated using ultrasound to offer an additional phase within
gel-type food products thereby accommodating novel functional and textural
needs [60, 120]. To mimic the rheological properties of their counterparts, the
structure of foams and emulsions ought to be the end goal that the rheological
characteristics, responsible for release characteristics of flavors and mouth-
feel, must match the final food product [60, 90].

1. **Energy Reduction by Increasing Water Content:** By increasing
    water content, the energy value of food products can be decreased.
    By way of gel or emulsion structuring, this increased water is bound
    to a food matrix in the manufacturing of fat-reduced food products
    (e.g., dressings, or mayonnaise, double emulsions). The incorpora-
    tion of biopolymers, along with polysaccharides and proteins into
    inner and outside aqueous stages of double emulsions to enhance
    yield and stability of model systems, has been successfully achieved
    [29]. These disperse systems should be effectively stabilized by the
    addition of surface-active molecules [86].

2. **Energy Density Reduction by Reducing Sugar Content:** The
    energy density of a food product is brought down by bringing down its
    sugar content and their perceived sweetness is supplanted by natural
    or non-caloric sweeteners. The sweetness of steviol glycoside extracts,
    for example, is approximately 350 times the sweetness of sugar and
    is being increasingly used as a healthy and natural sugar alternative.
    However, the replacement of sugars by non-caloric sweeteners may
    change physicochemical properties (volume and matrix structure) of
    such food products. This loss in volume or matrix structure could be

compensated by using a combination of bulking agents. However, it involves knowing the contribution of each bulking agent in building up the structure and subsequent influence on several other sensory characteristics [60]. Hence, a blend of bulking agents might be utilized to make amends for the loss in network structure or volume.

3. **Energy Density Reduction by Reducing Fat Content:** Energy density of a food product could be diminished by bringing down or controlling of their fat. An example of energy density reduced food product is frozen low-fat ice-cream. The texture of such an ice cream is viewed to be creamier than the ice cream produced conventionally (Figure 1.4). This creamier texture is due to a notable change in the recipe i.e., diminishing fat, balanced adjustment, and an additional processing step (twin-screw extrusion). The shear forces in the twin extrusion system are 3 times higher than in conventional scraped heat exchangers, which provide small-sized air bubbles and ice crystals and increase the functionality of destabilized fat droplets [34]. The type of emulsifier determines the increase in fat destabilization. This increase in the fat destabilization thus advances fat organizing and hence consequently improves meltdown behavior and overall stability [11]. The adjustments in the structure can impact the nature of food products, which can be surveyed by estimating their rheological properties. For example, studies on the evaluation of sensory properties have demonstrated the high correlation between loss moduli (G'') and other typical quality parameters (e.g., creaminess and scoop ability) [90, 115].

A typical example of a good emulsion is Mayonnaise (O/W), portrayed by high energy density value containing 80–95% oil. It is prepared with the aid of an emulsification of oil into a consistent aqueous phase. The consistency of such an O/W emulsion relies on the interplay between droplets of oil. The fat content of such emulsions may be decreased by lowering the amount of oil droplets, which in turn results in the reduction of viscosity (which is undesirable). However, the desirable viscosity can be achieved by the addition of starch solution. Starch solution is made by suspending starch in water followed by heating to allow its swelling and gelling followed by blending with an emulsion to obtain a viscous low-fat emulsion. The rheological properties of such a solution are like nearer to that of full-fat food products [90].

Microparticulated proteins (MPP) and hydrocolloid structures that mimic full-fat food products can also be used to provide fat-reduced textures [60].

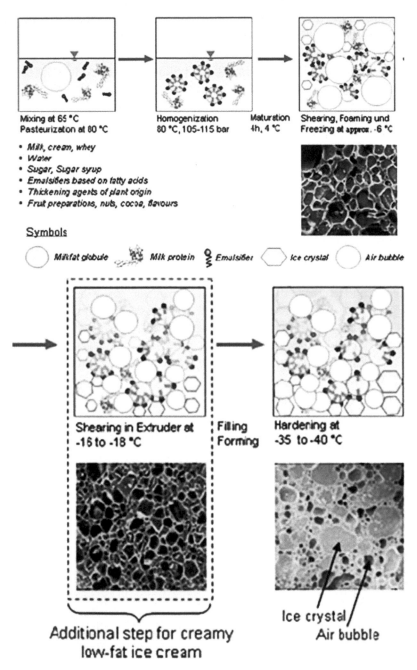

FIGURE 1.4   Creation of microstructure for fat-reduced ice cream.

Source: Reprinted with permission from Ref. [90]. © 2009 Elsevier.

Another example of the low fat product is low-fat spread that consists primarily of water; and therefore, to get an acceptable viscosity, the structuring of the water phase is to be done through liquid phases [1]. In addition, there are various commercially available fat replacers for developing low-fat products. These fat replacers provide similar textural and rheological properties as are being provided by fats and oils and also consist of lower calories than the fats/oil itself.

Reducing fat in tablet structures, powders, or solid bars is a challenging task in such products. Fat not only provides sensorial properties but also keeps the particles together. It is, therefore, obvious that replacing fat in such products should be accompanied by using an alternative binder system. Hydrophilic amorphous food components could be utilized to cohere generally incoherent food components together. As low-fat seasoning tables are compressed, the amorphous particles, which have been added externally, are deformed; and therefore, particle sintering could be observed at the contact points (Figure 1.5), resulting in the production of stable tablet structure [90].

FIGURE 1.5   Electron microscopic images of low-fat seasoning tablet with added amorphous binder.

Source: Reprinted with permission from Ref. [90]. © 2009 Elsevier.

1.9 DESIGNING LOW SALT FOODS

Food industries are developing and designing food products with reduced salt content [8, 60]. Besides its natural presence in many food products, sodium is also present in processed and prepared foods or other sodium-containing ingredients, e.g., sodium bicarbonate (leavening agent). Consumers want sodium content to be reduced in food and the same time to be tasty. It is of utmost importance for the food technologists to gain an understanding of the basic taste of salty. There is no other salt, which is capable mimicking the saltiness of sodium chloride by 100%. Thus, it is quite arduous to find an acceptable substitute that brings about the same taste characteristics. As different colors are mixed, a new shade is formed; similarly, if different substances are mixed, a new taste is created.

Salt also possesses an enhancing effect on other food flavors. There are two major types of taste receptors associated with taste buds, which include G-protein receptors and ion channels. G-protein coupled receptors help in sensing the sweet, umami, and bitter taste; while salty and sour taste is sensed via ion-channels [8]. Ion channels for sodium are very specific and further knowledge about ion channels is needed. There are several other attributes associated with saltiness, such as mouth feel, body, and enhanced flavor, and there exists a second mechanism to account for it. However, this mechanism is not fully clear, thus there is a need to fully understand this mechanism so that effective salt replacers could be developed.

Reducing or replacing sodium chloride poses an extra challenge because salt contributes more to the flavor than it contributes to salty taste. Some of the studies have indicated that salt helps in the rapid release of some flavoring compounds while as it selectively filters some unpleasant tastes, such as the taste of metallic ions and bitterness. Salt also increases the volatility of some compounds, and, therefore, they get released easily. Salt is an inherent flavor on our tongue, therefore, its addition to food brings out this flavor.

Sodium affects the mouth-feel and viscosity of a food product. Viscosity of a food product assumes a key part with decrease of sodium content in different food products, such as sauces, soups, or dressings. It is, therefore, of paramount importance to select the right kind of emulsifier system or hydrocolloid to ensure the effective and proper release of salty taste. Flavors are not, however, properly released in an over-stabilized food product and more sodium is, therefore, required to achieve the right taste. On the other hand, under-stabilized food product may have less viscosity thus may not remain inside the oral cavity long enough for the tongue to detect or process the flavors.

Potassium chloride (KCl) comes closer to mimic the saltiness of sodium chloride (NaCl); however, its saltiness is perceived more slowly and often has a hint of bitterness. Therefore, taste modifiers are needed when KCl is used to simulate the saltiness of NaCl [8]. The most common approach to lower the sodium content is the replacement of NaCl by KCl by certain proportion only. For several other applications, other ingredients (herbs, flavors, and minerals) may be considered when replacing NaCl by KCl. Some alternative cooking processes may also be considered, such as grilling, and caramelizing.

Some formulators advise the use of specialty sea salts as these contain potassium (K) naturally. Other minerals present in sea salt include magnesium and sulfur, which may enhance the flavors of food products, the amount of which that may be used to replace the refined salt and is unique to each application. This concept of the replacement of sodium with sea salt is because the higher mineral content in sea salt adds to the flavor, as a result of which less salt is used. A 50% reduction in added salts could be achieved by carefully blending the sea or refined salt with KCl and various other flavors.

1.10 STRUCTURING FIBERS FOR IMPROVING NUTRITIONAL PROFILE OF FOOD PRODUCTS

Dietary fibers are partially digested inside the small intestine and protect against several chronic diseases and exhibit prebiotic effects [69, 90, 96]. Whole grains, vegetables, and fruits are natural and most common sources of dietary fiber [60]. Therefore, by increasing an overall content of whole grains or dietary fibers of different food products, their nutritional profile could be improved. However, specific methods are often required for food structuring for inclusion of whole grains or dietary fibers. Process conditions also need to be adjusted during food production. It is quite arduous to incorporate the fibers derived from whole grains into food products without compromising its taste and texture [60, 94]. The incorporation of fiber into food products tends to increase the viscosity [90]. The addition of wheat bran to extruded cereal products exert a negative effect on physicochemical properties of starch, such as change in glass transition temperature, sorption isotherm, and melting temperature which results in reduction of expansion properties [99].

The natural food structure of raw ingredients is often less stable, both physically as well chemically. This could be improved by way of protecting layers and structuring [107] and adding natural stabilizing components to the recipes. Further, through physical processing, supramolecular structure of an individual component could be changed [60]. It is, therefore, important to increase the fiber content of food products; however, this tends to increase the viscosity [90] affecting its taste and texture [60, 94]. The incorporation of insoluble fibers (e.g., cellulose) and whole grains to raw dough or batter results in a denser structure [90]. These changes in the physicochemical properties result in the change of expansion properties of extruded cereal products.

Robin et al. [99] reported similar results after adding whole grain flours obtained from different sources (such as tritordeum, rye, barley, wheat, triticale) to the cake batter. A significant correlation between firmness and symmetry and water absorption and specific volume was observed. Thus, structuring of dietary fibers and whole grains should be carried out in such a way to minimize its negative effect on the texture of food products. To develop the desired texture, modification of the structure of cell wall and fibers is done through the enzymatic process [96]. Insoluble fibers can be made soluble by enzymatic hydrolysis, which in turn could be used in structuring the aqueous phase of low-fat products. Alternatively, mechanical treatment could be employed to modify the properties of the cell wall and fibers. These fibers display a higher viscosity or loss modulus compared to native ones [95].

1.11 DESIGNING FOOD STRUCTURE FOR IMPROVING PLEASURE

An ever-increasing demand exists for high quality and cost-effective food products possessing improved health benefits within the emulsion-based food formulations. Since the structures of food emulsions play a vital role in determining their functional attributes, it is, therefore, essential to understand the microstructure-function relationship of food emulsions [88].

1.11.1 MICROSTRUCTURAL APPROACH TO DESIGN EMULSIONS

Microstructural approach can be used to manipulate the microstructure of food emulsions. Emulsions are dispersions of two immiscible phases, typically oil, and water and can be either water-in-oil (W/O) or oil-in-water (O/W) type. The use of various ingredients in the formulation of food emulsions assumes an essential role in determining their physicochemical properties and subsequent resultant functions. Such ingredients in use are water, edible oils, emulsifiers, and other additional ingredients as stabilizers, texture enhancing agents, or flavors [88].

The demand for proteins, in comparison to the conventionally used low molecular weight surfactant type emulsifier (typically fat-derived), as emulsifying agents is increasing in the food industry. This is due to the ability of this emulsifier to stabilize the oil-water interface. In addition to this, demand for proteins is increasing for being of natural origin and providing nutritional value to the formulation [88].

In addition to these ingredients, energy is also utilized for the fabrication of emulsions. This energy is provided by the disruption of large volumes of oil into discrete emulsion droplets thereby yielding a homogenous product. Typically, the droplet size of emulsions is <1 μm, which is resistant to gravitational separation from creaming. The emulsions possessing such droplet size are considered stable. Therefore, sufficient energy with appropriate formulation is necessary to produce stable emulsion droplets [75]. The rationale for the creation of such emulsions is microstructural approach to food designing. The desired microstructures of food emulsions aim to possess certain functions, such as long-term stability and ideal sensory attributes for given applications. This ultimately results in better consumer acceptance.

Fundamentally, the visual properties (opacity and color) within an emulsion system are determined by the droplet size of an emulsion, concentration of droplets within an emulsion and refractive index differential between oil and water phases. Emulsions possessing smaller droplet sizes and/or higher oil contents have lighter optical characteristics, manifesting whiter color. Unlike conventional macroemulsions, nano-sized emulsion droplets do not scatter light and can appear optically transparent. This provides a potential for product novelty while allowing for the discrete incorporation of lipophilic components, such as oil-soluble vitamins, within specific applications [44].

Based on the appearance of a product and its behavior when manipulated such as with cutlery or being poured, consumers will generate an opinion on the food's sensory properties, based on previous consumption experiences of the same or similar foods. For example, consumers have been shown to visually assume liquids that pour slower are thicker, and products with shiny, even yellowish coloring to have a creamier profile. Evidently, visual behavioral characteristics are result of the system's physical properties, such as viscosity and adhesion, which can be varied by the system's microstructure. Physical properties are primarily responsible for the product's texture and mouth-feel [44]. The presence of ingredients such as biopolymers, particularly those having thickening or gelling properties especially high molecular weight polysaccharides within an aqueous continuous phase will ultimately affect the overall physical properties of emulsions [88].

The dispersed phase also has a role in deciding the perception of attributes of emulsion-based food products. Higher concentrations of oil droplets have been shown to increase the perception of attributes, such as 'creaminess,' 'richness,' 'thickness' and 'fattiness,' a familiar example of this is the comparison in sensory properties between skimmed and full-fat milk. Smaller oil droplet sizes have shown to increase the perception of sensory attributes including 'creaminess,' 'thickness,' 'smoothness' and 'slipperiness.' These smaller oil droplets increase the viscosity of an emulsion system and improve its lubrication capacity. Furthermore, sensory perception is significantly affected by colloidal interactions between droplets. Depletion, flocculation, interactions between emulsion droplets, and the aggregation of emulsion droplets have improved perception of attributes, such as 'thickness' and 'fattiness' through increasing the system's viscosity. During oral processing, the coalescence of droplets has been related to an increase in the perception of fat-released attributes, such as fattiness, and creamy mouthfeel through increasing lipophilic flavor release [88].

Another sensory characteristic is flavor, the intensity of which depends on the distribution of flavor molecules among emulsion phases and their release profile during oral processing. Structurally, increasing the oil droplet concentration increases polar flavors, while conversely decreases non-polar flavors. In addition, the size of emulsion droplets influences flavor-release kinetics, whereby larger droplets produce a more delayed and sustained release of non-polar flavor molecules, ascribed to greater diffusion path length to reach the aqueous phase than the tongue. A delayed diffusion of flavor molecules to the tongue's taste receptors is also observed if the continuous phase contains biopolymers, which promote thickening or gelling due to altered partitioning and mass transport of the flavor molecules. The larger surface area associated with nano-sized emulsion droplets allows for a more rapid and intense release of flavor compared with food systems possessing larger emulsion droplets [44].

Understanding of structure-sensory relationships and designing microstructures for the desired sensory profile is further complicated by the fact that the perception of sensory attributes may change throughout oral processing as structure changes due to the mechanical action of the teeth, tongue, and palate and the chemical mixing and dilution of the food by saliva. Research within this area is attempting to understand this complex phenomenon via further understanding of the mechanical and chemical breakdown of food and the application of novel time-dependent sensory techniques such as temporal dominance of sensations [21].

1.12 SUMMARY

This chapter suggests that the food microstructure modulates various physiological functions inside our body and also plays detrimental or beneficial roles in some diet-related diseases. It is also concluded that the microstructure affects the nutritional composition as well. Bridging food structuring and the physiology of the gastrointestinal tract (GIT), along with the development of the methods for non-obstructive estimation/measurement and interdisciplinary research will offer the basis for the design and development of novel and future food products with a tailor-made useful behavior within the human frame. Such food products may have the potential to overcome the problem of weight loss. While developing food products for specific health, wellness, and nutritional needs, it is important to adapt the food structure. From the harvesting of raw material to its consumption stage, food structure gets changed. Hence, an exhaustive understanding of these time-dependent changes is needed.

KEYWORDS

- **bioactive components**
- **bioavailability**
- **dietary bioactive lipids**
- **excipient foods**
- **functional foods**
- **medical foods**

REFERENCES

1. Aguiler, J., & Stanley, D., (1999). *Micro Structural Principles of Food Processing and Engineering* (p. 432). Gaithersburg, USA: Aspen Publishers.

2. Aguilera, J., (2005). Why food microstructure? *Journal of Food Engineering*, *67*(1–2), 3–11.
3. Amin, T., Mahapatra, H., Bhat, S. V., & Gulleria, S. P., (2012). Applications of nutrigenomics in food industry. *Indian Horticulture Journal*, *2*(3–4), 54–59.
4. Amin, T., Thakur, M., & Jain, S. C., (2013). Microencapsulation: The future of probiotics. *Journal of Microbiology, Biotechnology and Food Sciences*, *3*(1), 35–43.
5. Arnott, J. A., & Planey, S. L., (2012). The influence of lipophilicity in drug discovery and design. *Expert Opinion on Drug Delivery*, *7*, 863–875.
6. Aust, O., Agha, A. N., Zhang, L., Wollersen, H., Sies, H., & Stahl, W., (2003). Lycopene oxidation product enhances gap junctional communication. *Food and Chemical Toxicology*, *41*, 1399–1407.
7. Bauer, E., Jakob, S., & Mosenthin, R., (2005). Principles of physiology of lipid digestion. *Asian-Australasian Journal of Animal Science*, *2*, 282–295.
8. Berry, D., (2011). Flavorful sodium-reduced foods. *Food Products Design*, *21*(8), 1–7.
9. Bharat, S. S., & Bajaj, A. N., (2010). Interactions and incompatibilities of pharmaceutical excipients with active pharmaceutical ingredients: A comprehensive review. *Journal of Excipients and Food Chemistry*, *1*(3), 1–26.
10. Black, R. E., (2003). Zinc deficiency, infectious disease, and mortality in the developing world. *Journal of Nutrition*, *133*(5), 1485S–1489S.
11. Bolliger, S., Kornbrust, B., Goff, H., Tharp, B., & Windhab, E., (2000). Influence of emulsifiers on ice cream produced by conventional freezing and low-temperature extrusion processing. *International Dairy Journal*, *10*(7), 497–504.
12. Boon, C., Xu, Z., Yue, X., & McClements, D. J., (2008). Factors affecting lycopene oxidation in oil-in-water emulsions. *Journal of the Agricultural and Food Chemistry*, *56*, 1408–1414.
13. Brown, M. J., Ferruzzi, M. G., & Nguyen, D. A., (2004). Carotenoid bioavailability is higher from salads ingested with full-fat than with fat-reduced salad dressings as measured with electrochemical detection. *American Journal of Clinical Nutrition*, *80*, 396–403.
14. Burton, G., Joyce, A., & Ingold, K., (1983). Is vitamin E the only lipid-soluble, chain-breaking antioxidant in human blood plasma and erythrocyte membranes? *Archives of Biochemistry and Biophysics*, *221*, 281–290.
15. Cercaci, L., Estrada, R. M., Lercker, G., & Decker, E., (2007). Phytosterol oxidation in oil-in-water emulsions and bulk oil. *Food Chemistry*, *102*, 161–167.
16. Chee, C., Gallaher, J., Djordjevic, D., Faraji, H., & McClements, D. J., (2005). Chemical and sensory analysis of strawberry-flavored yogurt supplemented with an algae oil emulsion. *Journal of Dairy Research*, *72*, 311–316.
17. Chen, B., Han, A., Laguerre, M., McClements, D. J., & Decker, E., (2011). Role of reverse micelles on lipid oxidation in bulk oils: Impact of phospholipids on antioxidant activity of α-tocopherol and trolox. *Food and Function*, *2*, 302–309.
18. Chen, B., McClements, D. J., & Decker, E., (2010). Role of continuous phase anionic polysaccharides on the oxidative stability of menhaden oil-in-water emulsions. *Journal of Agricultural and Food Chemistry*, *58*, 3779–3784.
19. Chen, B., McClements, D. J., & Decker, E., (2013). Design of foods with bioactive lipids for improved health. *Annual Review of Food Science and Technology*, *4*, 35–56.
20. Chen, C., & Wagner, G., (2004). Vitamin-E nanoparticle for beverage applications. *Chemical Engineering and Research Design*, *82*, 1432–1437.

21. Chen, J., (2009). Food oral processing: A review. *Food Hydrocolloids*, *23*, 1–25.
22. Chen, L., Remondetto, G., & Subrirade, M., (2006). Food protein-based materials as nutraceutical delivery systems. *Trends in Food Science and Technology*, *17*, 272–283.
23. Clinton, S., (1998). Lycopene: Chemistry, biology, and implications for human health and disease. *Nutrition Reviews*, *56*, 35–51.
24. Colmenero, J. F., (2013). Potential applications of multiple emulsions in the development of healthy and functional foods. *Food Research International*, *52*, 64–74.
25. Crowe, K. M., (2013). Designing functional foods with bioactive polyphenols: Highlighting lessons learned from original plant matrices. *Journal of Human Nutrition and Food Science*, *1*(3), 10–18.
26. Crowe, K., & Francis, C., (2013). Position of the academy of nutrition and dietetics: Functional foods. *Journal of the Academy of Nutrition and Dietetics*, *113*, 1096–1103.
27. Davis, E., & Gordon, J., (1982). Food microstructure, an integrative approach. *Food Microstructure*, *1*, 11–12.
28. Decker, E., Ryan, J., & McClements, D. J., (2010). Oxidation in foods and beverages and antioxidant applications: Management in different industry sectors. In: *Understanding Mechanisms of Oxidation and Antioxidant Activity* (Vol. 2, p. 552). Cambridge, UK: Woodhead Publishers.
29. Dickinson, E., (2011). Double emulsions stabilized by food biopolymers. *Food Biophysics*, *6*(1), 1–11.
30. Dickinson, E., (2013). Stabilizing emulsion-based colloidal structures with mixed food ingredients. *Journal of the Science of Food and Agriculture*, *93*, 710–721.
31. Dickinson, E., (2012). Use of nanoparticles and micro particles in the formation and stabilization of food emulsions. *Trends in Food Science and Technology*, *24*, 4–12.
32. Donato, L., Kolodziejcyk, E., & Rouvet, M., (2011). Mixtures of whey protein micro gels and soluble aggregates as building blocks to control rheology and structure of acid induced cold-set gels. *Food Hydrocolloids*, *25*(4), 734–742.
33. Douaire, M., & Norton, I., (2013). Designer colloids structured food for the future. *Journal of the Science of Food and Agriculture*, *93*, 3147–3154.
34. Eisner, M., Wildmoser, H., & Windhab, E., (2005). Air cell micro structuring in a high viscous ice cream matrix. *Colloids and Surfaces: A Physicochemical and Engineering Aspects*, *263*(1–3), 390–399.
35. Etherton, K. P., Taylor, D., & Poth, Y. S., (2000). Polyunsaturated fatty acids in the food chain in the United States. *American Journal of Clinical Nutrition*, *71*, S179–S188.
36. Evans, M., Ratcliffe, I., & Williams, P., (2013). Emulsion stabilization using polysaccharide-protein complexes. *Current Opinion in Colloid and Interface Science*, *18*, 272–282.
37. Ezhilarasi, P., Karthik, P., Chhanwal, N., & Anandharamakrishnan, C., (2013). Nano encapsulation techniques for food bioactive components: A review. *Food Bio-Processing and Technology*, *6*, 628–647.
38. Fajardo, A., Akoh, C., & Lai, O., (2003). Lipase-catalyzed incorporation of n-3 PUFA into palm oil. *Journal of the American Oil Chemists' Society*, *80*, 1197–2000.
39. Faulks, R., & Southon, S., (2008). Assessing the bioavailability of nutraceuticals. In: Ottaway, P. B., (ed.), *Food Fortification and Supplementation* (p. 1–10). Cambridge, UK: Woodhead Publishing.
40. Fave, G., Coste, T. C., & Armand, M., (2004). Physicochemical properties of lipids: New strategies to manage fatty acid bioavailability. *Cellular and Molecular Biology*, *50*, 815–831.

41. Feinle, C., Christen, M., & Grundyetal, D., (2002). Effects of duodenal fat, protein or mixed-nutrient infusions on epigastric sensations during sustained gastric distension in healthy humans. *Neuro Gastroenterol. Motility*, *14*, 205–213.

42. Florence, A. T., & Attwood, D., (2011). *Physicochemical Principles of Pharmacy* (p. 664). London, UK: Pharmaceutical Press.

43. Foley, P. J., (2012). *Development of Reactive Polyanions for Encapsulation of Live Yeast Cells Within Polymer Hydrogel Films* (p. 213). PhD thesis, Hamilton, ON: McMaster University.

44. Frøst, M., & Janhøj, T., (2007). Understanding creaminess. *International Dairy Journal*, *17*, 1298–1311.

45. Grandois, J. L., Marchioni, E., Zhao, M., Giuffrida, F., Ennahar, S., & Bindler, F., (2010). Oxidative stability at high temperatures of oleyol and linoleoyl residues in the forms of phosphatidylcholines and triacylglycerols. *Journal of Agricultural and Food Chemistry*, *58*, 2973–2979.

46. Guo, M., Fox, P., Flynn, A., & Kindstedt, P., (1995). Susceptibility of beta-lactoglobulin and sodium caseinate to proteolysis by pepsin and trypsin. *Journal of Dairy Science*, *78*, 2336–2344.

47. Haham, M., Shalom, I. S., & Nodelman, M., (2012). Stability and bioavailability of vitamin D nanoencapsulated in casein micelles. *Food and Function*, *3*, 737–744.

48. Hamman, J., & Steenekamp, J., (2012). Excipients with specialized functions for effective drug delivery. *Expert Opinion on Drug Delivery*, *9*, 219–230.

49. Haraldsson, G., & Thorarensen, A., (1999). Preparation of phospholipids highly enriched with n-3 polyunsaturated fatty acids by lipase. *Journal of the American Oil Chemists' Society*, *76*, 1143–1149.

50. Henna, F. L., Nielsen, N., Heinrich, T. M., & Jacobsen, C., (2011). Oxidative stability of marine phospholipids in the liposomal form and their applications. *Lipids*, *46*, 3–23.

51. Honest, K., Zhang, H., & Zhang, L., (2011). Lycopene: Isomerization effects on bioavailability and bioactivity properties. *Food Reviews International*, *27*, 248–258.

52. Hossen, M., & Hernandez, E., (2005). Enzyme-catalyzed synthesis of structured phospholipids with conjugated linoleic acid. *European Journal of Lipid Science and Technology*, *107*, 730–736.

53. Hu, B., Ting, Y., Yang, X., Tang, W., Zeng, X., & Huang, Q., (2012). Nanochemoprevention by encapsulation of (-)-epigallocatechin-3-gallate with bioactive peptides/chitosan nanoparticles for enhancement of its bioavailability. *Chemical Communications (Camb)*, *48*(18), 2421–2433.

54. IamettiE, S., Donnizzelli, P., Pittia, P., & Rovere, N., (1999). Characterization of high-pressure-treated egg albumen. *Journal of Agricultural and Food Chemistry*, *47*, 3611–3616.

55. Iveta, G., Irina, G., Sue, P., James, T., & Duolao, W., (2007). A randomized cross-over trial in healthy adults indicating improved absorption of omega-3 fatty acids by pre-emulsification. *Journal of Nutrition*, *6*, 1–9.

56. Jones, P. J., & Jew, S., (2007). Functional food development: Concept to reality. *Trends in Food Science and Technology, 18*(7), 387–390.

57. Joye, I. J., & McClements, D. J., (2013). Production of nanoparticles by anti-solvent precipitation for use in food systems. *Trends in Food Science and Technology*, *34*, 109–123.

58. Kang, M., Dai, J., & Kim, J., (2012). Ethylcellulose microparticles containing chitosan and gelatin: pH dependent release caused by complex coacervation. *Journal of Industrial and Engineering Chemistry, 18*, 355–359.

59. Kanner, J., (2007). Dietary advanced lipid oxidation end products are risk factors to human health. *Molecular Nutrition and Food Research, 51*, 1094–1101.

60. Kaufmann, S. F., & Palzer, S., (2011). Food structure engineering for nutrition, health, and wellness. *Procedia. Food Science, 1*, 1479–1486.

61. Kim, I., Garcia, H., & Hill, C. J., (2007). Phospholipase A1-catalyzed synthesis of phospholipids enriched in n-3 polyunsaturated fatty acid residues. *Enzyme and Microbial Technology, 40*, 1130–1135.

62. Krasaekoopt, W., Bhandari, B., & Deeth, H., (2003). Evaluation of encapsulation techniques of probiotics for yoghurt. *International Dairy Journal, 13*, 3–13.

63. Lattanzia, V., Kroon, P., Linsalata, V., & Cardinali, A., (2009). Globe artichoke: A functional food and source of nutraceutical ingredients. *Journal of Functional Foods, 1*(2), 131–144.

64. Lesmes, U., & McClements, D. J., (2009). Structure-function relationships to guide rational design and fabrication of particulate delivery systems. *Trends in Food Science and Technology, 20*, 448–457.

65. Let, M. B., Jacobsen, C., & Meyer, A. S., (2007). Lipid oxidation in milk, yoghurt, and salad dressing enriched with neat fish oil or pre-emulsified fish oil. *Journal of Agricultural and Food Chemistry, 55*, 7802–7809.

66. Li, Y., Hu, M., Du, Y. M. H., Xiao, & McClements, D. J., (2011). Control of lipase digestibility of emulsified lipids by encapsulation within calcium alginate beads. *Food Hydrocolloids, 25*, 122–130.

67. Lundin, L., Golding, M., & Wooster, T. J., (2008). Understanding food structure and function in developing food for appetite control. *Nutrition Dietetics, 65*(3), S79–S85.

68. Madene, A., Jacquot, M., Scher, J., & Desobry, S., (2006). Flavor encapsulation and controlled release: A review. *International Journal of Food Science and Technology, 41*, 1–21.

69. Malkki, Y., (2004). Trends in dietary fiber research and development: A review. *Acta Alimentaria, 33*, 39–62.

70. Malone, M., Appelqvist, I., & Norton, I., (2003). Oral behavior of food hydrocolloids and emulsions: Part 1: Lubrication and deposition considerations, Part 2: Taste and aroma release. *Food Hydrocolloids, 17*, 763–784.

71. Marty, C., & Berset, C., (1990). Factors affecting the thermal degradation of all-trans β-carotene. *Journal of the Agricultural and Food Chemistry, 38*, 1063–1067.

72. Matalanis, A., Jones, O. G., & MClements, D. J., (2011). Structured biopolymer-based delivery systems for encapsulation, protection, and release of lipophilic compounds. *Food Hydrocolloids, 25*, 1865–1880.

73. Maurelli, S., Blasi, F., Cossignani, L., Bosi, A., Simonetti, M., & Damiani, P., (2009). Enzymatic synthesis of triacylglycerols with CLA isomers starting from sn-1,3-diacylglycerols. *Journal of the American Oil Chemists' Society, 86*, 127–133.

74. McClements, D. J., (2010). Edible nanoemulsions: Fabrication, properties, and functional performance. *Soft Matter, 6*, 2297–2316.

75. McClements, D. J., (2005). *Food Emulsions: Principles, Practices and Techniques* (p. 690). Boca Raton, FL: CRC Press.

76. McClements, D. J., (2014). *Nanoparticle- and Micro Particle-Based Delivery Systems: Encapsulation, Protection and Release of Active Components* (p. 572). Boca Raton, FL: CRC Press.

77. McClements, D. J., (2013). Utilizing food effects to overcome challenges in delivery of lipophilic bioactives: Structural design of medical and functional foods. *Expert Opinion on Drug Delivery*, *10*, 1621–1632.

78. McClements, D. J., Decker, E. A., & Park, Y., (2008). Controlling lipid bioavailability through physicochemical and structural approaches. *Critical Reviews in Food Science and Nutrition*, *49*(1), 48–67.

79. McClements, D. J., Decker, E. A., & Park, Y., (2009). Structural design principles for delivery of bioactive components in nutraceuticals and functional foods. *Critical Reviews in Food Science and Nutrition*, *49*(6), 577–606.

80. McClements, D. J., & Li, Y., (2010). Structured emulsion-based delivery systems: Controlling the digestion and release of lipophilic food components. *Advances in Colloid and Interface Science*, *159*, 213–228.

81. McClements, D. J., & Xiao, H., (2014). Excipient foods: Designing food matrices that improve the oral bioavailability of pharmaceuticals and nutraceuticals. *Food and Function*, *5*, 1320–1333.

82. Mezzenga, R., & Fischer, P., (2013). The self-assembly, aggregation and phase transitions of food protein systems in one, two, and three dimensions. *Reports on Progress in Physics*, *76*, 8, e-article ID 046601.

83. Mourao, D., Bressan, J., Campbell, W., & Mattes, R., (2007). Effects of food foam on appetite and energy intake in lean and obese young adults. *International Journal of Obesity and Related Metabolic Disorders*, *31*, 1688–1695.

84. Mullally, M., Mehra, R., & FitzGerald, R., (1998). Thermal effects on the conformation and susceptibility of beta-lactoglobulin to hydrolysis by gastric and pancreatic endoproteinases. *Irish Journal of Agricultural and Food Research*, *37*, 51–60.

85. Murakami, R., & Takashima, R., (2003). Mechanical properties of the capsules of chitosan-soy globulin polyelectrolyte complex. *Food Hydrocolloids*, *17*, 885–888.

86. Murray, B., Durga, K., Yusoff, A., & Stoyanov, S., (2011). Stabilization of foams and emulsions by mixtures of surface active food grade particles and proteins. *Food Hydrocolloids*, *25*(4), 627–638.

87. Nicolai, T., & Durand, D., (2013). Controlled food protein aggregation for new functionality. *Current Opinion in Colloid and Interface Science*, *18*, 249–256.

88. O'Sullivan, J., & Park, M., (*2016*). Applications of ultrasound for the functional modification of proteins and nano-emulsion formation: A review. *Food Hydrocolloids, e-article, p. 49.* doi: 10.1016/j.foodhyd.2016.12.037.

89. Ottino, J., (2005). New tools, new outlooks, new opportunities. *AIChE Journal*, *51*(7), 1839–1845.

90. Palzer, S., (2009). Food Structures for nutrition, health and wellness. *Trends in Food Science and Technology*, *20*(5), 194–200.

91. Parker, R., (1997). Bioavailability of carotenoids. *European Journal of Clinical Nutrition*, *51*, 86–90.

92. Patel, A., & Velikov, K., (2011). Colloidal delivery systems in foods: A general comparison with oral drug delivery. *LWT Food Science and Technology*, *44*, 1958–1964.

93. Raatz, S., Redmon, J., Wimmergren, N., Donadio, J., & Bibus, D., (2009). Enhanced absorption of n-3 fatty acids from emulsified compared with encapsulated fish oil. *Journal of the American Dietetic Association, 109*, 1076–1081.

94. Redgwell, R., (2010). Dietary fiber in food fabrication: A changing landscape for consumer and industry. *Food Science and Technology, 24*(4), 18–20.

95. Redgwell, R., Cutri, D., & Delval, G. C., (2008). Physicochemical properties of cell wall materials from apple, kiwifruit, and tomato. *European Food Research Technology, 227*, 607–608.

96. Redgwell, R., & Fischer, M., (2005). Dietary fiber as a versatile food component: An industrial perspective. *Molecular Nutrition and Food Research, 49*, 412–535.

97. Rein, M. J., Renouf, C. C., & Hernandez, A. L., (2013). Bioavailability of bioactive food compounds: A challenging journey to bio-efficacy. *British Journal of Clinical Pharmacology, 75*, 588–602.

98. Robertson, M., (2006). Food perception and postprandial lipid metabolism. *Physiology and Behavior, 89*, 4–9.

99. Robin, F., Théoduloz, C., Gianfrancesco, A., Pineau, N., Schuchmann, H., & Palzer, S., (2011). Starch transformation in bran-enriched extruded wheat flour. *Carbohydrate Polymers, 85*(1), 65–74.

100. Sabally, K., Karboune, S., St-Louis, R., & Kermasha, S., (2006). Lipase-catalyzed transesterification of dihydrocaffeic acid with flaxseed oil for the synthesis of phenolic lipids. *Journal of Biotechnology, 127*, 167–176.

101. Sagalowicz, L., & Leser, M., (2010). Delivery systems for liquid food products. *Current Opinion in Colloid and Interface Science, 15*, 61–72.

102. Salminen, S., Bouley, C., & Ruault, B. M., C., (1998). Functional food science and gastrointestinal physiology and function. *British Journal of Nutrition, 80*(1), 147–171.

103. Schmitt, C., Bovay, C., Rouvet, M., Rami, S. S., & Kolodziejczyk, E., (2007). Whey protein soluble aggregates from heating with NaCl: Physicochemical, interfacial, and foaming properties. *Langmuir, 23*(8), 4155–4166.

104. Scholten, E., Moschakis, T., & Biliaderis, C. G., (2014). Biopolymer composites for engineering food structures to control product functionality. *Food Structure, 1*, 39–54.

105. Shahidi, F., & Zhong, Y., (2010). Lipid oxidation and improving the oxidative stability. *Chemical Society Reviews, 39*, 4067–4079.

106. Shchukina, E., & Shchukin, D., (2012). Layer-by-layer coated emulsion microparticles as storage and delivery tool. *Current Opinion in Colloid and Interface Science, 17*, 281–289.

107. Shih, F., Daigle, K., & Champagne, E., (2011). Effect of rice wax on water vapor permeability and sorption properties of edible pullulan films. *Food Chemistry, 127*(1), 118–121.

108. Soupas, L., Huikko, L., Lampi, A., & Piironen, V., (2005). Esterification affects phytosterol oxidation. *European Journal of Lipid Science and Technology, 107*, 107–118.

109. Stieger, M., & Velde, F. V., (2013). Microstructure, texture, and oral processing: New ways to reduce sugar and salt in foods. *Current Opinion in Colloid and Interface Science, 18*, 334–348.

110. Tieken, S., Leidy, H., Stull, A., Mattes, R., Schuster, R., & Campbell, W., (2007). Effects of solid versus liquid meal-replacement products of similar energy content on hunger,

satiety, and appetite-regulating hormones in older adults. *Hormone and Metabolic Research, 39*, 389–394.

111. Tomoyori, H., Kawata, Y., Higuchi, T., & Ichi, I., (2004). Phytosterol oxidation products are absorbed in the intestinal lymphatics in rats but do not accelerate atherosclerosis in apolipoprotein E-deficient mice. *Journal of Nutrition, 134*, 2738.

112. Velikov, K. P., & Pelan, E., (2008). Colloidal delivery systems for micronutrients and nutraceuticals. *Soft Matter, 4*, 1964–1980.

113. Villeneuve, P., Barouh, N., & Barea, B., (2007). Chemoenzymatic synthesis of structured triacylglycerols with conjugated linoleic acids (CLA) in central position. *Food Chemistry, 100*, 1443–1452.

114. Whelan, J., & Rust, C., (2006). Innovative dietary sources of n-3 fatty acids. *Annual Review of Nutrition, 26*, 75–103.

115. Wildmoser, H., Scheiwiller, J., & Windhab, E., (2004). Impact of disperse microstructure on rheology and quality aspects of ice-cream. *Lebensmittel-Wissenschaft Und-Technologie (Food Science and Technology), 37*(8), 881–891.

116. Williams, C., & Burdge, G., (2006). Long-chain n-3 PUFA: Plant v. marine sources. *Proceedings of the Nutrition Society, 65*, 42–50.

117. Woodall, A., Lee, S., Weesie, R., Jackson, M., & Britton, G., (1997). Oxidation of carotenoids by free radicals: Relationship between structure and reactivity. *Biochimica et Biophysica Acta, 1336*, 33–42.

118. Zasypkin, D. V., Braudo, E. E., & Tolstoguzov, V. B., (1997). Multicomponent biopolymer gels. *Food Hydrocolloids, 11*, 159–170.

119. Zou, L., Liu, W., Liu, C., Xiao, H., & McClements, D. J., (2015). Utilizing food matrix effects to enhance nutraceutical bioavailability: Increase of curcumin bioaccessibility using excipient emulsions. *Journal of Agricultural and Food Chemistry, 63*(7), 2052–2062.

120. Zúñiga, R., Kulozik, U., & Aguilera, J., (2011). Ultrasonic generation of aerated gelatin gels stabilized by whey protein β-lactoglobulin. *Food Hydrocolloids, 25*(5), 958–967.

CHAPTER 2

SECONDARY METABOLITES FROM *CLERODENDRUM INFORTUNATUM* L.: THEIR BIOACTIVITIES AND HEALTH BENEFITS

R. L. HELEN, K. JAYESH, S. SYAMA, and M. S. LATHA

ABSTRACT

Tribal communities use various parts of *Clerodendrum infortunatum* (Hill glory bower) to treat colic, scorpion sting, snake bite, tumor, and certain skin diseases. The leaves of the plant are used orally for curing fever and bowel complaints. Roots are used as a cure for diarrhea. The plant has been used in the traditional Indian medicinal systems like Ayurveda, Siddha, Unani, and Homeopathy. Besides the medicinal properties, the plant has been used as fumigants and to control louse. The presence of flavonoids, steroids, saponins, phenols, and fixed oils has been reported in this plant. Different parts of *Cleroendrum infortunatum* possess antioxidant, anti-inflammatory, antidiabetic, anticonvulsant, antimicrobial, anthelmintic, analgesic, and hepatoprotective activities. Therefore, *Clerodendrum infortunatum* is a plant with immense medicinal potential. This chapter discusses the medicinal properties of *Clerodendrum infortunatum* L.

2.1 INTRODUCTION

Plants have been used as medicines since the beginning of civilization. The earliest record on the use of plants as medicine appeared in *Rigveda*, which has been written between 1600–3500 BC. The scholastic works of Charaka and Susrutha contain knowledge about the use of preventive and curative medicines [23]. Ancient people depended on nature for their day to day medication needs. The document of Hippocrates has provided insights into the use of herbal medicines [60]. Scientific innovations and advancements in modern medicines, antibiotics, and synthetic drugs have revolutionized the medical field. However, the continuous and excessive use of synthetic drugs is associated with liver damage and carcinogenesis [25]. This turns the focus of attention to the traditional plant-based medicines. The past decade witnessed a shift towards revisiting natural products for healthcare needs and this turns the focus of interest of the pharmaceutical companies to nature.

Today herbal medicines provide a first-line defense to rural people, where the accessibility of herbs is plenty. Even in places where modern medicines are available, the dependence of traditional herbal remedies is on the rise due to the holistic approach to health problems, lack of adverse reactions, etc. Plants provide a safe, cost-effective, efficient remedy with fewer side effects compared to the synthetic drugs and this contributes to the upsurge interest in the usage of plants and plant-based drug sources [22].

The World Health Organization (WHO) recognizes herbal medicines as easily accessible remedial measures and thus encourages their felicitous utilization. WHO also provides initiatives to make systematic inventory and assessment of medicinal plants and instigate proper means for the management of plant-based medications to ascertain the standard of such medicines [88]. Herbal remedies remain the primary source of health care for about 80% of the population living in developing countries [21]. In developed countries like the UK and Europe, complementary and alternative medicines (CAM) have become a mainstream of research [10].

The market of herbal remedial measures has a tremendous increase and the global market for natural medicines is expected to progress because of the customer's growing preferences for herbal preparations in various forms [91]. There is a common belief that herbal remedies provide a longer and healthier life, and this has also diverted the attention of the public to natural herbal remedies. Plant drug constitutes about 25% of the total drugs in developed countries (such as United States), whereas in the developing country (such as India) the contribution of the plant-based drug is about 80%. Therefore, the economic importance of the medicinal plant is significantly higher in India [33].

India is regarded as the medicinal garden of the world due to its biodiversity richness. The geographical and climatic conditions prevailing in India make it a rich source of biological diversity [81]. Many recognized indigenous healthcare systems (such as Ayurveda, Yunani, Homeopathy, Yoga, Naturopathy, and Siddha) have been practiced for health care in India. Therefore,

India has a unique position in the world with regard to the traditional treatment systems and has been rightly described as the medicinal garden. These medicinal systems use drugs of herbal origin, which are derived either from the whole plant or from different parts of the plant. Ayurveda (https://www.ayurvedanama.org/) is the most developed and widely practiced medicinal system in India.

The term Ayurveda comes from two Sanskrit words *Au* (life) and *Veda* (Science), i.e., it is the science of life. The essence of Ayurvedic knowledge is contained in the four *vedas*. The oldest of the four, *Rigveda* contains a description of 67 herbs. Despite being a treatment system, the Ayurvedic system provides a healthy life involving the physical, psychological, ethical, and spiritual aspects of mankind [64]. *Siddha* system of medicine has proximity to Ayurveda having the five proto-elements and three *doshas* concept with differences in their interpretation [53]. This system has evolved from the Dravidian culture and is one of the oldest systems practiced in South India, especially in Tamil Nadu. The term *Siddha* originates from 'Siddhi' meaning achievement. The main aim of *Siddha* treatment is to impart a perfect and imperishable body and to promote longevity.

In Siddha, *Materia Medica* drugs are categorized into herbal products, metal and mineral products, and animal products. About 80% of the drugs are formulated from herbal plants and products [64]. The *Unani* system established by Hippocrates has its origin in Greece. It maintains and promotes positive health besides the prevention and cure of diseases. The *pharmacopeia* involves solely natural drugs. According to the *Unani* system, a balance is established between the mind and body so that all processes are in symmetry. The human body is composed of seven components called '*Umoor* e *Tabaiyah*' (fundamentals of physique), which are responsible for proper health [41].

Clerodendrum infortunatum is a plant with immense medicinal potential. This chapter discusses medicinal properties of *Clerodendrum infortunatum* L.

2.2 SECONDARY METABOLITES FROM MEDICINAL PLANTS

The forests in India are a rich repository of medicinal plants. About 200,000 medicinal plants have been recorded in India [19]. Of these, about 8000 plants have been utilized in the herbal industry and more than 3000 formulations have been registered [72]. Many plants remain unexplored in the wild and their wide potential is yet to be discovered. Compared to the modern pharmaceutics, plant-based medicines have several advantages. Plants contain a variety of compounds, which act in synergy. They can complement each other and neutralize the negative effects if any. Sometimes the whole plant extract or crude extract will be more fruitful in the treatment of a disease than purified components. The constituents may act on one another to improve the solubility or interact to improve the bioavailability of the phytoconstituents [84].

For example, *Cannabis sativa* is a therapeutic agent for curing rheumatoid arthritis, multiple sclerosis, and AIDS (acquired immunodeficiency syndrome), where the phytoconstituents cannabidiol increases the level of tetrahydro cannabidiol (THC) in the brain. THC alone induces anxiety but the presence of cannabidiol causes attenuation of anxiety. Therefore, the herbs are preferred than isolated compounds for the treatment of multiple sclerosis [86]. Also, plants can be used as preventive medicine because of fewer side effects. This reduces the dependence of synthetic drugs for the treatment of chronic diseases [30]. For example, the consumption of tea is linked with reduced likelihood of diseases and boosts immunity [71]. Medicinal plants are generally considered as safe but there exist evidences that point to the toxicity regarding the use of herbals as medicines [55].

The public health issues and concerns regarding the safety become more recognized with increase in the global use of herbal products. The information regarding traditional medicinal herbs relies mostly in the hands of traditional healers and is transferred through generations orally. The systematic and proper identification, validation, and screening are necessary to impart its proper usage in modern therapeutics. Considering the diversity exists in the Indian sub-continent about medicinal plants, plant products, and traditional healthcare systems, there is an urgent need to establish these traditional strategies realizing the developmental trends underway.

Plants produce a vast array of compounds that besides producing nutrients for their growth and development are involved in the defense and signaling mechanisms. Normal cell processes do not depend on these secondary metabolites, but they perform various cellular processes [5]. For example, flavonoids offer protection from the reactive species produced during photosynthesis. Ethanopharmacology deals with the study of traditional cultures using medicinal plants and the observation, identification, description, and experimental investigation of the active ingredients in plants and the effect of such drugs. This leads to the identification of new bioactive compounds that form the source of new drugs.

Plants are natural factories, which form sources of pharmaceuticals, agrochemicals, food additives, and pesticides, etc. [59]. These bioactive compounds known as secondary metabolites elicit specific pharmacological or toxicological effects in humans and animals. They possess immense variability in the structure and exhibit diverse biological properties. The secondary metabolites are grouped into phenylpropanoids, terpenoids, and alkaloids based on their biosynthetic origin. About 200,000 secondary metabolites have been estimated in the plant kingdom [29]. Secondary metabolites may serve as pheromones, toxins (ex. Digoxin from Foxglove), and drugs (ex. Quinine from Cinchona, Morphine from Poppy).

2.2.1 PHENYLPROPANOIDS

Phenylpropanoids are the most diverse group of organic compounds, which include simple low molecular weight phenolic acids, coumarins, and benzoic acid derivatives to complex flavonoids, stilbenes, tannins, and lignans [45]. They are obtained from six carbon phenyl group and the three-carbon propene tail of cinnamic acid, produced during the initial step in the phenyl-propanoid pathway [89]. They provide protection against biotic and abiotic stresses, UV radiation, herbivore, and pathogen attack and serve as signaling molecules to mediate bio-interactions [83].

2.2.2 TERPENOIDS

Terpenes are a diverse group containing one or more five-carbon isoprene units. They are synthesized from common precursor isopentenyl pyrophosphate (IPP) and dimethylallyl pyrophosphate (DMAPP) via two independent pathways: the cytosolic mevalonic acid (MVA) pathway and the plastid localized methylerythritol phosphate (MEP) pathway [89]. According to the isoprene units present, terpenoids are categorized into: hemiterpenes (C_5), monoterpenes (C_{10}), sesquiterpenes (C_{15}), diterpenes (C_{20}), triterpenes (C_{30}), etc. Terpenoids function as plant and animal hormones, membrane lipids, insect attractants, antifeedants, mediators of the electron transport system, and play a major role in the plant-environment interaction [35].

2.2.3 ALKALOIDS

Alkaloids are compounds containing nitrogen and are derived from amino acids. They are heterocyclic compounds with a bitter taste [5]. About 12,000 alkaloids have been identified, which display a vast degree of bioactivities. A major portion of drugs are contributed by alkaloids, of which the isolation of morphine from opium poppy forms the breakthrough in the history of pharmacy [89]. Alkaloids constitute most of poisons, neurotoxins, traditional psychedelics (atropine), and social drugs (nicotine). They serve as feeding deterrents and toxins to insects and herbivores [40]. Based on the chemical structure they are grouped into true alkaloids, proto alkaloids and pseudo alkaloids [87]. Alkaloids also serve as a source of antibiotics and analgesics in medicine. For example, the alkaloid berberine is used in ophthalmic and sanguinarine in toothpastes [65].

In this context, the present chapter is a rational approach to discuss the beneficial effects of *Clerodendrum infortunatum*, which has been used in a number of Indian folk medicines.

2.3   HERBAL MEDICINES: PHARMACOLOGICALLY BIOACTIVE COMPOUNDS

Plants serve as an important source of pharmacologically active compounds for the manufacture of new drugs, which are obtained either directly or indirectly from plants. Out of 252 fundamental drugs recognized by WHO, 11% were solely obtained from angiosperms [82]. Recent reports indicate that herbal preparations are more effective than isolated and purified compounds. The presence of interacting substances makes the crude extract more active. Also, pure compounds are difficult to synthesize and are costly [63]. This turns the focus of attraction to herbal preparations. Plants form the main ingredients in both modern medicines and traditional medicinal preparations.

Numerous drugs were obtained from plants or plant products, e.g., vincristine, and vinblastine from *Catharanthus roseus*, codeine from *Papaver somniferum*. Drugs are prepared either from the whole plant or from the roots, leaves, flowers, stem, bark, fruits, seeds, etc. The excretory products of the plant (such as resins, gum, and latex) also serve as raw materials for the drugs. Sometimes crude extract of the plant is used as a drug. Isolation and purification of the active principle responsible for the medicinal property is important in some other cases. Thus, modern research is ongoing on crude extracts and active principles. Research studies are now focused on exploring the pharmacological profile of traditionally used plants and identification of bioactive principles from these for the development of new therapeutics.

2.4   *CLERODENDRUM INFORTUNATUM* (HILL GLORY BOWER)

Clerodendrum infortunatum is a medicinal plant described in Puranas as *'Ghantakarna'* and is used as a remedy for many diseases in Ayurveda, Yunani, and Homeopathy. *Clerodendrum infortunatum* L. is a perennial shrub. The generic name of the plant was obtained from two words: Kleros meaning "Chance" and Dendron meaning "a tree" [52]. Around 508 species have been reported in the genus *Clerodendrum* and among them, *Clerodendrum infortunatum* is an Indian species. It belongs to the family Verbenaceae. The plant is commonly known as "Hill Glory Bower" (Figure 2.1).

FIGURE 2.1   *Clerodendrum infortunatum* Linn.

Clerodendrum L. is a genus extensively dispersed in the tropical and subtropical regions of the world. In 1753, Linnaeus made the first description of the genus with the identification of *Clerodendrum infortunatum*. Later in 1763, the Latin name Clerodendrum was changed to its Greek form "Clerodendron" by Adanson. And Moldenke readopted the Latin name Clerodendrum in 1942 and it has now been commonly used by the taxonomists for the classification and description of the species [74]. Many of the species have been described by more than one author and they have been denoted in different ways [44, 46], e.g., *Clerodendrum infortunatum* Linn., *Clerodendrum infortunatum* Gaertn., etc.

Traditional medicinal systems practiced in India, China, Korea, Thailand, and Japan use various species of Clerodendrum as folk medicine [39, 66, 74]. Most members of the family possess aromatic odor [56]. A representative of the family Verbenaceae contains phytoconstituents such as anthraquinones, terpenes, steroidal saponins, alkaloids, and flavonoids [75]. The family is closely related to Lamiaceae [13, 62].

2.5 TRADITIONAL USES OF *CLERODENDRUM INFORTUNATUM*

Clerodendrum infortunatum is an important ethnomedicine as a remedial measure for vast spectrum of ailments in the traditional healthcare systems. The ethnomedicinal properties of *Clerodendrum infortunatum* has been exploited by various tribes in India. The plant is used in preparation of "Shuktani," a recipe used by Bengali community of Barak Valley, Assam. [54]. The Kuki and Rongmai Naga tribes of North-East Asia use leaves of *Clerodendrum infortunatum* orally for the treatment of fever and bowel complaints.

Certain tribe communities of North Bengal (such as Rabha, Rajbanshi, and Lepcha) use root bark of the plant to cure diarrhea. Khumis tribe in Thanchi district of Bangladesh use this plant as a remedy for burning sensation in the chest, salt taste in mouth, flatulence, and gastric pain [8]. Paste of tender leaves is used to cure cut-wounds and leprosy by the medical practitioners of Bhadra wildlife Sanctuary in Karnataka [24]. Santhals of Bihar, Jharkhand, and Orissa use various parts of the plant against colic, snake bite, and skin diseases [23].

The roots, leaves, flowers, seeds, and tender branches possess medicinal properties in Ayurveda [7]. The plant is used to cure common ailments, such as fever, cough, and cold, asthma, bronchitis, inflammation, diseases of the skin infections and blood, liver, and spleen, malarial fever, non-malignant tumors, indigestion, etc. [23]. The leaf extract is bitter and is used as a vermifuge, laxative, and cholagogues. Juice of fresh leaves is inserted into the rectum for the removal of ascarids. Extract from the leaf is a good appetizer. Juice of fresh leaves is given to anemic patients to increase the hemoglobin content and to treat piles. The leaves are slightly warmed and applied along with edible oil to get relief from body pain [47]. In diabetic patients, the leaves reduce the sugar percentage [62]. A paste prepared from the leaf tissue is effective against scorpion sting.

Extracts from the leaf and root of the plant are used for curing fresh wounds caused by a sharp knife. The leaves and roots are applied externally over the affected area for the treatment of tumors and skin infections and employed internally as tonics. Root bark juice is used to treat indigestion and abdominal pain. Root bark paste is spread as a bandage in swelling [4]. Paste of root is given to children in mother's milk for treatment of helminthiasis [70].

The plant is found to possess applications in the Homeopathy, Siddha, and Unani medicinal systems. It is employed as a medication for diarrhea, post-natal complications and to dress wounds in Indian Homeopathy. In the Unani system of Indian medicine, *C. infortunatum* is used to remove worms from the intestinal tract and to treat rheumatism [77]. Siddha medicine prepared with *C. infortunatum* along with other medicinal herbs is used for the treatment of all types of fever [90]. The decoction prepared from the root and stem bark is used for treatment of respiratory diseases, fever, cough, bronchial asthma, etc. The plant also has antidiarrheal, hepatoprotective, antirheumatic, and antimicrobial benefits in Siddha system.

2.6   OTHER USES OF *C. INFORTUNATUM*

Besides medicinal properties exhibited by various parts of *C. infortunatum* [42], it has implications in other fields also. Aqueous extract of the leaf is applied to the control louse. The plant extract is used as fumigant for bed bugs. Farmers of certain tribes use twigs of *C. infortunatum* on rice fields to repel rice bugs [6]. Pharmacological properties exhibited by different parts of *Clerodendrum infortunatum* are listed in Table 2.1.

TABLE 2.1   Pharmacological Properties Exhibited by Different Parts of *Clerodendrum infortunatum*

Plant Parts	Properties
Flowers	Used for scorpion sting.
Leaves	Bitter, acrid, thermogenic, laxative, antiseptic, demulcent, anti-inflammatory, depurative, vermifuge, expectorant, antipyretic.
Roots	Diuretic, laxative, analgesic, anti-inflammatory, antitumor, relieves congestion and torpidity of bowels, cramps, rheumatism, employed for skin diseases and alopecia.
Stem	Toothache, snake bite antidote.

2.7   PHYTOCONSTITUENTS IN *CLERODENDRUM INFORTUNATUM*

The phytoconstituents present in the plant contribute to its medicinal properties. The initial phytochemical screening of the leaf and root extracts of *Clerodendrum infortunatum* reported the presence of flavonoids, alkaloids, tannins, phenols, steroids, saponins, and glycosides [32]. Fixed oils containing glycerides of linoleic, oleic, stearic, and lignoceric acid have been reported in the leaves of *Clerodendrum infortunatum* [38]. The leaves contain saponin, alkyl sterols, enzymes, and 2,-(3,4-dehydroxyphenyl) ethanol-1-O-α-2 rhamnopyranosyl (1→3)-β-D-(4-O-caffeoyl) glycopyranoside (acteoside) [51]. Leaf also showed the presence of diterpene, clerodin, riboflavin, ascorbic acid, and thiamine [70].

Ethanolic extract of the powdered root bark contains carbohydrates, such as fructose, galactose, glucose, lactose, maltose, raffinose, and sucrose, steroids, tannins, flavonoids, saponins, and alkaloids [2]. Roots contain β-sitosterol, lupeol, and steroidal glycosides [79], clerosterol (5, 25-sigmastadien-3β-ol), clerodolone (lup_20(30)-en-3β-diol-12-one) and clerodone (3β-hydroxy-lupan-12-one). Chemical analysis of the flower revealed the existence of β-sitosterol, lupeol, cleridine, hentricontane, and fumaric acid esters of caffeic acid [14]. GC-MS (gas chromatography-mass spectrometry) analysis of the leaf, stem, root, flower, and seed of *Clerodendrum infortunatum* identified the presence of several bioactive compounds, such as limonene, phytol, catechol, hexadecenoic acid, dodecanoic acid, vitamin E, squalene, hydroxymethylfurfural, and stigmasterol. Several phenolics and phenolic acid derivatives were also identified by GC-MS analysis [20]. Table 2.2 indicates phytochemical constituents isolated from different parts of *Clerodendrum infortunatum*.

2.8 PHARMACOLOGICAL STUDIES

Extensive investigation on medicinal plants have revealed a broad spectrum of pharmacological properties. High intake of diet rich in fruits, vegetables, berries, and whole grains and low intake of red meat and junk food can reduce the incidence of cancer, cardiovascular diseases (CVD), and other degenerative diseases. Excessive free radicals produced in the cell during the physiological and biochemical process cause damages to the cell wall and DNA (deoxyribonucleic acid) and result in chronic diseases [9]. Dietary intake of antioxidant-rich foods decreases the impact of free radical-induced damage [58].

Free radical is a molecular fragment containing one or more unpaired electrons and is capable of independent existence. Free radicals and reactive oxygen species (ROS) from internal or external origin can impair signal transduction pathways and cause damage of macromolecules, such as lipids, proteins, and DNA. Hepatocytic proteins, lipids, and DNA are primarily affected by ROS and cause abnormalities in the structural and functional status of liver [12]. It ultimately results in the impaired function of organelles and leads to diseases, such as cancer, inflammation, atherosclerosis, CVDs, neurodegenerative diseases, rheumatoid arthritis, etc. [61].

TABLE 2.2 Phytochemical Constituents Isolated from Different Parts of *Clerodendrum infortunatum* [14, 58, 80, 100]

Group	Medicinal Properties	Chemical Constituents Reported
Fixed oils	Antioxidants.	Glycerides of linoleic, oleic, stearic, and lignoceric acid.
Flavonoids	Anticancer, anti-inflammatory, antiallergy, protection from UV radiation.	Apigenin, acacetin, and methyl esters of acacetin-7-O-glucuronide, cabruvin, quercetin, scutallaren, scutellarein-7-O-β-D-glucuronide, hispidulin, quercetin.
Phenolics	Antioxidant, anticancer, anti-inflammatory, lowered cardiovascular effects.	Acetoside, fumaric acid, methyl, and ethyl esters of caffeic acid.
Steroids	Offset the build-up of cholesterol.	Clerodolone, clerodone, clerodol, and clerosterol, β-sitosterol.
Terpenoids	Antioxidants, antimalarial, antibacterial, insect attractants and antifeedant, immune-modulatory, natural preservative.	Clerodin, oleanolic acid, clerodinin A, lupeol.
Vitamins	Antioxidant, anticancer.	Riboflavin, ascorbic acid, thiamine, vitamin E.

ROS initiated DNA damage contributes to the onset and progression of carcinogenesis. Certain oncogenes may increase the production of ROS, which leads to genomic instability. Therefore, ROS triggers tumorigenesis by activating cellular proliferation, metabolic alteration, and angiogenesis [78]. The progression of inflammatory responses is also favored by ROS. At the site of inflammation, polymorph nuclear neutrophils inflict ROS generation causing endothelial dysfunction and tissue injury by oxidation of cellular signaling proteins, such as tyrosine phosphatases [48]. Precise molecular mechanisms exist in cells to curb the ill-effects elicited by ROS and thus maintain the oxidative balance. The sources of ROS and its impact on our body are shown in Figure 2.2.

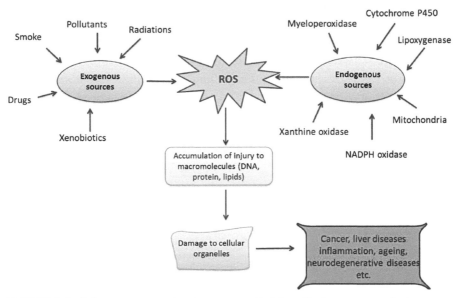

FIGURE 2.2   Diagram representing the sources of ROS and its impact on our body.

The enzymatic and non-enzymatic antioxidants residing within our body counteracts the effect of oxidants. In addition to these inbuilt-antioxidants to fight against chronic diseases, fruits, vegetables, and herbal medicines form rich source of exogenous antioxidants. High intake of fruits and vegetables slinked with low rate of cancer, heart diseases, and neurological disorders [34]. The bioactive constituents may act as free radical scavengers and convert these radicals to less reactive forms. Thus, drugs with antioxidant potential may minimize the damages caused by ROS and prevent the incidence of chronic diseases.

2.8.1   TOXICITY STUDY

The acute and sub-chronic studies revealed that the methanolic extract of *Clerodendrum infortunatum* was safe up to dose of 2000 mg/kg body weight [15]. The physical, biochemical, and histological parameters were normal, and the plant extract did not show any indications of toxicity. Aqueous acetone extract of the root bark also did not show any change in the antioxidant status, liver function enzymes, and liver microarchitecture, which points to the non-toxic nature of the plant [37].

2.8.2 ANTIOXIDANT ACTIVITY

Antioxidants protect the cell from free radical-induced damages and thus avert the pathogenesis of chronic ailments like cancer, inflammation, and CVDs. Plants contain many compounds with antioxidant potential and this account for their beneficiary effects [73]. The *in vitro* assays (such as DPPH radical scavenging assay, reducing power assay, and hydrogen peroxide scavenging assay) exhibited antioxidant activity in different parts of *C. infortunatum* [24, 51, 67].

2.8.3 ANTI-INFLAMMATORY ACTIVITY

Inflammation is the normal protective reaction to the tissue injury caused by chemicals, biological agents, etc. This tissue injury results in the release of inflammatory mediators, like, cytokines, interleukins, tumor necrosis factor by the activation of transcription factor NFκB [36]. Phytochemical constituents may inhibit these molecular targets and reduce inflammatory processes. The methanolic leaf extract showed significant and dose-dependent anti-inflammatory activity against the carrageenan-induced acute inflammatory model in Wistar albino rats. The extract also resulted in significant inhibition against the inflammation mediated by histamine and dextran [18].

2.8.4 CYTOTOXICITY AND ANTICANCER ACTIVITY

Drugs with ability to kill cancer cells without causing damage to the normal cells are promising alternatives in the cancer therapeutics. Exploration of plants in the folklore medicine has resulted in the identification of principles with antitumor activity. Different extracts of *C. infortunatum* exhibit significant anticancer activity in- *in vitro* and *in vivo* models. Methanolic extract of the plant exhibits anticancer activity against Ehrlich's ascites carcinoma in mice. Oleanolic acid and clerodenin-A from the plant was responsible for the increased life span, antioxidant status, and decreased tumor volume [69].

The antitumor activity of the hydroalcoholic extract on murine tumor cells was reported [11]. The extract induced apoptosis and mortality in Dalton's lymphoma ascites cells with minimum toxicity. The expression of anti-apoptotic gene *bcl-2* and pro-apoptotic gene *bax* confirmed the antitumor activity. Hexane and chloroform extract of root of the plant showed significant *in vitro* anticancer activity against prostate (PC-3), lung (A549) and colon (HCT-116) cancer cell lines [28]. The cytotoxic effect may be due to the presence of phytoconstituents, such as alkaloids, terpenoids, and flavonoids. Treatment with chloroform and hexane root extract resulted in diminished colony formation and wound closure of A549-cells and thus inhibits the migration in human lung cancer [27].

2.8.5   ANTIDIABETIC ACTIVITY

The diabetes mellitus condition characterized by high blood sugar is associated with disrupted protein, carbohydrate, and fat metabolism [49]. Diabetes is a multifactorial disease and thus medicinal plants with hypoglycemic effect have been promising antidiabetic agents. Methanolic root extract of *Clerodendrum infortunatum* produced marked dose-dependent decline in glucose level in the blood of streptozotocin (STZ) induced diabetic rats [37].

Methanolic extract of the plant exhibited anti-hyperglycemic potential due to the outcome of an augmented endogenous antioxidant mechanism [16]. Chloroform extract of the leaf also revealed significant anti-hyperglycemic effect in STZ-induced diabetic rats [3]. Administration of 200 and 400 mg/kg chloroform extract resulted in a prominent and dose-dependent reduction in the blood glucose levels and restored the biochemical parameters to normal levels.

2.8.6   ANTIMICROBIAL ACTIVITY

Leaf, root, and stem of *C. infortunatum* have antibacterial activity against Gram-positive and Gram-negative bacteria. Antifungal activity of the plant parts was observed against *Aspergillus Niger*, *Aspergillus flavus,* and *Candida albicans* by ethyl alcohol and ethyl acetate extract [65]. The leaf extract manifested greater activity than the root and stem. In another study, the ethanolic leaf extract exhibited significant antifungal activity against *A. Niger*, *P. freuentance, P. notatum*, and *B. cinera*, when tested by turbidity and spore germination method [43].

2.8.7   HEPATOPROTECTIVE ACTIVITY

Drug-induced toxicity, alcohol abuse; environmental toxicants, chemicals, etc. contribute to liver diseases. Methanolic extract of the leaf revealed significant hepatoprotection against carbon tetrachloride (CCl_4) induced hepatic toxicity [68]. The extract significantly replenished the antioxidant levels and decreased the malondialdehyde (MDA) levels suggestive of hepatoprotection by increased antioxidant status.

2.8.8   ANTHELMINTIC ACTIVITY

Alcohol and aqueous leaf extracts of *C. infortunatum* manifest significant anthelmintic activity against *Pheretima prosthuma*, which can be perceived from the death and paralysis of worm in a shorter time that is comparable to the reference piperazine citrate [50].

2.8.9   ANALGESIC AND ANTICONVULSANT ACTIVITY

Methanolic extract of the leaf of *Clerodendrum infortunatum* offered protection from pentylenetetrazole (PTZ)-induced seizures in Swiss albino mice [17]. The anticonvulsant activity exhibited may be due to the depressant effect produced by the phytoconstituents present in the plant which in turn increase the pentobarbitone induced sleeping time. Saponins obtained from the leaf also revealed significant anticonvulsant potential against leptazol induced seizures [57].

The writhing induced by acetic acid in Swiss albino mice was inhibited by saponins isolated from the leaf of *C. infortunatum*. The results are comparable with the reference standards paracetamol, morphine sulfate, and aspirin [57]. In the hot plate method, also saponins induce analgesia and potentiated the activity of the reference drugs pentazocine and aspirin.

2.8.10   OTHER ACTIVITIES

Leaf extract of plant exhibited nootropic effect (memory-enhancing activity) in Wistar albino rats at a dose of 100 and 200 mg/kg body weight [26]. The 200 mg/kg methanolic extract exhibited the greatest activity, which is approximated to the standard drug Brahmi.

Powdered leaf and stem of *C. infortunatum* mixed with cow dung induced larval and pupal mortality and deformed adult formation in the grubs of *Oryctes rhinoceros* [77]. Clerodin, 15-methoxy-14,15-dihydroclerodin, and 15-hydroxy-14,15-dihydroclerodin isolated from the compound exhibited significant antifeedant activity on cotton bollworm (*Helicoverpa armigera*), which was compared with the commercial pesticide azadirachtin [1].

2.9 FUTURE PERSPECTIVES

The present century is facing an exponential increase in the acceptance and usage of herbal drugs. This provides an opportunity for us to scrutinize therapeutically products from ancient medicinal systems, which can be utilized for the development of new drugs. Issues related to drug safety is a major concern in the present scenario and thus regulatory policies of herbal products and functional foods are required to be standardized to avoid misuse. The increased use of herbal medicine is to be checked and remedial measures must be undertaken to preserve the natural habitat of the plants.

2.10 SUMMARY

The literature provides an insight into the potential benefits of various parts of *Clerodendrum infortunatum*. The properties of the plant are boundless, and it is very effective and safe to be used as a medicine. Different parts of the plant can be efficiently utilized for the development of potent therapeutics to cure cancer, inflammation, and other oxidative stress-related disorders using the reverse pharmacological approach. Most of the research now focuses on the identification and isolation of phytoconstituents and it does not reach the market. Thus, rational approaches require the development of new and novel products must be reckoned so that the fruits of the developments ultimately reach society.

KEYWORDS

- **alkaloids**
- **beta-sitosterol**
- ***Clerodendrum infortunatum***
- **oleanolic acid**
- **phenylpropanoids**
- **quercetin**
- **reactive oxygen species**
- **terpenoids**

REFERENCES

1. Abbaszadeh, G., Srivastava, C., & Walia, S., (2014). Insecticidal and antifeedant activities of clerodone diterpenoids isolated from the Indian *bhant* tree (*Clerodendron infortunatum*) against the cotton bollworm. *Helicoverpa Armigera. J. Insect. Sci.,* *14*(29), 1–13.

2. Azad, A. K., Azizi, W. S., Syafiq, T. M. F., Mahmood, S., Almoustafa, H. A., & Labu, Z. K., (2013). Isolation, characterization, and cytotoxic effect exploration of methanolic extract of local medicinal plant *Clerodendrum viscosum* vent. *Aust. J. Basic and Appl. Sci.,* *7*(4), 641–647.

3. Baid, S. S., (2013). Evaluation of antihyperglycemic and hypolipidemic activities of *Clerodendrumn infortunatum* Linn. leaf extracts. *Asian J. Comp. Alt. Med.,* *1*(1), 1–8.

4. Barbhuiya, A. R., Sharma, G. D., Arunachalam, A., & Deb, S., (2009). Diversity and conservation of medicinal plants in Barak valley, Northeast India. *Indian Journal of Traditional Knowledge,* *8*(2), 169–175.

5. Bernhoft, A., (2010). A brief review on bioactive compounds in plants. In: Bernhoft, A., (ed.), *Bioactive Compounds in Plants-Benefits and Risks for Man and Animals* (pp. 11–17). The Norwegian Academy of Science and Letters, Oslo.

6. Bhattacharjee, P. P., & Ray, D. C., (2010). Pest management beliefs and practices of Manipuri rice farmers in Barak Valley, Assam. *Indian Journal of Traditional Knowledge,* *9*(4), 673–676.

7. Bhattacharya, S., (2001). *Chiranjeev Vanaushadhi* (in Bengali) (p. 215). Ananda Publishers, Kolkata.

8. Bijoy, S., Fatema, A., Umma, A., & Rolee, S., (2012). Ethnomedicinal investigations among the Sigibe clan of the khumi tribe of thanchi sub-district in Bandarban district of Bangladesh. *American Eurasian Journal of Sustainable Agriculture,* *6*(4), 378–386.

9. Birben, E., Sahiner, U. M., Sackesen, C., Erzurum, S., & Kalayci, O., (2012). Oxidative stress and antioxidant defense. *World Allergy Organ. J.,* *5*(1), 9–19.

10. Braun, L. A., Tiralongo, E., & Wilkinson, J. M., (2010). Perceptions, use and attitudes of pharmacy customers on complementary medicines and pharmacy practice. *BMC Complement Altern. Med.,* *10*(1), 38.

11. Chacko, T., Menon, A., Nair, S. V., Al Suhaibani, E., & Nair, C. K. K., (2015). Cytotoxic and antitumor activity of the extract of *Clerodendrum infortunatum*: A mechanistic study. *American Journal of Phytomedicine and Clinical Therapeutics,* *3*(2), 145–158.

12. Cichoz-Lach, H., & Michalak, A., (2014). Oxidative stress as a crucial factor in liver diseases. *World J. Gastroenterol.,* *20*(25), 8082–8091.

13. Cronquist, A., (1981). *An Integrated System of Classification of Flowering Plants* (p. 110). Columbia University Press: Columbia.

14. Das, B., Pal, D., & Haldar, A., (2014). Review on biological activities and medicinal properties of *Clerodendrum infortunatum* Linn. *International Journal of Pharmacy and Pharmaceutical Sciences,* *6*(10), 41–43.

15. Das, S., Bhattacharya, S., Biswas, M., & Kar, B., (2011). Acute and sub-chronic toxicity study of *Clerodendrum infortunatum* leaf in adult male albino rats. *Am-Euras. J. Sci. Res.,* *6*(4), 189–191.

16. Das, S., Bhattacharya, S., Prasanna, A., Kumar, R. B. S., Pramanik, G., & Haldar, P. K., (2011). Preclinical evaluation of antihyperglycemic activity of *Clerodendron*

infortunatum leaf against streptozotocin-induced diabetic rats. *Diabetes Ther.*, *2*(2), 92–100.

17. Das, S., Haldar, P. K., Pramanik, G., Panda, S. P., & Bera, S., (2010). Anticonvulsant activity of methanolic extract of *Clerodendron infortunatum* Linn. in Swiss albino mice. *Thai J. Pharm. Sci.*, *34*(4), 129–133.

18. Das, S., Haldar, P. K., Pramanik, G., & Suresh, R. B., (2010). Evaluation of anti-inflammatory activity of *Clerodendrum infortunatum* Linn extracts in rats. *Global J. Pharmacol.*, *4*(1), 48–50.

19. Dev, S., (1997). Ethno therapeutics and modern drug development: The potential of Ayurveda. *Current Sci.*, *73*(11), 909–928.

20. Dey, P., Dutta, S., & Chaudhuri, T. K., (2015). Comparative phytochemical profiling of *Clerodendrum infortunatum* L. using GC-MS method coupled with multivariate statistical approaches. *Metabolomics*, *5*(3), 1–10.

21. Dey, Y. N., Ota, S., Srikanth, N., Jamal, M., & Wanjari, M., (2012). A phytopharmacological review on an important medicinal plant: *Amorphophallus paeoniifolius*. *Ayu.*, *33*(1), 27–32.

22. Ekor, M., (2014). The growing use of herbal medicines: Issues relating to adverse reactions and challenges in monitoring safety. *Frontiers in Pharmacology*, *4*(177), 1–10.

23. Ghosh, D., (2012). Bhantweed having multifarious medicinal properties. *Sci. Cult.*, *78*, 174–176.

24. Gouthamchandra, K., Mahmood, R., & Manjunatha, R. H., (2010). Free radical scavenging, antioxidant enzymes and wound healing activities of leaves extracts from *Clerodendrum infortunatum* L. *Environ. Toxicol. Pharmacol.*, *30(1)*, 11–18.

25. Grice, H. P., (1988). Enhanced tumor development by butylated hydroxyl anisole (BHA) from the prospective of effect on fore-stomach and oesophageal squamous epithelium. *Food Chem. Toxicol.*, *26*, 717–723.

26. Gupta, R., & Singh, H. K., (2012). Nootropic potential of *Alternanthera sessilis* and *Clerodendrum infortunatum* leaves on mice. *Asian Pac. J. Trop. Dis.*, *2*, S465–S470.

27. Haris, M., Mahmood, R., Rahman, H., Nazneen, & Rah, B., (2015). Inhibition of wound closure and decreased colony formation by *Clerodendrum infortunatum* L. in lung cancer cell line. *Int. J. Curr. Res. Biosci. Plant Biol.*, *2*(9), 66–73.

28. Haris, M., Mahmood, R., Rahman, H., & Rahman, N., (2016). In vitro cytotoxic activity of *Clerodendrum infortunatum* L. against T47D, PC-3, A549 and HCT-116 human cancer cell lines and its phytochemical screening. *Int. J. Pharm. Pharm. Sci.*, *8*(1), 439–444.

29. Harvey, A. L., Edrada-Ebel, R., & Quinn, R. J., (2015). The re-emergence of natural products for drug discovery in the genomics era. *Nature Reviews Drug Discovery*, *14*(2), 111–129.

30. Hassan, B. A. R., (2012). Medicinal plants: Importance and uses. *Pharmaceut Anal. Acta*, *3*, E139–E145.

31. Helen, L. R., Jayesh, K., & Latha, M. S., (2016). Safety assessment of aqueous acetone extract of *Clerodendrum infortunatum* L. in Wistar rats. *Comp. Clin. Pathol.*, *25*(6), 369–375.

32. Helen, L. R., Jyothilakshmi, M., & Latha, M. S., (2014). Phytochemical and antioxidant screening of various extracts of the root bark and leaf of *Clerodendrum infortunatum* Linn. *Pharmacophore*, *5*(2), 343–351.

33. Hossein-zadeh, S., Jafarikukhdan, A., Hosseini, A., & Armand, R., (2015). The application of medicinal plants in traditional and modern medicine: A review of *Thymus vulgaris*. *International Journal of Clinical Medicine, 6*(9), 635–642.

34. Husain, N., & Kumar, A., (2012). Reactive oxygen species and natural antioxidants: A review. *Advances in Bioresearch, 3*(4), 164–175.

35. Irchhaiya, R. A., Kumar, A., Yadav, N., & Gupta, S., (2015). Metabolites in plants and its classification. *World Journal of Pharmacy and Pharmaceutical Sciences, 4*(1), 287–305.

36. Iwalewa, E. O., McGaw, L. J., Naidoo, V., & Eloff, J. N., (2007). Inflammation: The foundation of diseases and disorders. A review of phytomedicines of South African origin used to treat pain and inflammatory conditions. *Afr. J. Biotechnol., 6*(25), 2868–2885.

37. Kalita, P., Dey, B. K., & Talukdar, A., (2014). Antidiabetic activity of methanolic root extract of *Clerodendrum infortunatum. J. Adv. Pharm. Res. Biosci., 2*(3), 68–71.

38. Kapoor, L. D., (2001). *Handbook of Ayurvedic Medicinal Plants* (1ˢᵗ edn., pp. 124–125). CRC Press, New Delhi—India.

39. Kar, P., Goyal, A., Das, A., & Sen, A., (2014). Antioxidant and pharmaceutical potential of *Clerodendrum* L.; An overview. *International Journal of Green Pharmacy, 8*(4), 210–216.

40. Kennedy, D. O., & Wightman, E. L., (2011). Herbal extracts and phytochemicals: Plant secondary metabolites and the enhancement of human brain function. *Adv. Nutr., 2*(1), 32–50.

41. Khaleefathullah, S., (2002). Unani medicine. In: Roy, C. R., & Muchatar, R. U., (eds.), *Traditional Medicine in Asia* (pp. 31–46). WHO-Regional Office for South East Asia, New Delhi.

42. Khare, C. P., (2007). *Indian Medicinal Plants: An Illustrated Dictionary* (p. 160). Springer, London.

43. Kharkwal, H., & Joshi, D. D., (2012). Antifungal activities of the leaf extract of *Clerodendrum infortunatum* Retz. *World Appl. Sci. J., 20*(11), 1538–1540.

44. Kirthikar, K. R., & Basu, B. D., (1987). *Indian Medicinal Plants* (Vol. 2, p. 1479). Lalit Mohan Basu Publisher, Allahabad, India.

45. Korkina, L., Kostyuk, V., De Luca, C., & Pastore, S., (2011). Plant phenylpropanoids as emerging anti-inflammatory agents. *Mini Rev. Med. Chem., 11*(10), 823–835.

46. Leena, P. N., & Aleykutty, N. A., (2013). Pharmacognostic, phytochemical studies on the root of *Clerodendrum infortunatum* Linn root. *International Journal of Research in Pharmacy and Chemistry, 3*(2), 182–187.

47. Maisch, J. M., (1885). On some useful plants of the natural order of *Verbenaceae. American Journal of Pharmacy, 57*(7), 189–199.

48. Mittal, M., Siddiqui, M. R., Tran, K., Reddy, S. P., & Malik, A. B., (2014). Reactive oxygen species in inflammation and tissue injury. *Antioxidants and Redox Signaling, 20*(7), 1126–1167.

49. Modak, M., Dixit, P., Londhe, J., Ghaskadbi, S., & Devasagayam, T. P. A., (2007). Indian herbs and herbal drugs used for the treatment of diabetes. *J. Clin. Biochem. Nutr., 40*(3), 163–173.

50. Modi, A. J., Khadabadi, S. S., & Deore, S. L., (2010). *In vitro* anthelmintic activity of *Clerodendrum infortunatum. International Journal of Pharm. Tech. Research, 2*(1), 375–377.

51. Modi, A. J., Khadabadi, S. S., Deore, S. L., & Kubde, M. S., (2010). Antioxidant effects of leaves of *Clerodendrum infortunatum* (Linn.). *International Journal of Pharmaceutical Science and Research*, *1*(4), 67–72.

52. Moldenke, H. N., (1971). *Fifth Summary of Verbenaceae, Avicenniaceaę, Sillbaceae, Dicrastylidaceae, Nymphoremaceae, Nyctanthaceae and Eriocaneaceae of the World as to Valid Taxa, Geographic Distribution and Synonymy* (Vol. 1(M), p. 312). Brawn-Brumfield Inc., Michigan.

53. Narayanaswamy, V., (1975). *Introduction to the Siddha System of Medicine* (p. 217). Institute of Siddha Medicine, Tulsi Nagar, Madras (Chennai), India.

54. Nath, A., & Hati, G. G., (2012). *Shuktani*: A new ethnomedico recipe among the Sylheti Bengali valley, Southern Assam, India. *Indian Journal of Traditional Knowledge*, *11*(1), 156–166.

55. Newall, C. A., Anderson, L. A., & Philipson, J. D., (1996). *Herbal Medicines: A Guide for Health Care Professionals* (p. 218). The Pharmaceutical Press, London E1W 1AW, UK, online E-book.

56. O'Leary, N., Calvino, C. I., Martínez, S., Lu-Irving, P., Olmstead, R. G., & Mulgura, M. E., (2012). Evolution of morphological traits in Verbenaceae. *American Journal of Botany*, *99*(11), 1778–1792.

57. Pal, D., Sannigrahi, S., & Mazumder, U. K., (2009). Analgesic and anticonvulsant effects of saponin isolated from the leaves of *Clerodendrum infortunatum* Linn in mice. *Indian J. Exp. Biol.*, *47*(9), 743–747.

58. Pandey, K. B., & Rizvi, S. I., (2009). Plant polyphenols as dietary antioxidants in human health and disease. *Oxidative Medicine and Cellular Longevity*, *2*(5), 270–278.

59. Patil, U. H., & Gaikwad, D. K., (2011). Phytochemical screening and microbicidal activity of stem bark of *Pterocarpus marsupium*. *International Journal of Pharma: Sciences and Research*, *2*(1), 36–40.

60. Paulsen, B. S., (2010). Highlights through the history of plant medicine. In: Bernhoft, A., (ed.), *Bioactive Compounds in Plants-Benefits and Risks for Man and Animals* (p. 18–29). The Norwegian Academy of Science and Letters, Oslo.

61. Rahman, T., Hosen, I., Islam, M., & Shekhar, H., (2012). Oxidative stress and human health. *Advances in Bioscience and Biotechnology*, *3*(7A), 997–1019.

62. Rajurkar, B. M., (2010). Morphological study and medicinal importance of *Clerodendrum infortunatum* Gaertn. (Verbenaceae), found in Tadoba National Park, India. *Asian Journal of Pharmaceutical Research and Health Care*, *2*(2), 216–220.

63. Rasoanaivo, P., Wright, C. W., Willcox, M. L., & Gilbert, B., (2011). Whole plant extracts versus single compounds for the treatment of malaria: Synergy and positive interactions. *Malaria Journal*, *10*(1), S4–S10.

64. Ravishankar, B., & Shukla, V. J., (2007). Indian systems of medicine: Brief profile. *Afr. J. Tradit. Complement. Altern. Med.*, *4*(3), 319–337.

65. Robert, M. F., & Wink, M., (1998). *Alkaloids, Biochemistry, Ecology and Medicinal Applications* (p. 1–7). Plenum Press, New York.

66. Rueda, R. M., (1993). *Annals of the Missouri Botanical Garden* (p. 870–890). Missouri Botanical Garden Press, St. Louis, MO.

67. Sannigarhi, S., Mazumder, U. K., Pal, D. K., & Parida, S., (2009). *In vitro* antioxidant activity of methanol extract of *Clerodendrum infortunatum* Linn. *Oriental Pharmacy and Experimental Medicine*, *9*(2), 128–134.

68. Sannigrahi, S., Mazumder, U. K., Pal, D., & Mishra, S. L., (2009). Hepatoprotective potential of methanol extract of *Clerodendrum infortunatum* Linn. against CCl_4 induced hepatotoxicity in rats. *Phcog Mag.*, *5*(1), 394–399.

69. Sannigrahi, S., Mazumder, U. K., Pal, D., & Mishra, S. L., (2012). Terpenoids of methanol extract of *Clerodendrum infortunatum* exhibit anticancer activity against Ehrlich's ascites carcinoma (EAC) in mice. *Pharmaceutical Biology*, *50*(3), 304–309.

70. Saroj, S., (2016). Comprehensive overview of a traditional medicinal herb: *Clerodendrum infortunatum* Linn. *Journal of Pharmaceutical and Scientific Innovation*, *5*(3), 80–84.

71. Sharangi, A. B., (2009). Medicinal and therapeutic potentialities of tea (*Camellia sinensis* L.): A review. *Food Research International*, *42*(5), 529–535.

72. Sharma, A., Shanker, C., Tyagi, L. K., Singh, M., & Rao, C. V., (2008). Herbal medicine for market potential in India: Overview. *Academic Journal of Plant Sciences*, *1*(2), 26–36.

73. Sharma, S. K., Singh, L., & Singh, H., (2013). A review on medicinal plants having antioxidant potential. *Indian Journal of Research in Pharmacy and Biotechnology*, *1*(3), 404–409.

74. Shrivasatava, N., & Patel, T., (2007). Clerodendrum and healthcare: An overview. *Medicinal and Aromatic Plant Science and Biotechnology*, *1*(1), 142–150.

75. Siegler, D. S., (1998). *Plant Secondary Metabolism* (pp. 337–405). Kluwer Academic Publishers, New York - United States.

76. Singh, V. K., Ali, Z. A., & Siddiqui, M. K., (1997). Medicinal plants used by the forest ethnics of Gorakhpur district (Uttar Pradesh). *Int. J. Pharm.*, *35*(3), 194–206.

77. Sreeletha, C., & Geetha, P. R., (2011). Pesticidal effects of *Clerodendron infortunatum* on the fat body of *Oryctes rhinoceros* (Linn.) male. *Journal of Biopesticides*, *4*(1), 13–17.

78. Sullivan, L. B., & Chandel, N. S., (2014). Mitochondrial reactive oxygen species and cancer. *Cancer Metab.*, *2*(1), 17–21.

79. Suman, & Gupta, R., (2012). Detection and quantification of quercetin in roots, leaves and flowers of *Clerodendrum infortunatum* L. *Asian Pacific Journal of Tropical Diseases*, *2*(2), S940–S943.

80. Thoppil, R. J., & Bishayee, A., (2011). Terpenoids as potential chemo preventive and therapeutic agents in liver cancer. *World J. Hepatol.*, *3*(9), 228–249.

81. Umadevi, M., Kumar, K. P. S., Bhowmik, D., & Duraivel, S., (2013). Traditionally used anticancer herbs in India. *Journal of Medicinal Plant Studies*, *1*(3), 56–74.

82. Veeresham, C., (2012). Natural products derived from plants as a source of drugs. *Journal of Advanced Pharmaceutical Technology and Research*, *3*(4), 200–201.

83. Vogt, T., (2010). Phenylpropanoid biosynthesis. *Molecular Plant*, *3*(1), 2–20.

84. Wagner, H., & Ulrich-Merzenich, G., (2009). Synergy research: Approaching a new generation of phytopharmaceuticals. *Phytomedicine*, *16*(2), 97–110.

85. Waliullah, T. M., Yeasmin, A. M., Ashraful, A., Wahedul, M. I., & Parvez, H., (2014). Antimicrobial potency screening of *Clerodendrum infortunatum* Linn. *Int. Res. J. Pharm.*, *5*(2), 57–61.

86. Williamson, E. M., (2001). Synergy and other interactions in phytomedicines. *Phytomedicine*, *8*(5), 401–409.

87. Wonjo, J., Bogatek, R., Sampietro, D. A., & Sgariglia, M. A., (2009). Alkaloids. In: Narwal, S. S., Bogatek, R., Zagdanska, B. M., Sampietro, D. A., & Vattuone, M. A., (eds.), *Plant Biochemistry* (pp. 315–335). Stadium Press LLC, New Delhi.

88. World Health Organization (WHO), (2004). *Guidelines on Safety Monitoring of Herbal Medicines in Pharmacovigilance Systems* (p. 34). World Health Organization, Geneva, Switzerland.

89. Yang, L., Yang, C., Li, C., Zhao, Q., Liu, L., Fang, X., & Chen, X. Y., (2016). Recent advances in biosynthesis of bioactive compounds in traditional Chinese medicinal plants. *Science Bulletin, 61*(1), 3–17.

90. http://siddham.in/clerodendrum-infortunatum (accessed on 18 May 2020).

91. https://www.hexaresearch.com/research-report/global-herbal-medicine-market/ (accessed on 18 May 2020).

92. http://www.biologyreference.com/Re-Se/Secondary-Metabolites-in-Plants.html (accessed on 18 May 2020).

CHAPTER 3

FUNCTIONAL BENEFITS OF
FICUS HISPIDA L.

D. SUMA, A. VYSAKH, R. N. RAJI, NINAN JISHA, and M. S. LATHA

ABSTRACT

The present review discussed the various important pharmacological activities of *Ficus hispida Linn.*, which is used as a folklore medication for the healing of various diseases, because of its phytochemical properties, such as antidiarrheal, antipyretic, antimicrobial, astringent, vulnerary, hepatoprotective, anti-inflammatory, antineoplastic, hemostatic, and anti-ulcer medicine. Diverse parts of these plant parts have been used for healing of jaundice, vitiligo, hemorrhage, diabetes, ulcers, psoriasis, anemia, piles, convulsion, hepatitis, dysentery, biliousness, lactagogue, and purgative. The present review revealed the importance of isolation and identification of active principles and it is crucial that to explain the unknown potential of this plant with potent researches in conjunction with the organization of superior quality assurance should be established. The current review forms the source for the collection of plant species for more examination in the possible finding of new innate biocompounds.

3.1 INTRODUCTION

Since ancient times, traditional herbal medicines are key remedial measures among rural dwellers throughout the world. Treatment with medicinal plants is as old as humankind itself. Preserved monuments, written documents, and even original plant medicines are some of the ample evidences, which have shown the relation between man and his look for new effective drugs from the past. Understanding of phytochemical treatment is a product of the many years of knowledge against diseases due to which man learned to follow drugs in barks, fruit bodies, seeds, and other parts of the plants.

This chapter discusses the important pharmacological activities of *F. hispida*. Additionally, it is crucial to explain the unknown potential of this plant with potent researches in conjunction with the organization of superior quality assurance.

3.2   HISTORY OF PLANT-BASED MEDICINES

The curative use of plants originated during remote times. One of the earliest authors recorded about 450 diverse species of plants for therapeutic uses and mentioned plant- and animal-based natural products. Prehistoric conventional medicine systems (such as Chinese, Ayurvedic, and Egyptian) used plant-based natural products. Vedas mentioned health-care therapies with medicinal plants, which are surplus in India. Ayurveda, Homeopathy, Sidda, and Unani are some of the age-old conventional systems of medicine [12].

The rich diversity of plant species is available in the Indian sub-continent. Tribal village communities use approximately 8,000 species of medicinal plants among the 17,000 species of higher plants, as traditional medicinal systems, such as the Ayurveda. Countless plants used as spices originated from India, such as nutmeg, ginger, etc. Spices and herbs have presumed pivotal role in the modern era as the natural origin of chemotherapeutical agents and explore substitute sources of drugs. Conventional plant-based medicines are relied by most of the people in the developing countries. About 88% of the earth's population depends on the conventional remedy for their predominant health maintenance. Many of the wild plants are endemic and are found only in specific ecological niches.

3.3   CURRENT SCENARIO

According to the report of the World Health Organization (WHO), about 80% of the inhabitants in many third world countries even today use conventional medicine for their curative care, owing to poverty and lack of entry into modern medicines. More than 8000 species of plants have been evaluated to be used in the endemic health system and medicinal plants play a pivotal role in the primary life-supporting system of rural and tribal communities. Since about 80% of the 6.1 billion people of the world live in developing countries, 3.9 billion people are frequently using medicinal plants.

Therefore, there is an urgent need to search for invaluable medicinal compounds from which innovative therapeutic drugs may be generated and to investigate medicinal plants for their efficacy, quality, and safety.

The discovery of medicinal plants as sources of novel drugs to cure cancer, acquired immunodeficiency syndrome (AIDS), etc. needs to explore as many solutions as possible. Moreover, the side effects of natural curatives are comparatively very low.

3.4 PHYTOCONSTITUENTS

According to World Health Organization (WHO), any plant (one or more of its parts) is considered a medicinal plant, which constitutes components that can be utilized for curative basis, or act as predecessors for the synthesis of natural medicinal products. Leaves, barks, roots, flowers, fruits, rhizomes, grains, stems, or seeds of medicinal plants are utilized in the healing of a pathological condition because of medically active chemical constituents. These non-nutritious bioactive compounds are known as phytoconstituents or phytochemicals or secondary metabolites and are responsible to keep the plant resistant to microbial infestations or infections by various insects and pests. It is understood that plants construct these phytochemicals to defend themselves. However, latest research studies illustrate that these phytoconstituents also provide protection against human diseases [19].

The pharmacological activity has been investigated in a few members of the plant population till date. Only 5–10% of the plants have been explored so far among 250,000 species of plants that exist on this earth. Phytochemistry is the study of natural products (chemicals) collected from the plants. Phytochemicals or phytoconstituents are biologically dynamic, innately available chemical constituents available in plants, which can deliver health welfare to human beings. Phytoconstituents are mainly classified as primary and secondary chemicals based on chemical reactions related to plants.

Sugars, proteins, amino acids, pyrimidines, and purines of nucleic acids, chlorophyll, etc. are considered as primary phytoconstituents. Plants require these substances for reproduction, normal growth, and development of plants.

Secondary phytoconstituents are compounds that have no significant role in the perpetuation of fundamental life activities in plants, but they have a remarkable position in the communication of plants with its surroundings [9]. They safeguard plants from damage and diseases and give color, flavor, and aroma to the plant. Plants synthesize an inexplicable variety of phytochemicals, but most are derivatives of a few biochemicals, e.g., alkaloids, flavonoids, phenolics, terpenoids, and glycosides [2]. Plants produce thousands of secondary metabolites, and most of them have potent physiological effects in humans and are used for medicinal purposes. Traditionally used medicinal plants produce various phytoconstituents of known therapeutic properties [7].

Alkaloids, flavonoids, Phenolics, terpenoids, glycosides, tannins, and saponins are major groups of secondary metabolites present in plants. Leaves, flowers, roots, stems, fruits, or seeds are parts of plants, where phytoconstituents are accumulated. The outer layer of plant tissues is occupied with surplus phyto-constituents, such as pigment molecules. Potency of phytochemicals as defense metabolites mainly relies on their several molecular targets, both in prokaryotic and eukaryotic cells, including receptors, enzymes, structural macromolecules, ion channels, and nucleic acids. As a result, secondary metabolites intercede with number of cell processes, such as cell cycle, mitosis, programmed cell death, signaling pathways, protein synthesis and folding, and so on.

TABLE 3.1   Bioactive Phytocompounds in Medicinal Plants and Herbs

Classification	Major Classes of Compounds	Biotic Functions
Antibacterial and Antifungal	Terpenoids, alkaloids, phenolics.	Microbial inhibitors diminish the possibility of fungal infection.
Anticancer	Carotenoids, polyphenols, curcumine, flavonoids.	Tumor inhibitors, retarded progression of lung cancer, anti-metastatic effect.
Antioxidants	Polyphenolic compounds, tocopherols, ascorbic acid flavonoids, carotenoids.	Oxygen-free radical quenching, lipid peroxidation inhibitors.
Detoxifying agents	Reductive acids, coumarins, flavones, carotenoids, tocopherols, indoles, phenols, retinoids, cyanates, aromatic isothiocyanates phytosterols.	Procarcinogen activation inhibitors, drug binding inducers of carcinogens, suppressors of tumorigenesis.
NSA (polysaccharides non-starch)	Cellulose, gums, hemicellulose, mucilages, lignins, pectins.	Water holding ability, interruption in absorption of nutrients, binding of toxins and bile acids.
Others	Alkaloids, biogenic amines, terpenoids, volatile flavor compounds.	Neuropharmacological agents, anti-oxidants, cancer chemoprevention.

Because plant tissues have hundreds of natural compounds and unique nature of phytochemicals, there exists a low risk of budding phytochemical-resistant pathogens or insects compared with conventional commercial agrochemicals or pharmaceuticals. The use of herbal drugs is once again becoming more popular in the form of dietary supplements, nutraceuticals, and complementary and alternative medicines. Phytoconstituents have been isolated from fruits, spices, beverages as well as many other sources. Bioactive and infection resisting phytocompounds available in the plant are listed in Table 3.1.

3.5 FICUS HISPIDA L.

Ficus hispida (FH) *Linn.* is grown for its comestible fruits and traditional value and is found throughout the year. Conventionally, diverse plant parts have been included in the healing of jaundice, vitiligo, hemorrhage, diabetes, ulcers, psoriasis, anemia, piles, convulsion, hepatitis, dysentery, biliousness, lactagogue and as purgative. Alkaloids, carbohydrates, phenols, flavonoids, amino acids and proteins, sterols, gums and mucilage, saponins, glycosides, and terpenes are bioactive phytochemical groups found in FH. To substantiate the scientific reason of customary medicinal values endorsed to FH, numerous experimental studies have been reported [20]. Pharmacological activities (like antineoplastic, antimicrobial, cardioprotective, anti-inflammatory, and neuroprotective) are also described in this chapter.

It is reported that traditional herbal practitioners for the treatment of diabetes use leaves and fruits of *F. hispida*. FH has significant hypoglycemic effect. Pratumvinit et al. [24] studied *in vitro* antineoplastic efficacy of the stem of FH sequentially extracted with crude methanol, ethyl acetate, methanol, water, and ethanol, and water fractions against human breast cancer cell lines, such as SKBR3, MCF7, MDA-MB435 and T47D [24]. Anti-neoplastic activity against T47D cells was shown only in the ethanolic fraction. The presence of constituent 6-O-methyltylophorinidine was the key factor, which contributed to the mechanism behind this activity, and it has been investigated to display cytotoxicity for prostrate, nasopharynx, lung, and colon cancer cell lines. Inhibition of cell growth and apoptosis initiation is due to the cytotoxic effect. Apart from the action of current known anticancer agents, recent studies have revealed the unique action of the tylophorine analogs. One of its unique actions is the suppressive result on NF-KB binding arbitrated transcription, guiding to apoptosis. An effective anticancer agent present in the bark of FH is hispidin [13].

Research studies confirm that many plants (e.g., *Moringa oleifera*) show multiple medicinal properties like antibacterial [15], antioxidant [28], and anticancer activities [11]. And bioactive compounds in *Ficus hispida* can fight against different diseases. Recent study related to anticancer activity on human lung cancer cell lines by *Scutellaria barbata* extract implied that the treatment of cell lines with plant extract change the expression profile of around twenty cell cycle genes and also showed downregulation (102-folds) of CD209, a dendritic cell-specific ICAM-grabbing, non-integrin. Inspired from this work, the authors of this chapter anticipate that the extracts of *Ficus hispida L.* plants could change the gene expression profile in different breast cancer cell lines. Indian conventional practitioners utilize most of the plant parts as a folk remedy for various ailments. The leaves and bark among other parts of this plant are used specially as an antitussive [22], antidiarrheal, astringent, antihepatotoxicity, anti-inflammatory [21], vulnerary, hemostatic, febrifuge, and anti-ulcer drug [31]. All these studies give valuable information about the efficacy of the compounds isolated from this plant in cancer treatment.

Ficus hispida L. originated from Latin words: *FIK*-us and *HISS*-pih-duh. Its common names are hairy fig, devil fig, FigTree with rough-leaves [5, 27]. In diverse states of India, it is identified by dissimilar names. It belongs to the family Moraceae and genus *Ficus*.

3.6 EXTRACTION METHODS

In the therapy of jaundice [3] and diabetes, tribal communities of Assam and Manipur have utilized the extracts of leaves, bark, and roots. The plant extraction methods have been confirmed by various scientists and recorded in the literature. Methanolic extract was prepared by extracting powdered dried leaves of FH with methanol keeping in water bath (50°C). New solvent was added to the plant sample after removing the solvent by filtration. This procedure was repeated two times. The collected extract was kept at 0–4°C for the further study [10].

A somewhat altered technique was adopted by some workers in which petroleum ether was used to defeat the dried leaves, and then further extraction was done with methanol. Ghosh et al. [10] extracted ground dried bark and isolated the water-soluble portion after double distillation with ethanol. The same extraction method was approved for leaves.

3.7 PHYTOCHEMISTRY AND BIOLOGICAL ACTIVITIES OF THE SELECTED MEDICINAL PLANTS

Initial phytochemical screening of FH has shown the occurrence of most of the phytochemicals [1]. Acharya et al. [1] reported that the FH bark constitutes β-amyrine acetate, lupeol acetate, β-sitosterylcapriate, lupeol, and α-amyrin [13]. In another study, purification of the dried bark powder extracted in petroleum ether was carried out with acetates of *n*-triacontanol, glucanol, and β-amyrin [23].

The 10-ketotetracosyl arachidate, a novel, and atypical compound, was discovered with petroleum extract. From stem and leaves of FH, two significant phenanthroindolizidine alkaloids (such as 6-O-methyltylophorinidine, 2-demethoxytylophorine) and bi-phenylhexahydroindolizinehispidine were also identified by *Venkatachalam* et al. *[30]*. Recently, it was shown that hispidin has anticancer activity [14].

Peraza-Sanchez et al. identified alkaloids (such as *n*-alkanes, coumarins, triterpenoid and ficustriol) from the methanolic extract of leaves and twigs of FH. Recently, hispidin, β-amyrine, oleanolic acid, bergaptine, and β-sitosterol were obtained from the leaves of FH by Huong et al. [13]. Linalool, linalool oxide, terpeneol, and 2,6-dimethyl-1,7-octadiene-3,6-diol are volatile chemicals isolated from the FH fruits. These compounds act as insect-attractants, which help in pollination of FH. The structures of selected significant compounds in FH are given in Figure 3.1.

3.7.1 HYPOGLYCEMIC AND THROMBOLYTIC BENEFITS

In diabetic albino rats, the antidiabetic effect of FH bark was established by Ghosh et al. [10]. Ethanolic extract of the bark showed considerable decrease in blood glucose level. It amplifies the intake of glucose and glycogen matter in liver, skeletal, and cardiac muscles [30]. Because of its reasonable antimicrobial activity, *F. hispida* extracts can be used as anti-thrombolytic agents. The thrombolytic activity as assessed in human blood samples; and the highest thrombolytic activity were shown by the methanolic soluble fraction (MSF).

3.7.1.1 MECHANISM OF ACTION

Lupeol acetate, β-sitosterol, and β-amyrine acetate are antioxidant constituents of FH bark. These can reduce oxidative stress and fight against destruction of β-cells of islets of Langerhans through various mechanisms [13]. Moreover, the increased uptake of glucose is due to improved expression of receptors of glucose on the cell membrane. The mechanism behind the thrombolytic activity is due to its moderate antimicrobial activity.

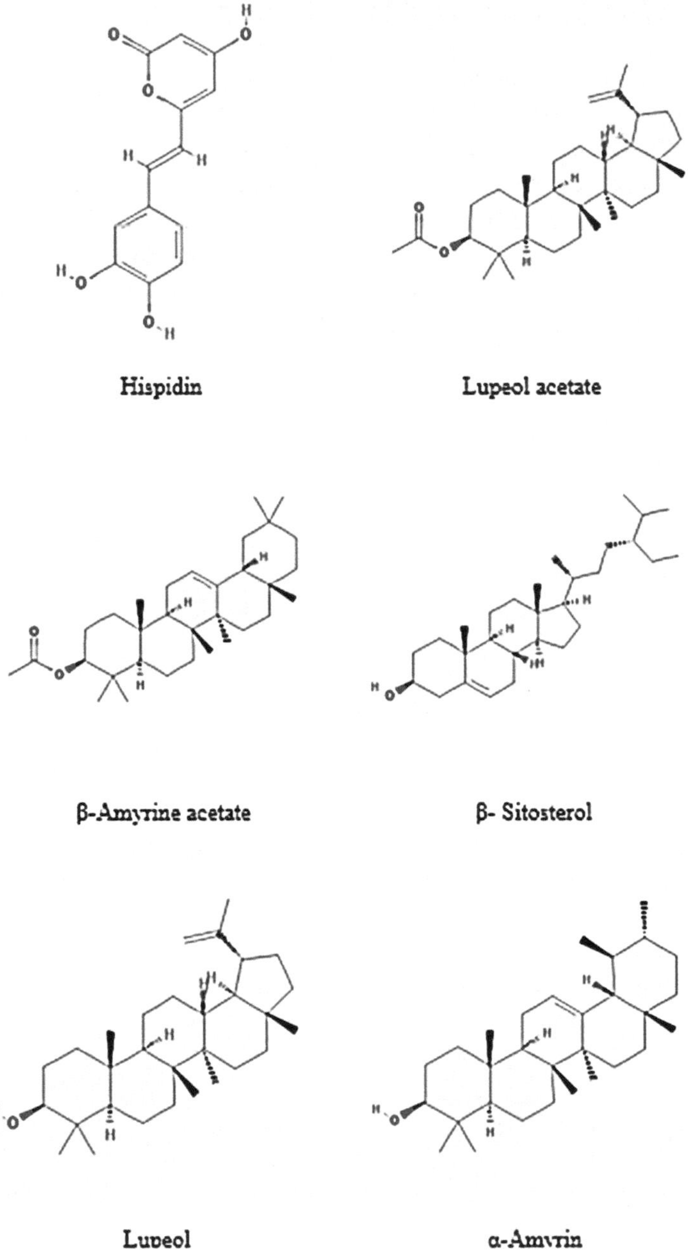

FIGURE 3.1 Some significant bioactive compounds from *Ficus hispida* L.

3.7.2 ANTINEOPLASTIC BENEFITS

A variety of Thai traditional remedies use stems of the *Ficus hispida L.* to prescribe for cancer therapy. The antineoplastic result of FH stem was illustrated during sequential extraction with methanol, water-crude ethanol, water-methanol-ethyl acetate fractions on SKBR3, MCF7, T47D, and MDA-MB435 cell lines *in vitro* by Pratumvinit et al. [24]. Anti-neoplastic activity against T47D cells was shown by only ethanol fraction [24]. MTT assay, cell cycle analysis, and colony-forming assay were also analyzed. According to MTT assay, the IC50 of this extract against T47D cell was $110.3+/-9.63$ µg/ml and cell growth inhibition in a dose-dependent manner was confirmed by colony-forming assay. The percentage increase of apoptotic cell population in herbal treated cells was confirmed by cell-cycle analysis. Therefore, *F. hispida L.* has valuable role in the healing of breast cancer by traditional medicine.

3.7.2.1 MECHANISM OF ACTION

The presence of constituent O-methyltylophorinidine is accountable for the cytotoxicity, which has been exhibited for nasopharynx, lung, colon, and prostate cancer cell lines. Prevention of cell growth and stimulation of apoptosis are major cytotoxic effects [24]. Recent studies have indicated that tylophorine analogs have a distinctive means of action. The inhibitory effect on transcription, which is mediated through binding of NF-KB, is one of its specific unique actions guiding to apoptosis. According to a fascinating report by Huong et al. [13], the bark of FH contains a strong anticancer agent, hispidin [25].

3.7.3 ANTINOCICEPTIVE AND NEUROPHARMACOLOGICAL BENEFITS

The neuropharmacological and antinociceptive properties of the ethanolic extract obtained from the fruits of *F. hispida L.* offer the systematic base for the customary use of this plant as a medicine for reducing pain and sadness. The results confirm the purpose of FH plant in conventional remedy. Fruit extracts of *Ficus hispida L.* show analgesic and dose-dependent activities. A significant peripheral antinociceptive effect is shown by oral administration of fruits of *F. hispida L.* on a writhing test induced with acetic acid. The intensity of CNS excitability is measured in terms of loco motor activity and diminution of this outcome may be directly linked to despair of the CNS resulting from sedation. Diminished locomotor activity specifies the CNS depressant activity with extracts of *F. Hispida L*. Major inhibitory neurotransmitter of the CNS is GABA (gamma-aminobutyric acid). Extracts of *F. hispida L*. may act via membrane hyperpolarization by potentiating GABAergic suppression in the CNS. The extract reduces the firing rate of critical neurons in the brain through direct activation of GABA receptors. Ethanol extract of the fruits of FH showed neuropharmacological and antinociceptive activities and thus furnishes the proof for the customary uses of FH plant parts as a remedy for pain and sadness.

3.7.3.1   MECHANISM OF ACTION

Phytoconstituents with possess sedative effects are tannins, glycosides, triterpenes and saponins. Various flavonoids and steroids of neuroactive substances act as benzodiazepine-like molecules and ligands for the $GABA_A$ receptors in the central nervous system (CNS). This explained the behavioral effects, such as anxiety, sedation, and convulsion in animal models. The actions of alkaloids are responsible for the depressant effect on the cerebral mechanism in sleep regulation.

3.7.4   CARDIOPROTECTIVE BENEFITS

The cardio-protective outcome of FH leaf extract was assessed by Shanmugarajan et al. [26] by injecting cyclophosphamide (CP) that is related to myocardial damage because of oxidative stress generated in rat heart. This study illustrated that the extract of leaf showed a considerable reticence of oxidative degradation of lipid and augmented the intensity of superoxide dismutase, glutathione reductase, catalase (CAT), glutathione peroxidase (GPx), glutathione-S-transferase; and diminished the glutathione activity in heart tissues established by CP [1].

3.7.4.1   MECHANISM OF ACTION

Cardio-protective activity is due to the antioxidant potential of the phytoconstituents. The compounds of FH (such as hispidin, β-amyrin, β-sitosterol, and bergapten) can inhibit oxidation [14]. Additionally, glutathione mediated the antioxidant property of the oleanolic acid and α-tocopherol giving defense against the injury caused by myocardial ischemia-reperfusion. This ability is due to the re-establishment of antioxidant stage and avoidance of oxidative degradation of lipids. It has been recently investigated that β-amyrinis an additional component that also donates to the expansion affect because of which there is a GSH-restoration effect [27].

3.7.5 ANTIOXIDANT BENEFITS

Polyphenols help in the neutralization and absorption of free radicals, quenching singlet, and triplet oxygen, or decomposing of peroxides. Numerous phytocompounds have considerable antioxidant capabilities that are connected with lower occurrence and lesser death rates of number of human diseases.

Antioxidative status phenomenon of human diseases shows an inverse connection. Antioxidants are used for the avoidance and healing of free radical-linked ailments. As a result, research related to identification of antioxidants is a significant subject. Based on new findings concerning their biological activities, polyphenols have recently acquired increasing attention. Free radical scavenging and retardation of lipid peroxidation are antioxidant properties of polyphenols that play a central role related to pharmacological and curative outlook.

However, the studies on isolation and evaluation of polyphenols and their antioxidative benefits are limited. Azathioprine induced oxidative stress can be assessed because of the antioxidant efficacy of the *Ficus hispida* leaves [26]. DPPH and the nitric oxide radical inhibition assays revealed strong antioxidant activity of methanolic extract of *F. hispida*. Liver disorders can be protected through the action of free radical scavengers. These effects support the curative efficacy of *F. hispida* extracts against free radical-mediated oxidative changes.

3.7.5.1 MECHANISM OF ACTION

Oleanolic acid, hispidin, bergapten, and β-sitosterol are the phytoconstituents present in FH and these act as antioxidants/anti-lipid peroxidants. Their function is to protect membrane lipids from generating oxidative damage. It is done through cessation of peroxyl radical dependent reactions. SOD, CAT, GPx, and GST are the enzymes of antioxidant nature, which have vital function in the removal of reactive oxygen species. GSH, vitamin C, and vitamin E are intimately related non-enzymatic antioxidants and play an important role in defending the cell from damage due to oxidation.

F. hispida extracts possess some bioactive antioxidant phytoconstituents, which afforded the antioxidant activity and lessened the utilization of endogenous antioxidants, which could be accountable for the decline of AZA activated oxidative stress probably through hepatic GSH reparative effect. Recently it has recognized that tylophorine counterpart as definite action compared to the available anticancer agents. Repressive action on NF-KB binding arbitrated transcription, guiding to apoptosis is one of its actions.

3.7.6   HEPATOPROTECTIVE EFFECT

Shanmugarajan et al. conducted oxidative liver wound study in Wistar rat with CP and studied the anti-hepatotoxic effect on the methanol extract of the *ficus* leaves [26]. It is reported that azathioprine induced liver damage was effective with methanolic leaf extract in Wistar rat model [6]. Mandal et al. investigated the hepatoprotective effect by generating acute liver injury by paracetamol in rats [20, 21]. The analysis was like that of a standard Liv-52 formulation [8].

3.7.6.1   MECHANISM OF ACTION

Phytoconstituents (such as oleanolic acid and β-sitosterol) in *F. hispida* extract are responsible for the hepatoprotective property by exerting an effect on the membrane-stabilizing effect. Preliminary phytochemical studies established the existence of flavonoids and triterpenoids in the methanolic extract of FH leaves. It may be hypothesized that this antioxidant may also be answerable for the experiential hepatoprotective property.

Liu et al. [16, 17] verified the function of triterpenoid in glutathione renovation; Glutathione is an enzyme liable for cell membrane defense and control of cellular function. The oleanolic acid, hispidin, and β-sitosterol act as free radical scavengers and thus might be connected with antihepatotoxic effects. Additionally, it has been reported that oleanolic acid, phenanthroin-dolizidine [18], and β-amyrin [27] have powerful antihepatotoxicity effects.

3.7.7   ANTI-INFLAMMATORY AND ANTIPYRETIC

Significant activity is observed in Albino rat model in which paw edema was generated with carrageenan. Here *F. hispida* extract showed 64% reduction in inflammation compared with 45.13% with the control [31].

3.7.7.1   MECHANISM OF ACTION

Anti-inflammatory and antipyretic effects are due to its antioxidant activity and synthesis of protective prostaglandins. Generation of free radical or inflammatory mediators reduces the excessive oxidative damage.

3.7.8   ANTIDIARRHEAL ACTIVITY

Antidiarrheal activity in rats was evaluated with extracts from the leaves of FH against castor-oil generated diarrhea and PEG_2-originated inter-pooling in rats was studied by Mandal et al. [20, 21]. Akey and dose-dependent antidiarrheal activity was shown by the methanolic extract of leaves. They also recognized the dosage of the extract and also established that 600 mg/kg had the same effect as by 5 mg/kg of diphenoxylate [27].

3.7.8.1   MECHANISM OF ACTION

It has been revealed that denaturation of proteins is caused by tannins and protein tannate may be accountable for the activity to stop diarrhea by cutting down the permeability on the intestinal mucosa.

3.7.9   ANTIULCEROGENIC EFFECT

Sivaraman et al. [27] organized an experiment on aspirin ulcerated rats with methanolic extract of FH roots and exhibited that a dosage of 200 and 400 mg/kg significantly diminished the occurrence of ulcers, improved ulcer healing and appreciably diminished free and total acidity [13].

3.7.9.1   MECHANISM OF ACTION

Antiulcer activity is due to the cytoprotective action of the antioxidant constituent of the methanolic extract of FH roots. It arrests deprivation of mucous membrane, produces renovation of carbohydrates with carbohydrate/protein percentage, and intensifies the production of cytoprotective prostaglandins like PGI [26].

3.7.10   ANTIMICROBIAL BENEFITS

The potential antimicrobial activity of the different parts of FH was evaluated against medically important bacterial and fungal strains by disc diffusion. Diameter of zone of inhibition expressed in mm determines the antimicrobial activity of test agents. Notable bacterial growth inhibition was obtained against the tested organisms. Presence of activity at concentration of 100 µg/disc for extracts and 10 µg/disc for isolated compounds indicated antibacterial activities.

3.7.10.1   MECHANISM OF ACTION

The extracts of *F. hispida* have moderate activity against microbes. A variety of secondary metabolites are responsible for its antimicrobial activity. Hence, these plants can be utilized for the progress of new pharmaceutical investigations. Isolation of the secondary metabolites needs additional research and the mechanism of metabolic transaction in bacterial metabolic pathways also should be revealed on the extract. This study confirmed the importance of folk medicines like modern medicine to compact pathogenic microorganisms. Therefore, this can be a safe alternative to treat against infectious diseases. Plants comprise antimicrobial properties are rich source of secondary metabolites, such as tannins, terpenoids, alkaloids, flavonoids, glycosides, etc.

3.7.11   AMELIORATING BENEFITS

Neuroprotective effect of methanolic extract of *Ficus* leaves was studied on β-amyloid induced oxidative stress and cognitive deficits in mice [27]. Reduction in cognitive behavior and memory deficit were shown by *Ficus* leaves. In brain, thiobarbituric acid reactive species level is reduced. During the study, antioxidant enzymes showed their increased activity. Alzheimer's disease and age-associated memory disorders can be treated by applying these activities [32].

3.7.11.1   MECHANISM OF ACTION

Oxidative stress-induced death of the neurons is the major cause of the neurodegenerative diseases of brain-like mental illness and Alzheimer's disease.

Decreased stage of acetylcholine in connection with the over expression of acetyl cholinesterase is the reason behind this diseases [29]. Therefore, the inhibition of the cellular damage in neuron along with decline in amyloid plug content might be due to the oxidation protective bioactive compounds. Depression of enzymatic activity on acetylcholine enhances the activity of acetylcholine has been reported [4, 27].

3.7.12   WOUND HEALING AND MATERNAL BENEFITS

In traditional medicines, *Ficus hispida* is used in curing wounds and as a medicine in post-and prenatal care (maternal remedies). Individual or collective action of phytoconstituents (like tannins, alkaloids, and sapo-nins) provides effective wound healing property of the extract of *Ficus hispida*.

Incision, excision, and dead space wound models were created in Albino rats and were treated with dexamethasone depressed curing conditions and normal with ethanol extract of roots of *Ficus hispida*. As contrast to control groups, rate of epithelium formation and wound healing in excision model was improved. Granulation tissue mass and hydroxyproline content were significantly increased in dead space model compared to control group. Significant wound healing activity is established with administration of *Ficus hispida* at the injury site. Regeneration of curing tissue with distinct aridness of wound-ends was observed in wounds and reduced wound area was correlated to control signifying the curative efficacy of *Ficus hispida*.

3.7.12.1   MECHANISM OF ACTION

Good wound healing properties exhibited by the extract of *Ficus hispida* is due to the individual or combined action of phytoconstituents like tannins, alkaloids, and saponins. Reduction in wound size is also stimulated by the plant extract due to enhanced epithelization. Therefore, curing at the various stages of tissue renovations enhances the post-healing ability of the extract.

3.8   TRADITIONAL FOODS: *FICUS HISPIDA*

The fruits of *Ficus hispida* safe to eat and are consumed locally in candied form. The fruits of FH can be made as a coolant and stimulant. Along with jaggery, the juice of fig is formed as a mild purgative. The juice of fig with jaggery is used in the curation of vitiligo [16]. Unripe fruit is used for the preparation of curry by tribes of Meghalaya. The fruit juices of fig mixed with honey can act as good anti-blood extravasations. Different varieties of FH species present in the Lembah Lenggong region are integral part of the livelihood of the locals.

3.9   FUTURE PROSPECTS

In developing countries, over 80% of the inhabitants depend on herb medicines, mainly plant drugs, for their primary healthcare. Ayurvedic products are practically cost-effective and well established by patients. They are easily obtainable and have lower side effects. These herbal drugs and Indian medicinal plants are also a wealthy source of valuable compounds including antioxidants and components that can be used in functional foods. Traditional medicinal properties and curative potentials of most herbs are well-documented. About 61% of new drugs particularly in transmittable illness and cancer have been developed in recent years. It has been observed that study rate of active new chemical principles are diminishing recently. The current review will form the source for collection of plant species for more examination in the possible finding of new innate biocompounds. Additional investigation is expected on the isolation and structure elucidation of active components from the plant.

3.10   SUMMARY

Drug-induced toxicity is a common side effect of many medicines, which constitute considerable risks to patient health and has a massive monetary impact on health care. Discovery of new drugs with low side effects has become an important goal in the field of pharmacology. Therefore, there is an increased demand for plant-based medicaments. *Ficus hispida Linn.* is used as a folklore medication for the healing of various diseases due to its properties, such as antidiarrheal, antipyretic, antimicrobial, astringent, vulnerary, hepatoprotective, anti-inflammatory, antineoplastic, hemostatic, and anti-ulcer. Different ailments can be treated with various parts of this plant. The present review revealed the importance of isolation and identification of active principles and further research is necessary to clarify specific properties of *Ficus hispida Linn.*

KEYWORDS

- **anti-inflammatory**
- **antineoplastic effect**
- **antinociceptive**
- **antipyretic**
- **apoptosis**
- **cytoprotective**
- ***Ficus hispida***
- **neurodegenerative disorders**
- **neuropharmacological properties**
- **prostaglandins**

REFERENCES

1. Acharya, B. M., & Kumar, K. A., (1984). Chemical examination of the bark of *Ficus hispida* Linn. *Curr. Sci.*, *53*, 1034–1035.
2. Arunkumar, N., & Deecaraman, M., (2009). Nanosuspension technology and its applications in drug delivery. *Asian J. Pharm.*, *3*, 168–173.
3. Arunsunder, M., & Shanmugarajan, T. S., (2010). *Ficus hispida* modulates oxidative-inflammatory damage in a murine model of diabetic encephalopathy. *Ann. Biol. Res.*, *1*, 90–97.
4. Augusti, K. T., Anuradha, S. P., & Prabha, K. B., (2005). Nutraceutical effects of garlic oil its non-polar fraction and a ficus flavonoid as compared to vitamin E in carbon tetrachloride induced liver damage in rats. *Indian J. Exp. Biol.*, *43*, 437–344.
5. Ayurvedic Pharmacopoeia of India, (2005). New Delhi: Government of India, Ministry of Health and Family Welfare Department of Ayush. *Part 1: Book of Standards* (Vol. 1, pp. 149–151). http://www.ayurveda.hu/api/API-Vol-1.pdf (accessed on 30 May 2020).
6. Bai, X., Qiu, A., Guan, J., & Shi, Z., (2007). Antioxidant and protective effect of an oleanolic acid enriched extract of *A. deliciosa* root on carbon tetrachloride induced rat liver injury. *Asia Pac. J. Clin. Nutr.*, *16, 169*–173.
7. Bruneton, J., (1995). *Pharmacognosy, Phytochemistry, Medicinal Plants* (pp. 265–380). France: Lavoisiler Publishing Co.
8. Buniatian, N. D., Chikitkina, V. V., & Lakavleva, L. V., (1998). The hepatoprotective action of ellagotannins. *Eksp. Klin. Formakol.*, *61, 53*–55.
9. Caldentey, O. K. M., & Inza, D., (2004). Plant cell factories in the post genomic era: New ways to produce designer secondary metabolites. *Trends Plant Sci.*, *9*, 433–440.

10. Ghosh, R., Sharathchandra, K. H., Rita, S., & Thokchom, I. S., (2010). Hypoglycemic activity of *Ficus hispida* (bark) in normal and diabetic albino rats. *France J. Pharma. Pharmacol.*, *4*, 72–82.

11. Guevara, A. P., & Vargas, C., (1999). An anti-tumor agent from *Moringa oleifera* Lam. *Mutation Res.*, *144(2)*, 181–188.

12. https://en.wikipedia.org/wiki/Medicinal_plants (accessed on 18 May 2020).

13. Huong, V. N., & Trang, V. M., (2006). Hispidin-strong anticancer agent isolated from the leaves of *Ficus hispida* L. *Tap Chi Hoa Hoc.*, *44*, 345–349.

14. Kirtikar, K. R., & Basu, B. D., (1956). *Indian Medicinal Plants* (pp. 23–22). Allahabad, India: Lalit Mohan Basu.

15. Koteswara, P., (2011). *In vitro* antibacterial activity of *Moringa oleifera* against dental plaque bacteria. *Journal of Pharmacy Research*, *4*(3), 695–697.

16. Liu, J., (1995). Pharmacology of oleanolic acid and ursolic acid. *J. Ethnopharmacology*, *49*, 57–68.

17. Liu, J., Liu, Y. P., & Klaassen, C. D., (1995). Effect of oleanolic acid on hepatic toxicant-activating and detoxifying systems in mice. *J. Pharmacol. Exp. Ther.*, *275*, 768–774.

18. Malathi, R., & Gomez, M. P., (2007). Hepatoprotective effect of methanolic leaf extracts of *Tylophoraasthamatica* against paracetamol induced liver damage in rats. *J. Pharmacol. Toxicol.*, *2*, 737–742.

19. Mamta, S., & Jyoti, S., (2013). Phytochemistry of medicinal plants. *J. Pharm. Phytochem.*, *1*, 168–170.

20. Mandal, S. C., & Ashok, C. K., (2002). Studies on anti-diarrheal activity of *Ficus hispida* leaf extract in rats. *Fitoterapia*, *73*, 663–667.

21. Mandal, S. C., Saraswathi, B., & Kumar, C. K., (2000). Protective effect of leaf extract of *Ficus hispida* Linn. against paracetamol-induced hepatotoxicity in rats. *Phytother Res.*, *14*, 457–459.

22. Nadkarn, A. K., (1976). *Indian Materia Medica* (pages 1031–1035). Mumbai: Popular Prakashan.

23. Peraza-Sánchez, S. R., Chai, H., & Shin, Y., (2002). Constituents of the leaves and twigs of *Ficus hispida*. *Planta. Med.*, *68*, 186–188.

24. Pratumvinit, B., Srisapoomi, T., & Worawattananon, P., (2009). *In vitro* antineoplastic effect of *Ficus hispida* L. against breast cancer cell lines. *J. Med. Plants Res.*, *3*, 255–261.

25. Rastogi, R., & Mehrotra, B. N., (1993). *Compendium of Indian Medicinal Plants* (p. 27). New Delhi: CDRI-Publication and Information Directorate.

26. Shanmugarajan, T. S., & Arunsundar, M., (2008). Ameliorative effect of *Ficus hispida* Linn leaf extract on cyclophosphamide-induced oxidative hepatic injury in rats. *J. Pharmacol. Toxicol.*, *3*, 363–372.

27. Sivaraman, D., & Muralidaran, P., (2009). Sedative and anticonvulsant activities of the methanol leaf extract of *Ficus hispida* Linn. *Drug Invent Today*, *1*, 23–27.

28. Sunday, E., Atawodi, C., & Atawodi, A., (2010). Evaluation of the polyphenol content and antioxidant properties of methanol extracts of the leaves, stem, and root barks of *Moringa oleifera* Lam. *Journal of Medicinal Food*, *6*, 710–716.

29. Tang, X. H., Gao, J., Fang, F., Chen, J., Xu, L. Z., & Zhao, X. N., (2005). Hepatoprotection of oleanolic acid is related to its inhibition on mitochondrial permeability transition. *Am. J. Chir. Med.*, *33*, 627–637.

30. Venkatachalam, S. R., & Mulchandani, N. B., (1982). Isolation of phenanthroindolizidine alkaloids and a novel biphenylhexahydroindolizine alkaloid from *Ficus hispida*. *Naturwissenschaften (Natural Sciences)*, *69*, 287–288.
31. Vishnoi, S. P., & Jha, T., (2004). Evaluation of anti-inflammatory activity of leaf extracts of *Ficus hispida*. *Indian J. Nat. Prod.*, *20*, 27–29.
32. Yokota, J. D., Takuma, A., & Hamada, M., (2006). Scavenging of reactive oxygen species by *Eriobotrya Japonica* seed extract. *Biol. Pharm. Bull.*, *29*, 467–471.

CHAPTER 4

ANTI-INFLAMMATORY PROPERTIES OF BIOACTIVE COMPOUNDS FROM MEDICINAL PLANTS

MUHAMMAD IMRAN, ABDUR RAUF, ANEES AHMED KHALIL, SAUD BAWAZEER, SEEMA PATEL, and ZAFAR ALI SHAH

ABSTRACT

Consumption of medicinal plants exerts a promising role in drug discovery that has proven very effective against various human syndromes. Diallyl trisulfide (DATS) suppresses naphthalene-induced oxidative injury and inhibits inflammatory responses, such as TNF-α, IL-6, IL-8, and nuclear factor-kappa B (NF-κB). DATS also suppresses the production of serum nitric oxide (NO) and myeloperoxidase (MPO). The 6-Gingerol can enhance the anti-inflammatory cytokine interleukin-10 and reduce the expression of monocyte chemotactic protein-1, TNF-α, interleukin-1β, and interleukin-6. Lupeol inhibits the stimulation and production of osteoclastogenesis of macrophages in a concentration-dependent manner. They momentously decrease the formation of pro-inflammatory cytokines (IL-6, IL-1β, IL-12, and TNF-α), and lead to an increase in the production of p38 mitogen-activated protein kinase (MAPK).

4.1 GARLIC (*ALLIUM SATIVUM*)

This chapter focuses properties of bioactive compounds from selected medicinal plants for health care. Emphasis is on the anti-inflammatory activities of these phytochemicals.

Naphthalene-induced oxidative injury is inhibited by diallyl trisulfide (DATS: an organosulfur compound), which is responsible for the inflammatory response *in vitro* in A549 cells. DATS application in low dose groups has shown good free radical scavenging potential mainly due to enhanced activities of SOD. Studies revealing the mechanism of action for DATS-mediated inhibition in induced oxidative injuries and formation of pro-inflammatory responses (IL-8, TNF-α, and IL-6) were ascribed to inhibition of NF-κB [56]. SAC (S-allyl cysteine) is another promising organosulfur compound present in garlic possessing anti-inflammatory and redox modulating potentials, along with pro-energetic and anti-apoptotic activities [24].

DATS have been scientifically explored to suppress the formation of NO (nitric oxide) and PGE2 (prostaglandin-E2) due to inhibitory action on the expression of COX-2 (cyclooxygenase-2) and NO synthase in LPS (lipopolysaccharides)-activated murine macrophage cell model (RAW 264.7). Evidently, DATS inhibited the release of TNF-α (tumor necrosis factor-α) and IL-1β (interleukin-1β) by attenuating the expression of mRNA. It also retarded LPS-activated DNA-binding potential of NF-κB along with NF-κB p65 nuclear translocation, which is associated with the inhibition of IκB (inhibitor κB) degradation [64]. Additionally, DATS inhibits LPS-induced TLR-4 (toll-like receptor 4) and expression of MYD88 (myeloid differentiation factor 88) and attachment of LPS to macrophages resulting in the antagonistic potential of DATS against TLR4. Moreover, the use of TLR4 signaling inhibitor (CLI-095) caused blockage of TLR4 signals and elevated the anti-inflammatory properties of DATS [55].

Likewise, treatment of allicin (@ 10 and 50 mg kg^{-1}) noteworthily decreased brain edema and deficits of motor functionality, and apoptotic neuronal cell death in the damaged cortex. These shielding effects were noticed in injured cortex that was subjected to allicin treatment even post four hours of injury. Allicin administration also minimized the MDA and protein carbonyl expression levels, elevated the concentration of antioxidative enzymes (AOE), and inhibited the inflammatory cytokines. Outcomes from analysis of Western blotting revealed increment in phosphorylation of Akt and eNOS (endothelial nitric oxide synthase) due to the administration of allicin. To understand the mechanistic anti-inflammatory properties of allicin, specific inhibitors like L-NIO (0.5 mg kg^{-1}) and LY294002 (10 μ L and 10 mmol/L) were used, which resulted in activation of blocked Akt/eNOS pathways. Antioxidative properties of allicin were reversed partly due to usage of LY294002; however, this effect was not noticed in the case of L-NIO [29].

The effect of allicin (100 μmol/L) on LPS-stimulated 3T3-L1 adipocytes was studied by Western blotting, RT-PCR, and microarray analysis of genes (22,000). The incubation of 100 ng/mL LPS-exposed 3T3-L1 adipocytes for 24 hours with allicin retarded the increment of expression of pro-inflammatory genes, MCP-1, Egr-1, and IL-6. In adipocytes, LPS-induced inflammation resulted in phosphorylation of ERK1/2, which was reduced due to the treatment of allicin. Additionally, profiling of gene expression through microarray revealed downregulation of cancer related genes, and upregulated immune responsive genes [2].

The phytochemicals (such as uracil, caffeic acid, and S-allyl cysteine (SAC)) from garlic modulate UVB-stimulated formation of wrinkles and influence the propagation of matrix-metalloproteinase (MMP) and NF-κB signaling. *In vivo*, these phytomolecules suppressed MMPs expression and type-I pro-collagen degradation and reduced histological collagen fiber disorder and oxidative stress conditions. Moreover, SAC, and caffeic acid diminished inflammatory and oxidative stress conditions due to modulation of NF-κB activity. Indirect antioxidant potential of uracil was also noticed, as it inhibited the expression of COX-2 and iNOS and downregulated transcriptional factors [53].

DADS (diallyl disulfide) reduced the production of nitric oxide as well as decreases IL-6 and IL-1β levels in RAW 264.7 cells and ovalbumin-induced model of allergic asthma. They also minimized the expression of pro-inflammatory proteins i.e., MMP-9, iNOS, and NF-κB and elevated the expressions of antioxidative proteins (hemeoxygenase-1 and Nrf-2). *In vivo* studies have shown reduction in inflammatory cell count of BALF (bronchoalveolar lavage fluid) along with reduction in mucus hypersecretion and airway inflammation induced due to ovalbumin challenge. Additionally, expression of HO-1 (heme-oxygenase-1) and activation of Nrf-2 and reduction of MMP-9, NF-κB, and iNOS were noticed owing to DADS administration [50].

In LPS-activated macrophages, Ajoene-an organosulfur component have reported to suppress the expression of IL-6, IL-8, and TNF-α and production of PGE2 (prostaglandin E 2) and NO (nitric oxide). Furthermore, these constituents inhibited the transcriptional activity of NF-κB and degraded IκBα in experimented macrophages. Moreover, they (20 µM) noticeably reduced the phosphorylation of p38 MAPKs (mitogen-activated protein kinases) and ERK (extracellular signal-regulated kinases), which were induced by LPS [42].

4.2 GINGER (*ZINGIBER OFFICINALE*)

The treatment of 6-gingerol in a dose-dependent manner significantly minimized the cell viability in LoVo (human colon cancer cell) cells. Moreover, results of flow cytometric method reveal induction of G2/M phase arrest and have an impact on sub-G1 phase of LoVo cells mainly due to the application of 6-gingerol. A study was conducted to assess the concentrations of cyclins, CDKs (cyclin-dependent kinases), and their regulating proteins responsible for S-G2/M transition. Results from this study showed that the concentration of cyclin A, cyclin B1, and CDK1 were decreased; on the other hand, the concentration of p21 (Cipl) and p27 (Kip1) were elevated due to use of 6-gingerol. Additionally, intracellular ROS (reactive oxygen species) and phosphorylation contents of p53 were elevated in response to the treatment of 6-gingerol [51].

In LPS-stimulated BV2 microglia, zingerone, 6-gingerol, and shogaols each at a dose of 20 μM were able to suppress the formation of NO, IL-1, IL-6 and TNF-α along with their mRNA levels. Retardation of activated NF-κB might be the primary mode of action responsible for suppressing the expression of the pro-inflammatory gene. Further in 3T3-L1 cells administrated with 6-Gingerol, an increment in levels of IL-10 and reduction in expression of MCP-1 (monocyte chemotactic protein-1), IL-6, TNF-α, and IL-1β was noticed. Similarly, in the co-culture insert using fully differentiated 3T3-L1 cells with RAW 264.7 macrophages, reduction in IKK-β (I kappa B kinase beta), and c-JUN N-terminal kinase and downregulation of AP-1 expression were shown due to treatment with 6-gingerol. Its application also reduced the iNOS expression and production of NO in experimented diet-induced obese Zebrafish test [9]. The 6-Gingerol (@ 50 μM) reduced *Vibrio cholerae* infection-triggered concentrations of IL-8, IL-6, IL-1α, and IL-1β by 3.2-folds in the protein level and two-folds in RNA level after 3.5 hours. The level of MAP-kinases signaling molecules like p38 and ERK1/2 were also reduced by two and three-folds, respectively, after 30 minutes of treatment. Additionally, there was an increment in phosphorylated IκBα and down-regulation of p65 resulting in down-regulation of NF-κB pathway [12].

The 6-Gingerol application led to the reinstatement of reduced intestinal barrier functionality and reportedly suppressed pro-inflammatory response in DSS (dextran sodium sulfate)-administrated monolayers of Caco-2 cells. *In vitro* studies revealed the activation of AMPK (adenosine monophosphate-activated protein kinase) owing to treatment of 6-gingerol. Outcomes of animal studies illustrated the ameliorating potential of 6-gingerol on DSS-induced colitis through bodyweight loss restoration, reducing intestinal bleeding, and preventing the shrinkage of colon length. Additionally, 6-gingerol also inhibited the elevated levels of TNFα, IL-1 β, and IL-12 that were increased due to induction of DSS [7].

The 6-gingerol, 8-gingerol, and 10-gingerol repressed synthesis of IFN-γ (interferon-γ) and DNA by T-lymphocytes. Alternatively, inhibition of IL-2 synthesis and CD69 and CD25 expression was noticed due to treatment with 8-gingerol and 10-gingerol. Among all examined gingerols, no one affected the synthesis of IL-4. In the presence of 8 and 10-gingerol, exogenous IL-2 increased proliferation of T-lymphocytes. In accordance with these findings, 8 and 10-gingerol reduced IL-2 induced proliferation of cells (CTLL-2)-subclone of T-cells from mouse (C57bl/6), nonetheless no affect was observed on expression of CD25, showing retardation in signaling of IL-2 receptors [49].

The 6-shogaol revealed a noteworthy reduction in the score of neurological deficit and mean infarct area. It inhibited the MAPK signaling proteins and CysLT1R (cysteinyl leukotriene-1 receptor), therefore proving the probable mechanism. In LPS, 6-shogaol repressed production of IL-6 and NO along with MCP-1 induced gene expression but not in IL-1β- stimulated chondrocytes. In chondrocytes, it also retarded MMP induction and TLR4-mediated innate immune responses [3]. The 3-Ph-3-SG (3-phenyl-3-shogaol) inhibited invasion of cancerous cells in MCF-7 breast cancerous cells and MDA-MB-231 via suppression in expression of PMA (phorbol 12-myristate 13-acetate) stimulated MMP-9. It also reduced the expression of the inflammatory mediators (such as NO, iNOS, COX-2, and PGE2) in RAW 264.7 macrophage-like cells [52].

4.3 CRAB'S EYE (*ABRUS PRECATORIUS* L.)

The flavonoid C-glycosides retarded paw edema and ear edema, which were induced by carrageenan and xylene, correspondingly. In the acute hepatitis tests, *Abrus mollis* extracts (AME) decreased the elevated amount of ALT (alanine aminotransferase) and AST (aspartate aminotransferase) induced by CCl_4 and β-D-galactosamine-(D-GaIN-). Histopathological portions of liver in CCl_4-stimulated hepatic fibrosis model revealed that AME curtailed liver injury in the induced experimented model. Liver functionality were also enhanced as validated by reduced serum AST, ALT, ALP, liver index, TBIL (total bilirubin), and hydroxyproline contents and elevated serum glutathione and ALB levels [14].

The *C*-glycosides also reduced LPS-induced accumulation of lipids in primary hepatocytes of mouse. They also minimized the levels of fat accumulation, liver index, and serum aminotransferase contents in liver. RT-PCR and data of immuno-blotting determine that AME reversed LPS-mediated gene expression of lipid metabolism, for instance, SREBP-1 (sterol regulatory element-binding protein-1), FAS (fatty acid synthase), CHOP (C/EBP homologous protein), and ACC1 (acetyl-CoA carboxylase 1). Additionally, AME also suppressed the LPS-enhanced expression of COX-2 and IL-6 (interleukin-6) [57].

4.4   HOLLY-LEAVED ACANTHUS (*ACANTHUS ILICIFOLIUS* L.)

The methanolic crude extract from the leaves of *Acanthus ilicifolius* (Holly mangrove, sea holly) inhibited carrageenan-induced paw edema in rats, like BW755C (also known to be a synthetic LOX (lipoxygenase) and COX (cyclooxygenase)) inhibitors. Methanolic extract (M.E.) reduced migration of leukocyte and exudation of protein in peritoneal fluid, thus signifying its effectiveness in suppressing peritoneal inflammation. Additionally, M.E. also significantly inhibited the activity of COX (1 and 2) and 5-LOX. In LPS-activated peripheral blood mononuclear cells (PBMCs; TRAP), preincubation of the methanolic extract suppressed the formation of proinflammatory cytokines (TNF-α and IL-6) [43].

Lupeol (0–80 μM) exerted anti-inflammation activity in RAW 264.7 cells and BMDMs (bone marrow-derived macrophages) stimulated by LPS. In a dose-dependent manner, it might suppress the activation, formation, and migration of osteoclastogenesis of macrophages. In arthritic mice, lupeol ameliorated inhibition of propagation of inflammation-related cytokines, clinical symptoms, and bone erosion. A significant reduction was observed in the accumulation of (18)-F-fluorodeoxyglucose ((18)-F-FDG) in the joints of arthritic mice [44].

IBD (Inflammatory bowel disease) i.e., Crohn's disease and ulcerative colitis are chronic inflammatory diseases of lower GIT (gastrointestinal tract). Lupeol significantly decreased the formation of pro-inflammatory cytokines (IL-1β, TNFα, IL-6, and IL-12), and caused remarkable elevation in formation of p38 MAPK [39]. *In vivo, in silico, ex vivo* and *in vitro* studies revealed that Lupeol significantly prevented hemorrhage, oxidative stress, cleavage of collagen and CX3CR1 receptors, edema, myotoxicity, dermonectosis, and myonecrosis in inflammatory cell induced due to *Echiscarinatus* (saw-scaled viper) bite venom [34]. Intraperitoneal injection of Lupeol (@ 10, 25, or 50 mg/kg) had significant effect on the severity of pancreatitis, which was proven by decreased neutrophil infiltration and pancreatic edema.

Additionally, lupeol suppressed the elevated content of digestive enzymes (IL-6) and cerulean-induced acinar cell death [54].

4.5 TURMERIC (*CURCUMA LONGA*)

Extracts from turmeric (@ 1 or 10 µg/mL for 14 hours) reduced LPS (50 to 100 ngmL^{-1}), stimulated IL-1β, and TNF-α formation of THP-1 cells analyzed by ELISA. LPS-stimulated IRAk1, IκBα degradation, TLR4-MyD88 interaction, TLR4 expression, and MAPK activation were noteworthy decreased by turmeric. It also attenuated the IL-1β, aortic iNOS expression, TNF-α, vascular dysfunction, and nitrite. It also induced apoptosis and increased PARP-1 and caspase-3 activation in HL-60 cells [58].

In ovariectomy (OVX) rats, curcumin, and tetrahydrocurcumin (THC) have revealed alike effectiveness for skin tail temperature, while THC prohibited glucose intolerance involved in aggravating osteoarthritis. Both these experimented components helped in protection from symptoms of osteoarthritis and behavior related to pain more compared to 17β-estradiol treatment in estrogen-deficient rats. Along with this, they preserved lean body mass; and fat mass was reduced equal to that of 17β-estradiol treatment. Symptoms of osteoarthritis were improved due to reduced expressions of MMP3, MMP13, TNF-α, IL-1β, IL-6, and matrix metalloproteinases genes in the articular cartilage [59].

In the anti-inflammatory role, curcumin decreased the damage of skeletal muscle and fibrosis-related to ischemic injury. ELISA and immunohistochemical staining were performed to validate the mechanistic approach in curcumin-arbitrated protection of tissue. It led to reduced macrophage infiltration and decreased the responses of inflammation, as revealed by decreased IL-6, TNF-α, and IL-1 levels. Conclusively in macrophage, curcumin had inhibitory action on NF-κB activation stimulated by LPS [45].

In intracerebral hemorrhage (ICH)-induced secondary brain injury, curcumin elevated neurological scores and decreased brain edema. In the mouse brain, it also minimized the expression of IL-17, VCAM-1 (vascular cell adhesion molecule-1), and INF-γ, post 72 hours induction of ICH. The results proposed that administration of curcumin might have alleviated cerebral inflammation due to ICH, causing reduction in infiltration of T-lymphocytes into the brain [47].

Curcumin (@200 mg kg^{-1}) minimized the expression of inflammatory cytokine, reduced lipid peroxidation and oxidative stress, prohibited apoptotic conditions, and elevated activity of antioxidative defense mechanism compared to saline or MP (methylprednisolone) treatment. Ultrastructural and histopathological abnormalities were decreased in curcumin-administrated rats compared with saline-subjected and MP-subjected groups [32, 48].

4.6 ONION (*ALLIUM CEPA*)

In LPS-stimulated HT-29 cells, morin, vanillic acid, p-coumaric acid, and epicatechin from onion peels downregulated TNF-α mRNA expression, up-regulated the expression of HO-1 and GSTs (glutathione S-transferase) detoxification genes like GSTP1, GSTT1, and GSTM1 [44]. Quercetin lowered the mesenteric fat weights, increased the adiponectin mRNA levels, and lowered the IL-6 mRNA levels.

By influencing the adipokine expression, quercetin addressed obesity-induced inflammation [4]. Quercetin decreased level of tissue NO and TNF-α. Further, *in vivo* studies have provided the evidence for protective effect of Welsh onion green leaves (WOE's) due to the decrease in lipid-oxidation and elevation in concentration of AOE (i.e., CAT (catalase), SOD (superoxide dismutase), and GPx (glutathione peroxidase)). In mice models, it also reduced formalin-induced pains and number of acetic acid-induced writhing responses [11]. Quercetin affected expression of gene and formation of Th-1-derived IFN-γ, along with down-regulating the production of Th-2-derived IL-4 by means of normal peripheral blood mononuclear cells [57]. It is due to the anti-inflammatory potential of quercetin that prevents from formation of secondary infection trailed by disturbance in skin barriers [33].

The anti-inflammatory potential of quercetin is related to inhibitory action of lipoxygenase and retardation of inflammation-causing mediators. Quercetin boosts immunity and reduces inflammation by targeting various intracellular targeting phosphatases and kinases, membrane proteins and leukocytes. It suppresses the formation and migration of histamine and other inflammatory substances due to stabilization of cell membranes in mast cells [31]. Quercetin also inhibits activation of human mast cells via inhibiting the release of histamine, prostaglandins, leukotrienes, and Ca^{2+} influx. Mast-cells are dominant immunity-enhancing cells significant in autoimmune diseases and pathogenesis of allergic responses. Release of various cytokines like TNF and IL-8 responsible for inflammatory reactions in body are affected by the administration of quercetin. Therefore, quercetin is significant in the treatment of allergic inflammatory diseases (RA, asthma, and sinusitis) [17].

In clinical trials, quercetin @ 0.01 µM was incubated for fifteen minutes with human umbilical cord blood-derived cultured mast-cells grown in the presence of IL-6 and stem cell factor. The results indicated that inhibition (82%) of TNF-α, IL-6, and IL-8 occurred due to application of quercetin; however, histamine, and tryptase were reduced by 52–77% and 79–96%, respectively. Likewise, in human mast cells, the impact of quercetin on pro-inflammatory cytokines expression was also examined. Outcomes of this investigation revealed that quercetin reduced the expression of gene and formation of IL-1β, IL-8, IL-6, and TNF-α in experimented human mast-cells. Administration of quercetin reduced phorbol calcium ionophore-induced activation of p38 MAPK and NF-κB [40].

Research studies on the inhibiting potential of black seed, chamomile, caraway, anise, cardamom, and fennel on histamine content revealed that combination of various herbs controlled the release of histamine from immunological (85%) and chemical (81%) induction of cells. Comparatively, quercetin was more effective in preventing histamine release by 95% and 97%, respectively [39].

4.7 CINNAMON (*CINNAMOMUM* SP.)

Cinnamon comprises of numerous bioactive constituents like cinnamic acid, cinnamaldehyde, and cinnamate. Bioactive compounds of cinnamon improve with appearance of dark color due to aging of cinnamon. Various essential oils (α-thujene, cinnamyl acetate, L-bornyl acetate, eugenol, E-nerolidol, trans-cinnamaldehyde, L-borneol, terpinolene, caryophyllene oxide, α-terpineol, b-caryophyllene, and α-cubebene) are present in cinnamon extracts. The 2′-hydroxycinnamaldehyde is known to inhibit the formation of nitric oxide by retarding the initiation of NF-κB, signifying the anti-inflammatory potential of this compound. Several bioactive components present in *C. ramulus* revealed potent anti-inflammatory characteristics due to inhibitory action on expression of NO, iNOSandCOX-2 production in CNS (central nervous system). Therefore, *C. ramulus* could be a potent nutraceutical agent responsible for preventing neuro-degenerative ailments caused due to inflammation [36].

Pretreatment of TCA (Trans-cinnamaldehyde) reduced NO formation, release of IL-1β and other morphological changes that were LPS-induced. Reduced formation of NO was associated with enhanced deprivation of iNOS via suppressing impact on MEK1/2-ERK1/2 signaling pathways. In hippocampus of experimented mice subjects, Trans-cinnamaldehyde reduced the concentrations of phosphorylated ERK1/2 and iNOS that were induced due to treatment with LPS. In LPS-induced mice, trans-cinnamaldehyde decreased the memory defects and elevated synaptic plasticity [38].

Treatment of cinnamaldehyde in fructose-administrated rats minimized cardiac oxidative stress to inhibit the activation of NLPR3 inflammasome and signaling of TGF-β/smads by retarding CD36-mediated signaling of TLR4/6-IRAK4/1 in fructose-induced subjects. Due to the administration of cinnamaldehyde, retarded CD36-mediated signaling of TLR4/6-IRAK4/1 in fructose-induced subjects might be responsible for inhibition of activation of NLRP3 inflammasome [42]. Additionally, in AR rat models, the administration of cinnamaldehyde decreased the symptoms of allergy. Cinnamaldehyde minimized vascular congestion along with eosinophil, plasma cell, and inflammatory cell infiltration into the lamina propria in the mucosa [59].

4.8   SAFFRON (*CROCUS SATIVUS*)

In ischemia/reperfusion (I/R)-induced acute kidney injury (AKI), intraperitoneal administration of saffron extract (@ 5, 10 and 20 mg/kg) decreased these disorders without any variations in FRAP and urea-nitrogen levels among saffron extract-treated groups. Saffron extracts reduced the levels of creatinine, MDA (malondialdehyde), TNF-α, and expressions of ICAM-1 and infiltration of leukocytes in a concentration-dependent manner [12]. Extracts of saffron elevated antioxidative enzymatic activities and GSH content; however, a significant reduction in levels of pro-inflammatory cytokine due to administration of crocin was noticed in kidneys of old rats [41].

Crocin-saffron derived carotenoid elevates the antioxidative capability and possesses free radical scavenging properties. Crocin acts as an immune-modulating agent as it curbs cell-mediated immunity responses, inflammatory mediators, and humoral immunity [47]. This carotenoid induced apoptotic conditions and retarded inflammatory markers like NF-κB. Results of *in vitro* studies in HepG2 cells have confirmed anti-inflammatory impact of crocin due to cell cycle arrest at S and G2/M phases [34].

In TBI, (traumatic brain injury) mice model (C57BL/6); crocin elevated NSS (neurological severity score) and brain edema, reduced activation of microglial, minimized the release of various pro-inflammatory cytokines and decreased apoptosis of the cell. Crocin triggered the signaling of the Notch pathway, which improves proliferation during neurogenesis [15]. Safranal (a saffron aldehyde) inhibited immune-reactivity and inflammatory cytokines expression of p38 MAPK, IL-1β, and TNF-α and elevated IL-10 expression post-SCI (spinal cord injury) revealing an anti-inflammatory potential. Safranal administration reduced the expression of a water channel protein (aquaporin-4; AQP-4), which is associated to SCE (spinal cord edema), demonstrating an edema attenuating potential [16].

Crocin (@ 25, 50, and 100 mg/kg) and safranal (@ 0.5, 1, and 2 mg/kg) curbed edema, reduced the concentration of neutrophils, and suppressed inflammatory pain responses [25]. Treatment of crocin suppressed expression of iNOS and production of NO due to down-regulation of NF-kB activity. Crocin inducted phosphorylation of CAMK-4 (Ca^{2+}/calmodulin-dependent protein kinase 4) and intracellular transfer of Ca^{2+}. Further, crocin mediated suppression of iNOS expression was also retarded due to inhibition of CAMK4 [13].

4.9 JUJUBE (*ZIZIPHUS JUJUBE*)

ZJRB (*Ziziphus jujube* root bark) revealed protection against CCl_4-intoxicated HepG2 cell lines due to elevated cell viability and reduced concentration of lactate dehydrogenase (LDH), compared with the control. ZJME (@ 50, 100, 200, and 400 mg/kg) repressed the lipid peroxidation and reinstated the liver functioning biomarkers (CAT, AST, LDH, ALT, SOD, and ALP) and cytokine contents (IL-10, I1-1β, and TNF-α) in the damaged liver of rats induced by application of CCL_4 [23].

Saponins suppressed LPS-induced production of nitric oxide without resulting any toxicity in RAW 264.7 macrophage. Similarly, protein and gene over-expression of TNF and iNOS (induced by LPS) were significantly downregulated due to administration of saponins. Phosphorylation of MAPK (p38 MAPK) and activation of NF-κB were inhibited by administration of saponins. Moreover, saponins evoked inhibitory effect on HL-60 cells (promyelocytic leukemia) and T-lymphocyte (leukemic cells); however, no cytotoxic impact on normal human PBMC (peripheral blood mononuclear cells) was noticed [24].

Saponins revealed inhibitory action on LPS-stimulated formation of NO and expression of TNF-α and IL-1β proteins. The saponins inhibited the formation of NO by retarding iNOS expression. Additionally, LPS-activated formation of IL-1β and TNF-α was decreased due to administration of saponins [22, 27]. TSG (total saponins ginseng) pre-treatment stimulated the seizure of NF-κB in cytosol by inhibiting degradation of IκB. In LPS-stimulated RAW 264.7 cells, TSG also helped in down-regulation of MAPKs [37].

CA (Coussaric acid) and BA (betulinic acid) repressed PGE2 (prostaglandin E_2) and NO formation and inhibited IL-1 β, TNF-α, and IL-6 contents. Additionally, they reduced the expression of COX-2 and iNOS. Treatment of both triterpenoids suppressed NF-κB and inducted protein expression of HO-1 in a concentration-dependent manner. Application of both these phytochemicals retarded the LPS-activated NF-κB binding potential along with inhibition of IL-6, NO, TNF-α, PGE_2, and IL-1β, due to reversal effect by tin protoporphyrin (a HO-1 inhibitor) [35].

BA inhibited NF-κB activity and breakdown of IκBα in high glucose-induced mesangial cells and kidneys of diabetic rats causing a decline in expression of FN (fibronectin). Additionally, BA inhibited transcriptional activity of NF-κB in high glucose-induced GMCs (glomerular mesangial cells). Besides, in mesangial cells, BA elevated the interaction among β-arrestin 2 and IκBα [58]. BA (@ 10 and 30 mg/kg) had a significant effect on lung injury and sepsis-induced mortality, due to decreased infiltration of neutrophils and protein. BA also reduced cytokine, expression of NF-κB, ICAMPs (intercellular adhesion molecule-1), MCP-1 (monocyte chemoattractant protein), and MMP-9 (matrix metalloproteinase-9) contents [26].

4.10   MEADOW SAFFRON (*COLCHICUM LUTEUM*)

Colchicine reduced the transcoronary (coronary sinus-arterial) gradients for IL-1β, IL-18, and IL-6 [29]. AMPK (AMP-activated protein kinase: a metabolic bio-sensor) having anti-inflammatory characteristics. MSU (monosodium urate) inhibited phosphorylated AMPK-α in bone marrow-derived macrophage production (BMDMs). According to *in vitro* and *in vivo* studies, colchicine at low concentration (10 nM) confirmed AMPKα phosphorylation, polarization of macrophage M2, decreased caspase-1 activation, and release of IL-1β and CXCL1 due to MSU crystals in BMDMs [30].

4.11   HEMP (*CANNABIS SATIVA*)

Hemp (*Cannabis sativa*) phenolics curtailed IBD (inflammatory bowel disease) and significantly protected against inflammatory ailments [5]. In LPS-stimulated murine peritoneal macrophages, non-psychotropic phytocannabinoid Δ^9-tetrahydrocannabivarin (THCV) down-regulated the over-expression of iNOS, COX-2 and IL-1β proteins. Additionally, THCV countered LPS-induced progression of CB_1 receptors, without alteration in mRNA expression of TRPV2, CB_2, and TRPV4. In both unstimulated and LPS-challenged macrophages, other transient receptor potential (TRP) channels (specifically, TRPV1, TRPA1, TRPV3 andTRPM8) were unnoticeable. Conclusively, nitrite formation in macrophages was inhibited by THCV-via CB_2 receptor activation. This phytocannabinoid down-regulated CB_1, but not CB_2 or mRNA expression of transient receptor potential channel [57]. Cannabidiol (5 µM) had modest effectiveness on TNF-α, IL-10, nitrotyrosine, iNOS, and Nrf-2. Progression of Bcl-2 and down-regulation of Bax and cleaved caspase-3 was noticed in cannabidiol-treated cells, whereas no impact was prompted by cannabinoid receptor-1 and 2 antagonists [21].

Cannabidiol (@ 30 mg/kg/day for 3 or 7 days) in a murine model of cerebral malaria (CM), significantly suppressed the cytokines (Th1/Th2), BDNF (brain-derived neurotropic factor) and NGF (nerve growth factor). On 5th day-post-infection (dpi) in hippocampus, the concentration of IL-6 and TNF-α was elevated, though in pretrontal cortex portion just levels of IL-6 were amplified. Cannabidiol administration caused an elevation in expression of BDNF and reduced the concentration of TNF-α in the hippocampus and IL-6 levels in pre-frontal cortex [18].

Cannabinoids (CBD) and Δ-9-tetrahydrocannabinol (THC) markedly reduced the Th17 phenotype, which is known to be increased in inflammatory autoimmune pathologies, such as multiple sclerosis (a nerve damaging disease). Furthermore, it was noticed that THC and CBD (@ 0.1–5 µM) inhibited the formation and excretion of cytokines in a dose-dependent manner. Additionally, protein, and mRNA of IL-6 (a significant aspect in induction of Th17) were also reduced. Pre-treatment with CBD leads to elevated contents of anti-inflammatory cytokine (IL-10). Both CBD and THC had no effect on IFN-γ and TNF-α concentration [14].

In adult and adolescent mice, THC inducted decrease in macrophage pro-inflammatory cytokines and elevated IL-10 contents. Application of THC had no impact on brain cytokines in adult mice; nonetheless, decrease in pro-inflammatory cytokine was obvious in brains of adolescent mice. Same trend regarding the effect on hippocampus and hypothalamus was noticed post-ten days treatment of THC. On the other hand, an inverted effect was noticed in adult mice treated as adolescents for examined brain cytokines levels post-47 days of final THC administration, revealing neuroinflammation [28].

4.12   LAPACHO TREE (*HANDROANTHUS IMPETIGINOSUS*)

In LPS-stimulated microglia, β-LAP (β-lapachone) from lapacho tree suppressed the iNOS expression, pro-inflammatory cytokines and MMPs (MMP-3, -8, and -9) at protein and mRNA levels. In contrast, β-LAP elevated the expressions of HO-1, IL-10, and TIMP-2 (tissue inhibitor of metalloproteinase-2). In an LPS-activated systemic inflammatory mouse model, anti-inflammatory potential associated with β-LAP was validated. In the LPS-treated mouse brain, β-LAP retarded activation of microglial and iNOS expressions, MMPs, and pro-inflammatory cytokines levels. Additionally, in LPS-activated microglia, studies regarding mode of action of β-LAP revealed anti-inflammatory potential by retarding NF-κB/AP-1, MAPKs, and PI3K/AKT signaling pathways. The β-LAP aided in suppressing the production of ROS by inhibiting phosphorylation of NADPH oxidase subunit proteins. Antioxidative properties of β-LAP seemed to increase NAD(P)H:quinone oxidoreductase-1 (NQO1) and HO-1 through Nrf2/ARE and PKA-pathways [74]. In mice (C57BL/6), β-LAP decreased, the instant hypersensitive response tempted by Con A (concanavalin A). Therefore, Th2 inhibited IL-10, IL-4, IL-6, and IL-5 averting the incidence of various allergic-inflammations and allergies [2].

4.13   OREGANO (*ORIGANUM VULGARE*)

Certain phenolics, flavonoids, sesquiterpenes, and monoterpenes from oregano inhibited formation of NO and ROS and mitochondrial activity. Production of NO were reduced by terpenes of LG (*Lippia graveolens*), LGF2 fractions and LP (*Lippia palmeri*) fractions like LPF-1, LPF-2 and LPF-3, thus confirming that experimented monoterpenes and sesquiterpenes were biologically active components of oregano. Additionally, extracts of HP, LG, and LP showed non-selective inhibitory action on COX-1 and COX-2 activity [10]. In LPS-induced, mMECs (mouse mammary epithelial cell), thymol (@ 10, 20, and 40 μg ml^{-1}) evidently suppressed the formation of IL-6 and TNF-α. Further, iNOS and COX-2 expression was inhibited in a concentration-dependent manner. Likewise, in LPS-activated mMECs, thymol also inhibited phosphorylation of p38 MAPKs, IκBα, JNK, NF-κB, and ERK. These results revealed thymol anti-inflammatory potential in LPS-activated mMECs due to interference with activity of MAPK and NF-κB signaling pathways [19].

Propionic and cinnamic acids, carvacrol, and cinnamaldehyde suppressed ST (*Salmonella typhimurium*) attack other than bacterial and cell viabilities, and motility of flagella. Synergistic usage of cinnamaldehyde and propionic acid with cinnamic acid resulted in structural diminishment of fimbriae [32]. Carvacrol and thymol reversed the doxorubicin-induced cardiotoxicity through significant elevation of serum creatine kinase isoenzyme-MB, lactate dehydrogenase, aspartate transaminase, TNF-α, creatine kinase, and cardiac troponin levels. They also reduced MDA (malondialdehyde) and caspase-3 content in heart along with noteworthy elevation in superoxide dismutase activity, CAT, and reduced glutathione in the heart [5]. Carvacrol reduced pancreatitis-stimulated MDA and 8-OHdG (8-hydroxy-2'-deoxyguanosine) contents. Furthermore, the activities of liver GSH-Px, CAT, and SOD in AP (acute pancreatitis) +CAR (carvacrol) groups were higher compared with rats in AP groups. However, LDH, ALT, and AST were minimized in the treatment groups [20].

Likewise, carvacrol and thymol meaningfully decreased TNF-α and IL-1β at mRNA and protein levels. They also reduced NFAT-2, NFAT-1, and c-Fos expression, whereas c-jun expression were minimized by carvacrol administration. They retarded expression of p-NF-κB p65, reduced phosphor-STAT3, and inducible phosphor-SAPK/JNK contents, while just carvacrol caused elevation of p-p38 contents in total cell extracts. These both might cause reduction of response of inflammation, via modulatory expression of NFATs, JNK, AP-1, and STAT-3 [1].

4.14 ROSEMARY (*ROSMARINUS OFFICINALIS*)

RPAE (Rosemary polyphenols alcoholic extract) attenuated neuropathic behavioral changes in comparison to CCI (chronic constriction injured) animals. Elevation in the concentration of cleaved caspases-3 and caspases-9, Bax, TNF-α, TLR4, iNOS, and Ibal (on day 7th and 14th) were observed in the vehicle-treated CCI group. RPAE (@ 400 mg/kg) reduced the levels of glial markers, inflammatory, and mentioned apoptotic in comparison with vehicle-treated CCI animals [38].

Carnosol (another well-known diterpene) reduced formalin-activated inflammation. Additionally, both the carnosol and extract had no impact on corticosterone concentration compared to the control group. Carnosol suppressed the activity of COX-1 andCOX-2 and inhibited the inflammation and pain stimulated due to injection of formalin, which might be the reason in retardation of COX-1 andCOX-2 [40].

The anti-inflammatory potential of RAME derived from a mutant culture of *Perilla frutescens* (L.) Britton has been investigated. In RAW 264.7 cells, RAME retarded LPS-activated production of NO having IC_{50} of 14.25 µM. RAME also suppressed the LPS-stimulated expression of IL-6, IL-1β, and IL-10, interferon-β, iNOS, and monocyte chemo-attractant protein-1. Furthermore, RAME inhibited the NF-κB activation. These outcomes suggested down-regulation of expression in iNOS content due to application of RAME was due to MyD88 (myeloid differentiation primary response gene 88)-dependent and-independent pathways. Besides, RAME suppressed HO-1 expression via activation of NFE2L2 (nuclear factor-erythroid 2-related factor 2). Tin protoporphyrin's treatment-aHO-1 inhibitor proliferated the RAME-activated inhibition of NO production. In a nutshell, RAME extracted from *P. frutescens* suppressed production of NO via concurrent inhibition of HO-1 and suppression of MyD88-dependent and-independent pathways [7].

In the BALF, marked inhibition in elevated inflammatory cells and cytokine (Th2) was noticed due to application of rosmarinic acid. It also decreased ova-specific IgE and total IgE contents and AHR (ameliorated airway hyper-responsiveness) and reduced the content of mucus hypersecretion and inflammatory cells in the airway. Additionally, rosemarinic acid could be facilitated due to inhibition of JNK, ERk and phosphorylation of p38 along with NF-κB (nuclear factor-κB) activation. Moreover, in lung tissue, pre-treatment of RA caused significant decline in expression of CCL11, AMCase, Ym2, CCR3, and E-selection mRNA [43].

4.15 *ALOE VERA*

Natural extracts of *Aloe barbadensis* (@0.5%, 1%, and 2%) administrated in diet for a period of fifteen days resulted in noteworthy elevation of macrophages contents that are accountable for phagocytic action in tissues [10]. *Aloe vera*-based *Nerium oleander* extracts increased antioxidative protection, protected cell viability, and intracellular reduced glutathione and significantly decreased the formation of ROS [80]. Similarly in propagation of RAW 264.7 cells, Aloe-emodin markedly inhibited IL-6, NO, and IL-1β formation, without any cytotoxicity. Expression levels of mRNA for IL-6, iNOS, and IL-1β genes were retarded by administration of Aloe-emodin. Analysis of Western blotting revealed suppressing behavior of aloe-emodin against LPS-activated expression of iNOS protein, degradation of IκBα and JNK, p38, ERK, and Akt phosphorylation [29].

Aloe-based phenols decreased TNF-α, IL-6, IL-12, IL-1β, CX3CL1, CCL5 and protein along with reduction in mRNA and protein of PPARγ/ LXRα and 11β-HSD1 in WAT (white adipose tissue) and liver [13]. Emodin decreased the DEP (diesel exhaust particles)-stimulated increase in resistance of airway and prohibited influx of neutrophils in BALF. It also augmented the elevation in proinflammatory cytokines (TNF-α, IL-1β, and IL-6) of lungs [6]. It diminished the secretion of IL-1β through retardation of NOD-like receptor family, NLRP3 (pyrin domain containing 3) stimulation tempted due to nigericin, silica crystals, and ATP (adenosine tri-phosphate). Additionally in LPS-induced endotoxin mouse models, emodin improved the harshness of NLRP3 inflammasome-mediated symptoms [1].

Emodin (@ 100 μM) in ConA-induced cells decreased production of NO and the production of IFN-γ, IL-6, TNF-α, IL-2, and IL-17; nonetheless induced in IL-4 and IL-10 cells. It can be concluded that immunomodulating potential of emodin might be due to anti-proliferative effects on lymphocytes, along with a shift in balance of Th-1/Th-2 and Th-17/Treg [9]. In BALF, emodin reduced the lung injury scores, pulmonary pathological changes, MPO activity, neutrophils, total cells, TNF-α, pulmonary edema, macrophages, IL-6, and IL-β. In lungs, it retarded the phosphorylation in LPS-induced binding potential of NF-κB p65 DNA. Emodin-retarded inflammatory biomarkers and pulmonary edema were mediated due to NF-κB inactivation in mice [43].

4.16 COMMON MYRTLE (*MYRTUS COMMUNIS*)

Myrtus communis-based alkaloids and flavonoids have reportedly displayed inhibitory effect Extracts of *Myrtus communis* have shown marked inhibition in secretions of IL-8 in a concentration-dependent manner, due to presence of alkaloids and flavonoids [8].

Significant reduction in MPO activity along with ear edema was noticed due to administration of *Myrtus communis*-based essential oil. This essential oil retarded IL-6 and TNF-α and cotton pellet-stimulated granuloma [9]. *In vitro*, MC (Myrtucommulone) inhibited (IC$_{50}$ values ranging from 1.8 to 29 μM) the biosynthesis of eicosanoids by suppression of COX-1 and 5-lipoxygenase; and retarded the release of elastase and ROS formation in stimulated polymorphonuclear leukocytes. In a concentration-dependent manner, it abridged the progression of carrageenan-stimulated mouse paw edema [46]. MC and S-MC (semimyrtucommulone) prohibited Ca^{2+} mobilization in poly-morphonuclear leukocytes, arbitrated by G-protein signaling pathways, and retarded the ROS formation and elastase release at comparable concentrations. MC somewhat suppressed the formation of peroxide and unsuccessfully blocked mobilization of Ca^{2+} and release of elastase when polymorphonuclear leukocytes were induced with 1 ionomycin [1].

4.17 HEN-OF-THE-WOOD (*GRIFOLA FRONDOSA*)

Extracts of WB (White button) encouraged macrophage formation of TNF-α and prevented DSS-activated colonic injury. Diet of mice supplemented with 2% WB mushrooms for a period of four weeks revealed no potential impact on *ex vivo* immune sensitivity or related toxicity. DSS-stimulated mice, which were subjected with 1% of WB mushrooms, had induced weight loss. Furthermore, ingestion of WB (2%) protected experimented mice from DSS-stimulated colonic injury. TNF-α responses in colon, serum of the DSS-induced, and WB (2%)-administrated mice were higher compared with the control [6].

These phenolics retarded the human colon cancer cells (HT-29) with monocytic cells (U937) and suppressed adhesion of TNF-α-activated monocyte. They also inhibited IL-8 and MCP-1, chief IBD-linked chemokines. Furthermore, the administration of mushroom-based phenolics retarded TNF-α-stimulated production of ROS and transcriptional activity of NF-κB in HT-29 examined cells. In the colon tissues, they also retarded the TNBS (trinitrobenzene sulfonic acid)-activated loss in weight, ulcer of colon, activity of MPO and expression of TNF-α. It shows that mushroom-based phenolics amends the inflammation in colon by inhibiting TNF-α production along with their signaling via NF-κB, causing expression of MCP-1, inflammatory chemokines, and IL-8 [30]. Consistently, mushroom-based phenolics minimized the expression of ICAM-1, E-selectin, and VCAM-1 [3]. Mushroom extracts (@ 500 mg/kg) retarded LPS-activated proliferation of NF-κB activation and formation of TNF-α, IL-1β, ICAM-1, iNOS, and COX-2. In rats, they blocked the acetic acid-stimulated abdominal contractions and formalin-activated impulsive nociceptive behaviors [33].

4.18 NEEM (*AZADIRACHTA INDICA*)

The extracts of neem declined carrageenan-induced acute paw edema. Furthermore, to resolve edema, azadirachtin@6, 60, and 120 mg/kg also retarded the proliferation phase in the inflammatory response due to decreased development of fibrovascular tissue growth. Azadirachtin@120 mg/kg helped in suppressing nociceptive response in nociceptive model and inflammatory pains. The activity of azadirachtin@120 mg/kg was attenuated in nociceptive pain model due to non-selective opioid antagonist, naltrexone (@10 mgkg^{-1}) but not by cyproheptadine (a non-selective serotonergic antagonist) [1].

At allergic site, intraperitoneally-treated gedunin reduced the concentrations of leukotriene B(4), CCL2, IL-5, CCL5, CCL11, and CCL3. Pretreatment with gedunin unsuccessfully inhibited adhesion of T-lymphocyte and chemotaxis in the direction of pleural washes recovered from OVA-induced mice, signifying *in vivo* inhibitory impact of gedunin on migration of T-lymphocytes through *in situ* inhibiting mediators of chemotactic. On the other hand, *in vivo* studies revealed that pre-administration with gedunin decreased CD69(+) and CD25(+) T-lymphocytes contents in pleura and number of ICD25(+) cells in thoracic lymph nodes post-24-hourinduction of OVA. In conformity, treatment of T-lymphocytes with gedunin (*in vitro*) suppressed α-CD3 mAb-activated expression of CD25 and CD69, production of IL-2 production and NF-κB nuclear translocation and NFAT-cells (Nuclear factor of activated T). Particularly, post-administration of mice with gedunin decreased inflammation of OVA-activated lung allergy due to reduction in T-lymphocyte and eosinophil counts and the contents of eosinophilotactic mediators in BALF [6].

The polyphenols from this plant (such as 15-hydroxyazadiradione, 7-benzoyl-17-hydroxynimbocinol, 23-deoxyazadironolide, limocin E, 23-epilimocin E, and 7-α-acetoxy-3-oxoisocopala-1, 13-dien-15-oic acid) showed 74 to 91% decrease in levels of melanin at a concentration of 25 μg/mL in TPA (12-*O*-tetradecanoylphorbol-13-acetate)-activated inflammation (1 μg/ear) in mice [34].

Rats that were orally administered with nimbidin (@ 5–25 mg/kg) for 3-days suppressed the transfer of macrophages to their peritoneal cavities as a response to inflammation and suppressed the phagocytosis and PMA-stimulated respiratory burst in these cells. In LPS-stimulated macrophages, nimbidin also repressed production of PGE2 (prostaglandin E2) and NO, while it partly inhibited IL-1. Thus, inhibition of NO showed that nimbidin augmented iNOS-induction deprived of any specific retardation in its catalytic activity. Additionally, nimbidin also weakened neutrophils degranulation evaluated as release of myeloperoxidase, lysozyme, and β-glucuronidase [20].

4.19 CREAT OR GREEN CHIRETA (ANDROGRAPHIS PANICULATA)

Andrographolide and neo-andrographolide from create predominantly inhibited SnRK1 (SNF1-Related Protein Kinase-1) activity in terpenoid pathway and eliminated downregulation of HMGR (3-hydroxy-3-methylglutaryl-CoA reductase), causing increased buildup of andrographolide and neon-andrographolide in begomo-virus-infected plants. In BALF, SRS27 suppressed the airway mucus hypersecretion, hyper-responsiveness, pulmonary eosinophilia, IL-5, IL-4, IL-13, and iNOS. Reduction in BALF inflammatory cytokines levels, airway mucus hypersecretion, production of IgE, and pulmonary eosinophilia was noticed in ovalbumin-stimulated mouse asthmatic model [28].

Phenolics present in creat decreased the amplified retinal vessels STZ-stimulated mice and reduced the breakdown of BRB in experimented mice. They also reduced the increased serum and retinal mRNA expression of VEGF levels. Further, they nullified NF-κB p65 nuclear translocation and Egr-1 (early growth response-1) and minimized the increased phosphor-NF-κBp65, IκB, and κB kinase. Additionally, they declined the elevated retinal and serum mRNA expression of TF, TNF-α, serpine-1, IL-1β, and IL-6 [20]. Likewise, *in vitro*, they inhibited formation of IL-6 and IL-17, inhibited p-Stat3 expression, and retarded Th17 variation of PBMCs [23].

4.20 AMLA (EMBLICA OFFICINALIS/PHYLANTHUS EMLICA L.)

In B16 murine melanoma cells, the alcoholic extracts from amla retarded the activity of tyrosinase, by employing the mRNA expression of tyrosinase and TRP-1 and TRP-2 (tyrosinase-related proteins). However, in RAW 264.7 murine macrophage cells, it reduced the expression of LPS-stimulated pro-inflammatory genes in a concentration-dependent manner [13]. The moderating potential of methanolic extract on cerulean-induced pancreatitis rat models was revealed via momentous increase in serum L/A ration, IL-1β, IL-18, caspase-1, oxidative stress index (OSI), collagen, MPO activity through the intervention on their mRNA expression [50].

4.21 MAIDENHAIR TREE (GINKGO BILOBA)

In RAW 264.7 cells, Ginkgolides have reportedly attenuated the aggregation of inflammatory cells (such as macrophages, lymphocytes, and neutrophils) and amended the damage of lungs. Furthermore, they reduced the excretion of plasma protein up to the same extent; while in terms of alleviation of myeloperoxidase activities, GB (Ginkgolide-B) was higher compared to GM (ginkgolides mixture). They minimized LPS-activated mRNA for IL-1β and IL-10 protein levels equally, though; GM was extra effective as compared to GB in reducing the levels of IL-10 mRNA [58].

GBE (*G. biloba* extract) decreased fibrosis of liver by retardation of p38 MAPK andNF-κBp65 through suppression of IκBα degradation, along with inhibition of apoptosis in hepatocyte via retardation of caspase-3 activation and Bax, and proliferation of Bcl-2. Inflammation-related aspects and HSC (hepatic stellate cell)-activation biomarkers revealed that GBE is an effective extract in inhibiting activation of HSC and inflammation because of upregulation of p38 MAPK and NF-kB/IκBα signaling. GBE markedly elevated IL-10 and Adipo-R1 gene expressions and minimized the phosphorylation of IR and Akt [43].

GA (Ginkgolide-A) suppressed the endothelial formation of high glucose activated interleukin and phosphorylation of STAT-3 (signal transducer and activator of transcription-3). GA elevated high glucose-induced vascular inflammation of lower-grade that could be attained via regulation of STAT3-mediated pathway. These results confirm that GA could be a potential phytonutrient for attenuation of inflammation in diabetic vascular complications [28].

GBE leaves decreased activity of MPO (myeloperoxidase), lipid peroxidation and suppressed the activity of MMP-9. Additionally, it attenuated LPS-induced activation of AOE and inhibited NF-κB phosphorylation and IκB. These outcomes recommend the protective mechanism of EGb 761 to be inhibition of NF-κB activation, probably through increment in levels of AOE. Further investigations are required to validate if EGb 761 is a candidate suitable to minimize the occurrences of ALI (endotoxin-induced acute lung injury) [12].

4.22 NONI (*MORINDA CITRIFOLIA*)

Juice of *Morinda citrifolia* (@ 200 µg/ml) improved the rate of proliferation for BMSC and increased the osteogenic variation biomarker genes ALP, OCN, and Runx2, as analyzed through RT-PCR. *In vivo* implantation of collagen scaffolds loaded with BMSC pre-treated to MCE (*M. citrifolia* extract) revealed elevation in density of bone calculated through CT (computed tomography) and further histological measurements showed neo-angiogenesis for the formation of bone [29].

Likewise, intraperitoneally administrated MC fruit juice decreased inflammation in OVA-induced Brown Norway rat along with reduction of inflammatory cellular count in lungs. Administration of juice also scavenged in a concentration-dependent manner (@ 8.1 nmol nitrites/50 µl juice). Additionally, juice suppressed calcium and cholinergic induced spasms with a shift of concentration-response curve [39].

Monotropein (a glycoside derived from roots of *M. officinalis*) diminished the apoptotic conditions in a concentration-dependent manner, in response to IL-1-βstimulation [59].

In LPS-induced RAW 264.7 macrophages, monotropeins retarded expressions of IL-1β mRNA, TNF-α, iNOS, and COX-2. Monotropein treatment reduced the activity of DNA-binding for NF-κB. Its administration also inhibited IκB-α phosphorylation and degradation and therefore NF-κB translocations. In DSS-induced colitis model, monotropein decreased DAI (disease activity index), activity of MPO and protein expressions due to inflammation by overturning activation of NF-κB in colon mucosa [47].

4.23 GINSENG (*PANAX GINSENG*)

Gold nanoparticles (AuNPs) produced from leaf extract of *Panax ginseng* using Meyer extract method revealed anti-inflammatory potential via inhibitory action on NF-κB activation in macrophages. They decreased the expressions of NO, IL-6, PEG$_2$ (prostaglandin E2), and TNF-α. Besides, in RAW 264.7 cells. AuNPs inhibited activation of LPS-stimulated NF-κB signaling pathway thru MAPK (p38 MAPK) [15]. *Panax quinquefolius* aqueous extract resulted in elevation of expression of TNF-α, Ccl5, Nos2, and Mcpl, and reduced the expression of adipocyte PPAR-γ [44].

In OVX (ovariectomized)-mice (C57BL/6J) model, ginseng administration minimized the density of blood vessel, activity of MMP and mRNA levels of FGF-2 and VEGF-A. Similarly, in examined adipose tissues, ginseng also reduced the concentration of inflammatory cells and expression of MCP-1, TNF-α, and CD68. In OVX mice, administrated extract decreased triglycerides and free fatty acids (FAs); and stabilized hyperglycemia and hyperinsulinemia. Ginseng suppressed ovariectomy-induced adiposity, obesity, and adipocyte hypertrophy by moderating activity of MMP and angiogenesis. In OVX mice, it also inhibited insulin resistance, adipose inflammation, and hepatic steatosis [35].

Ginsenoside Rg1 @10, 20, and 40 mgkg^{-1} reduced mortality rate stimulated by alcoholic consumption and eased in the impairment of liver, as proven by a decline in serum parameters. Moreover, ultrastructural, and histological analysis in alcoholic treated groups revealed impairment of hepatic cells none the less it was restored due to administration of Ginsenoside Rg1 [10].

4.24 SUMMARY

This chapter focuses on the properties of bioactive compounds from selected medicinal plants for health care. Emphasis is on the anti-inflammatory activities of these phytochemicals. In total, 23 medicinal plants are presented, namely:

1. Aloe vera;
2. Amla;
3. Cinnamon;
4. Common myrtle;
5. Crab's eye;
6. Garlic;
7. Ginger;
8. Ginseng;
9. Green chiretta;
10. Hemp;
11. Hen of the wood;
12. Jujube;
13. Lapacho tree;
14. Maidenhair tree;
15. Meadow saffron;
16. Neem;
17. Noni;
18. Onion;
19. Oregano;
20. Rosemary;
21. Saffron;
22. Sea holly; and
23. Turmeric.

KEYWORDS

- anti-inflammatory role
- bioactive compounds
- lactate dehydrogenase
- mechanisms of action
- medicinal plants
- oxidative stress index

REFERENCES

1. Boskabady, M. H., & Farkhondeh, T., (2016). Anti-inflammatory, antioxidant and immunomodulatory effects of *Crocus sativus* L. and its main constituents. *Phytotherapy Research, 30*(7), 1072–1094.
2. Boyle, A. P., Hong, E. L., Hariharan, M., Cheng, Y., Schaub, M. A., Kasowski, M., Karczewski, K. J., Park, J., Hitz, B. C., Weng, S., & Cherry, J. M., (2012). Annotation of functional variation in personal genomes using Regulome DB. *Genome Research, 22*(9), 1790–1797.
3. Campos, G. S., Bandeira, A. C., & Sardi, S. I., (2012). Zika virus outbreak, Bahia, Brazil. *Emerging Infectious Disease, 21*(10), 1885–1886.
4. Chen, C., Liang, B., Ogino, A., Wang, X., & Nagatsu, M., (2009). Oxygen functionalization of multiwall carbon nanotubes by microwave-excited surface-wave plasma treatment. *The Journal of Physical Chemistry, C. Nanomaterials and Interfaces, 113*(18), 7659–7669.
5. Chouard, N., Caurant, D., Majeus, O., Guezi-Hasni, N., Dussossoy, J. L., Baddour-Hadjean, R., & Pereira-Ramos, J. P., (2016). Thermal stability of SiO_2-B_2O_3-Al_2O_3-Na_2O-CaO glasses with high Nd_2O_3 and MoO_3 concentrations. *Journal of Alloys and Compounds, 671,* 84–99.
6. Colin-Gonzalez, A. L., Ali, S. F., Tunez, I., & Santamaria, A., (2015). On the antioxidant, neuroprotective and anti-inflammatory properties of S-allyl cysteine: An update. *Neurochemistry International, 89,* 83–91.
7. Faria, N. R., Azevedo, R. D., Kraemer, M. U., Souza, R., Cunha, M. S., Hill, S. C., Theze, J., Bonsall, M. B., Bowden, T. A., Rissanen, I., & Rocco, I. M., (2016). Zika virus in the Americas, early epidemiological and genetic findings. *Science, 352*(6283), 345–349.
8. Farzadinia, P., Jofreh, N., Khatamsaz, S., Movahed, A., Akbarzadeh, S., Mohammadi, M., & Bargahi, A., (2016). Anti-inflammatory and wound healing activities of *aloe vera,* honey, and milk ointment on second-degree burns in rats. *The International Journal of Lower Extremity Wounds, 15*(3), 241–247.
9. Feißt, C., Franke, L., Appendino, G., & Werz, O., (2005). Identification of molecular targets of the oligomeric nonprenylated acylphloroglucinols from *Myrtus communis*

and their implication as anti-inflammatory compounds. *Journal of Pharmacology and Experimental Therapeutics*, *315*(1), 389–396.

10. Finn, R. D., Bateman, A., Clements, J., Coggill, P., Eberhardt, R. Y., Eddy, S. R., Heger, A., et al., (2014). Pfam: The protein families database. *Nucleic Acids Research*, *42*(D1), D222–D230.

11. Flicek, P., Ahmed, I., Amode, M. R., Barrell, D., Beal, K., Brent, S., Carvalho-Silva, D., Clapham, P., Coates, G., Fairley, S., & Fitzgerald, S., (2012). Ensembl 2013. *Nucleic Acids Research*, *41*(Database issue), D44–D55. Online: doi: 10.1093/nar/*gks1236*.

12. Frenkel, D., Mooij, G. C., & Smit, B., (1992). Novel scheme to study structural and thermal properties of continuously deformable molecules. *Journal of Physics, Condensed Matter*, *4*(12), 3053–3060.

13. Gadekar, L., Mane, S., Katkar, S., Arbad, B., & Lande, M., (2009). Scolecite as an efficient heterogeneous catalyst for the synthesis of 2,4,5-triarylimidazoles. *Open Chemistry*, *7*(3), 550–554.

14. Garbett, K., Darnall, D. W., Klotz, I. M., & Williams, R. J., (1969). Spectroscopy and structure of hemerythrin. *Archives of Biochemistry and Biophysics*, *135*, 419–434.

15. García-Lafuente, A., Guillamon, E., Villares, A., Rostagno, M. A., & Martinez, J. A., (2009). Flavonoids as anti-inflammatory agents, implications in cancer and cardiovascular disease. *Inflammation Research*, *58*(9), 537–552.

16. Garten, R. J., Davis, C. T., Russell, C. A., Shu, B., Lindstrom, S., Balish, A., Sessions, W. M., et al., (2009). Antigenic and genetic characteristics of swine-origin 2009-A (H1N1) influenza viruses circulating in humans. *Science*, *325*(5937), 197–201.

17. Gholijani, N., Gharagozloo, M., Farjadian, S., & Amirghofran, Z., (2016). Modulatory effects of thymol and carvacrol on inflammatory transcription factors in lipopolysaccharide-treated macrophages. *Journal of Immunotoxicology*, *13*(2), 157–164.

18. Gunny, A. A., Arbain, D., Nashef, E. M., & Jamal, P., (2015). Applicability evaluation of deep eutectic solvents-cellulase system for lignocellulose hydrolysis. *Bioresource Technology*, *181*, 297–302.

19. Hammond, A., Galizi, R., Kyrou, K., Simoni, A., Siniscalchi, C., Katsanos, D., Gribble, M., et al., (2016). CRISPR-Cas9 gene drive system targeting female reproduction in the malaria mosquito vector *Anopheles gambiae*. *Nature Biotechnology*, *34*(1), 78–83.

20. Hassan, H. T., (2004). Ajoene (natural garlic compound): A new anti-leukemia agent for AML therapy. *Leukemia Research*, *28*(7), 667–671.

21. Herbst, R. S., Baas, P., Kim, D. W., Felip, E., Perez-Gracia, J. L., Han, J. Y., Molina, J., et al., (2016). Pembrolizumab versus docetaxel for previously treated: PD-L1-positive, advanced non-small-cell lung cancer (Keynote-010), a randomized controlled trial. *The Lancet*, *387*(10027), 1540–1050.

22. Horne, H. N., Sherman, M. E., Pfeiffer, R. M., Figueroa, J. D., Khodr, Z. G., Falk, R. T., Pollak, M., et al., (2016). Circulating insulin-like growth factor-I: Insulin-like growth factor binding protein-3 and terminal duct lobular unit involution of the breast, a cross-sectional study of women with benign breast disease. *Breast Cancer Research*, *18*(1), 24–28.

23. Hosseinzadeh, L., Behravan, J., Mosaffa, F., Bahrami, G., Bahrami, A., & Karimi, G., (2011). Curcumin potentiates doxorubicin-induced apoptosis in H9c2 cardiac muscle cells through generation of reactive oxygen species. *Food and Chemical Toxicology*, *49*(5), 1102–1109.

24. Huang, X., Qian, W., El-Sayed, I. H., & El-Sayed, M. A., (2007). The potential use of the enhanced nonlinear properties of gold nanospheres in photothermal cancer therapy. *Lasers in Surgery and Medicine, 39*(9), 747–753.

25. Jeon, J. S., Bersini, S., Gilardi, M., & Dubini, G., (2015). Human 3-D vascularized organotypic microfluidic assays to study breast cancer cell extravasation. *Proceedings of the National Academy of Sciences, 112*(1), 214–219.

26. Junlatat, J., & Sripanidkulchai, B., (2014). Hair growth-promoting effect of *Carthamus tinctorius,* floret extract. *Phytotherapy Research, 28*(7), 1030–1036.

27. Kandimalla, R., & Reddy, P. H., (2016). Multiple faces of dynamin-related protein 1 and its role in Alzheimer's disease pathogenesis. *Biochimica et Biophysica Acta (BBA)-Molecular Basis of Disease, 1862*(4), 814–828.

28. Kang, H. Y., Moon, S. H., Jang, H. J., Lim, D. H., & Kim, J. H., (2016). Validation of allergy. *Asthma and Respiratory Disease, 4*(5), 369–373.

29. Kim, D., Pertea, G., Trapnell, C., Pimentel, H., Kelley, R., & Salzberg, S. L., (2013). TopHat2: Accurate alignment of transcriptomes in the presence of insertions, deletions, and gene fusions. *Genome Biology, 14*(4), 36.

30. Klionsky, D. J., Abdelmohsen, K., Abe, A., & Abedin, M. J., (2016). Guidelines for the use and interpretation of assays for monitoring autophagy. *Autophagy, 12*(1), 1–222.

31. Krantz, I. D., McCallum, J., DeScipio, C., Kaur, M., Gillis, L. A., Yaeger, D., Jukofsky, L., et al., (2004). *Cornelia de Lange* syndrome is caused by mutations in NIPBL: The human homolog of drosophila melanogaster nipped-B. *Nature Genetics, 36*(6), 631–635.

32. Krause, A. J., Cines, B., & Pogrebniak, E., (2016). Associations between adiposity and indicators of thyroid status in children and adolescents. *Pediatric Obesity, 11*(6), 551–558.

33. Lee, J. Y., Nagano, Y., Taylor, J. P., Lim, K. L., & Yao, T. P., (2010). Disease-causing mutations in Perkins impair mitochondrial ubiquitination, aggregation, and HDAC6-dependent mitophagy. *The Journal of Cell Biology, 189*(4), 671–679.

34. Lessa, J. A., Parrilha, G. L., & Beraldo, H., (2012). Gallium complexes as new promising metallodrug candidates. *Inorganica Chimica Acta, 393*, 53–63.

35. Leyva-Lopez, N., Nair, V., Bang, W. Y., Cisneros-Zevallos, L., & Heredia, J. B., (2016). Protective role of terpenes and polyphenols from three species of Oregano (*Lippia graveolens, Lippia palmeri and Hedeoma patens*) on the suppression of lipopolysaccharide-induced inflammation in RAW 264.7 macrophage cells. *Journal of Ethnopharmacology, 187*, 302–312.

36. Lingaraju, G. M., Bunker, R. D., Cavadini, S., Hess, D., Hassiepen, U., Renatus, M., Fischer, E. S., & Thoma, N. H., (2014). Crystal structure of the human COP9 signalosome. *Nature, 512*(7513), 161–165.

37. Liu, Y. Y., Wang, Y., Walsh, T. R., Yi, L. X., Zhang, R., Spencer, J., Doi, Y., et al., (2016). Emergence of plasmid-mediated colistin resistance mechanism MCR-1 in animals and human beings in China, a microbiological and molecular biological study. *The Lancet Infectious Diseases, 16*(2), 161–168.

38. Lv, B. Q., Weng, H. M., Fu, B. B., Wang, X. P., Miao, H., Ma, J., Richard, P., et al., (2015). Experimental discovery of Weyl semimetal TaAs. *Physical Review X, 5*(3), e-article ID: 031013, Open Access: https://journals.aps.org/prx/abstract/10.1103/PhysRevX.5.031013 (accessed on 18 May 2020).

39. Mahmoud, A. B., Tu, M. M., Wight, A., Zein, H. S., Rahim, M. M., Lee, S. H., Sekhon, H. S., et al., (2016). Influenza virus targets class I MHC-educated NK cells for immunoevasion. *PLoS Pathog.*, *12*(2), 1005446. https://journals.plos.org/plospathogens/article?id=10.1371/journal.ppat.1005446 (accessed on 18 May 2020).
40. Mani, S. A., Guo, W., Liao, M. J., Eaton, E. N., Ayyanan, A., Zhou, A. Y., Brooks, M., et al., (2008). The epithelial-mesenchymal transition generates cells with properties of stem cells. *Cell*, *133*(4), 704–715.
41. Muraro, A., Werfel, T., Hoffmann-Sommergruber, K., Roberts, G., Beyer, K., Bindslev-Jensen, C., Cardona, V., et al., (2014). EAACI food allergy and anaphylaxis guidelines, diagnosis, and management of food allergy. *Allergy*, *69*(8), 1008–1025.
42. Murray, C. J., Vos, T., Lozano, R., Naghavi, M., Flaxman, A. D., Michaud, C., Ezzati, M., et al., (2013). Disability-adjusted life years (DALYs) for 291 diseases and injuries in 21 regions: (1990–2010), a systematic analysis for the global burden of disease study 2010. *The Lancet*, *380*(9859), 2197–2223.
43. Nair, V., (2002). HIV integrase as a target for antiviral chemotherapy. *Reviews in Medical Virology*, *12*(3), 179–193.
44. Nemmar, A., Hoylaerts, M. F., Hoet, P. H., Dinsdale, D., Smith, T., Xu, H., Vermylen, J., & Nemery, B., (2002). Ultrafine particles affect experimental thrombosis in an *in vivo* hamster model. *American Journal of Respiratory and Critical Care Medicine*, *166*(7), 998–1004.
45. Ng, M., Fleming, T., Robinson, M., Thomson, B., Graetz, N., Margono, C., Mullany, E. C., et al., (2014). Global, regional, and national prevalence of overweight and obesity in children and adults during 1980–2013: A systematic analysis for the global burden of disease study. *The Lancet*, *384*(9945), 766–781.
46. Ntalli, N. G., Manconi, F., Leonti, M., Maxia, A., & Caboni, P., (2011). Aliphatic ketones from *Rutachalepensis* (Rutaceae) induce paralysis on root knot nematodes. *Journal of Agricultural and Food Chemistry*, *13*, 7098–7103.
47. Opitz, C. A., Litzenburger, U. M., Sahm, F., Ott, M., Tritschler, I., Trump, S., Schumacher, T., et al., (2011). An endogenous tumor-promoting ligand of the human aryl hydrocarbon receptor. *Nature*, *478*(7368), 197–203.
48. Ottesen, A. R., Pena, A. G., White, J. R., Pettengill, J. B., Li, C., Allard, S., Rideout, S., et al., (2013). Baseline survey of the anatomical microbial ecology of *Solanum lycopersicum* (tomato). *BMC Microbiology*, *13*(1), 114–120.
49. Qu, S., Song, W., Yang, X., Wang, J., Zhang, R., Zhang, Z., et al., (2015). Microarray expression profile of circular RNAs in human pancreatic ductal adenocarcinoma. *Genomics Data*, *5*, 385–387.
50. Samarghandian, S., Boskabady, M. H., & Davoodi, S., (2010). Use of in vitro assays to assess the potential antiproliferative and cytotoxic effects of saffron (*Crocus sativus* L.) in human lung cancer cell line. *Pharmacognosy Magazine*, *6*(24), 309–318.
51. Skrovankova, S., Sumczynski, D., Mlcek, J., Jurikova, T., & Sochor, J., (2015). Bioactive compounds and antioxidant activity in different types of berries. *International Journal of Molecular Sciences*, *16*(10), 24673–24706.
52. Smith, J. A., Ware, E. B., Middha, P., Beacher, L., & Kardia, S. L., (2015). Current applications of genetic risk scores to cardiovascular outcomes and subclinical phenotypes. *Current Epidemiology Reports*, *2*(3), 180–190.
53. Tamaddonfard, E., Farshid, A. A., Ahmadian, E., & Hamidhoseyni, A., (2013). Crocin enhanced functional recovery after sciatic nerve crush injury in rats. *Iranian Journal of Basic Medical Sciences*, *16*(1), 83–90.

54. Tao, R., Coleman, M. C., Pennington, J. D., Ozden, O., Park, S. H., Jiang, H., Kim, H. S., et al., (2010). Sirt3-mediated deacetylation of evolutionarily conserved lysine 122 regulates MnSOD activity in response to stress. *Molecular Cell, 40*(6), 893–904.

55. Tsai, S. Q., Zheng, Z., Nguyen, N. T., Liebers, M., Topkar, V. V., Thapar, V., Wyvekens, N., et al., (2015). GUIDE-seq enables genome-wide profiling of off-target cleavage by CRISPR-Cas nucleases. *Nature Biotechnology, 33*(2), 187–197.

56. Verstovsek, S., Mesa, R. A., Gotlib, J., Levy, R. S., Gupta, V., DiPersio, J. F., Catalano, J. V., et al., (2015). Efficacy, safety, and survival with ruxolitinib in patients with myelofibrosis: Results of a median 3-year follow-up of COMFORT-I. *Haematologica, 100*(4), 479–488.

57. Villalvilla, A., Gomez, R., Roman-Blas, J. A., Largo, R., & Herrero-Beaumont, G., (2014). The SDF-1 signaling: A promising target in rheumatic diseases. *Expert Opinion on Therapeutic Targets, 18*(9), 1077–1087.

58. Zhang, L., Wang, F., Wang, L., Wang, W., Liu, B., Liu, J., Chen, M., et al., (2012). Prevalence of chronic kidney disease in China, a cross-sectional survey. *The Lancet, 379*(9818), 815–822.

59. Zhou, L., Ivanov, I. I., Spolski, R., Min, R., Shenderov, K., Egawa, T., Levy, D. E., et al., (2007). IL-6 programs TH-17 cell differentiation by promoting sequential engagement of the IL-21 and IL-23 pathways. *Nature Immunology, 8*(9), 967–974.

CHAPTER 5

THERAPEUTIC POTENTIAL OF ANTHOCYANIN AGAINST DIABETES

TAWHEED AMIN, H. R. NAIK, BAZILA NASEER, and
SYED ZAMEER HUSSAIN

ABSTRACT

Diabetes is the fastest growing non-communicable disease that contributes a huge burden on the healthcare system due to its high prevalence and potentially deleterious effects. It is an epidemic in many economically developing nations and high-income newly industrialized nations. Diabetes is characterized by abnormally high blood glucose (hyperglycemia) because of either impairment or destruction of β-cells secreting insulin and the activity of insulin in target cells. Anthocyanin acts through the inactivation of α-amylase and α-glucosidase and thus improving the function of β-cells of islets of Langerhans. Despite health beneficial effects, synthetic drugs result in side-effects, such as flatulence, bloating, diarrhea, etc. This chapter focuses on role of anthocyanin (plant-derived natural colorant) in diabetes therapy. The chapter also deals with methods to increase the bioavailability and process stability of anthocyanins. The chapter further highlights the pleiotropic effects of anthocyanin to control diabetes.

5.1 INTRODUCTION

Type-2 diabetes (T2D) is a well-known metabolic disorder and it continues to be a major medical concern and multifunctional disease due to its high prevalence and potentially deleterious effects [3, 73, 95, 117]. The most important contributing factor for the development of diabetes is postprandial hyperglycemia [73, 85]. Diabetes complications (the major cause of morbidity, mortality, and healthcare costs) can be significantly reduced (or prevented) by appropriate control of blood glucose at a normal level after the meal [30, 48, 49, 91].

Any agent that increases insulin sensitivity and the inhibition of starch hydrolyzing enzymes (α-amylase and α-glucosidase) are, therefore, considered therapeutic strategies for the treatment of T2D [102, 103]. It has been reported that there is a positive correlation between human pancreatic α-amylase activity and the increase in postprandial glucose levels, thereby demonstrating the importance of the inhibition of α-amylase activity as a therapeutic agent against T2D. The α-amylase is a starch blocker inhibits the hydrolysis of starch [109]. It cannot be inhibited completely; rather a mild α-amylase inhibitory activity is suggested to prevent the abnormal fermentation of undigested carbohydrates by the bacteria in the colon, which otherwise results in diarrhea and flatulence [56].

Therapeutic intervention using some commonly available synthetic anti-diabetic drugs (acarbose, voglibose, miglitol, etc.) have shown encouraging results in the management and treatment of diabetes [69, 70]. Although efficient in preventing the upsurge of blood glucose levels, the intake of these synthetic drugs is associated with the adverse gastrointestinal (GI) disturbances, such as diarrhea, flatulence, cramping, etc. [69]. Due to drug toxicity associated with the intake of synthetic anti-diabetic drugs, the current studies have begun to focus on plant-derived phytomolecules due to their low cost, easy availability/adaptability, less adverse side effects.

The utilization of plants and their phytochemicals for the management and treatment of diabetes not only provides safer alternatives to otherwise unsafe synthetic drugs, which transitory lowers blood glucose, yet additionally improves the activity and secretion of insulin [94]. The plant-based molecules (such as flavonoids, pheophytins [95], quercetin [55], rutin [67], and anthocyanin [116] possess inhibitory action against starch hydrolyzing enzymes. Anthocyanin shows promising activity against T2D by impeding the assimilation of glucose [26, 89]. It also improves insulin sensitivity in diabetic patients by targeting adenosine monophosphate-protein kinase (AMPK), glucose transporter type-4 (GLUT4), and metabolic enzymes [102].

The health-promoting characteristics of anthocyanins are being continuously explored as dietary supplements and nutraceuticals [2]. The natural colorant, anthocyanin, possess remarkable biological activities, such as anti-inflammatory [83], antioxidative [72], anticancer [108], anti-obesity [57] and antidiabetic [102]. Among its remarkable effects against starch hydrolyzing enzymes and ability to improve insulin sensitivity, natural anthocyanin is nontoxic in humans. However, the toxicity risk from food is minute that may be attributed due to the low bioavailability of anthocyanins [106].

This chapter focuses on plant-derived anthocyanin that can inhibit starch hydrolyzing enzymes, improve insulin sensitivity; and their prospective contribution in the treatment of diabetes along with mechanisms of action.

5.2   STARCH HYDROLYZING ENZYMES: BRIEF GLIMPSE

Digestion of carbohydrates inside our body occurs in a successive way with α-amylase acting initially on starch trailed by α-glucosidase to produce dietary glucose. Once the food is ingested, starch is acted upon by α-amylases (both salivary and pancreatic) and four small intestinal mucosal α-glucosidase subunits [57], and at an inner α-1,4 glucosidic linkages via an endo mechanism thereby producing linear and branched malto-oligosaccharides [55]. Maltase-glucoamylase and sucrose-isomaltase (the two membrane-bound protein complexes), and mucosal α-glucosidases are exo-type starch hydrolyzing enzymes [65, 68] that produce glucose by hydrolyzingα-1,4 glucosidic linkages opposite to the reducing end of dextrins already degraded by α-amylase [6, 27, 29]. Apart from illustrious maltase activity, the C-terminal subunit (maltase-glucoamylase) is named as isomaltase because of its action on long-chain oligomers [54] whereas the N-terminal subunit (sucrose isomaltase) is named as isomaltase because of its debranching activity [28].

Out of the two enzymes, α-glucosidases (including maltase and sucrose) [115] are the indispensable enzymes required for digestion of starch since these act upon oligosaccharides to produce monosaccharides before being absorbed into the enteral epithelia [81]. Restraint of α-glucosidase may in this way back-off the glucose production [115] and reduce the postprandial hyperglycemia, a predisposing factor for T2D [41].

5.3   ANTHOCYANIN: CHEMICAL STRUCTURE AND DISTRIBUTION

Anthocyanins are water-soluble pigments [36] and belong to flavonoids (most common group of phenols) [57]. Chemically, anthocyanins are glycosides of polymethoxyl derivatives of 2-phenylbenzopyrillium or flavylium salt [45, 90]. As observed in Figure 5.1, seven different side groups (R_1–R_7) exist on the flavylium cation. These side groups could be either of the three groups: a hydrogen atom (H) or a methoxy (-OCH$_3$) or a hydroxide (-OH) group.

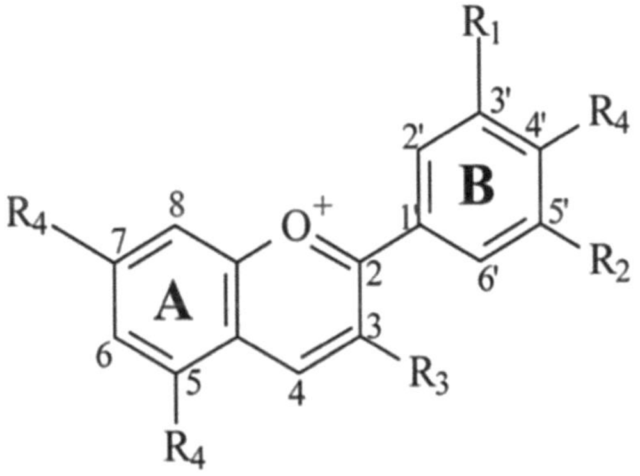

FIGURE 5.1   Flavylium ion: the basic structure of anthocyanin.

Legend: R_1 = H; R_2 = OH or OCH$_3$; R3 = OH or glycosyl; and R4 = OH or glycosyl.

Anthocyanins are derived from anthocyanidins; therefore, their structure consists of an aglycone or sugar-free anthocyanidin structure [32]. Anthocyanidins do not contain carbohydrate (glucose) esterified at the 3-position, while as it is present in anthocyanin [32]. Based on the number and position of OH⁻ groups on the flavan nucleus, six distinct anthocyanidins exist commonly in plants, such as pelargonidin (pg), cyanidin (cy), peonidin (pn), delphinidin (dp), petunidin (pt), and malvidin (mv) (Figures 5.2) [79]. The most predominant plant sources of anthocyanins are grapes (11%), berries (20%), red or purple vegetables (8%), yogurt (6%), wine (16%), and 100%-non-citrus juice (6%) [17].

Several experimental studies indicate that the consumption of polyphenol-rich vegetables and fruits decrease the incidence of diabetes [4, 5, 71]. Insulin resistance is a metabolic disorder in which the insulin incompetently stimulates the transport of glucose in fat and skeletal muscle and inadequately suppresses hepatic glucose production [45]. Oral hyperglycemic drugs (sulfonylurea-based) can increase the secretion of insulin from islets of Langerhans in type-2 diabetic patients sufficient to overcome insulin resistance, thus normalizing plasma glucose level [45]. However, the disadvantage with sulfonylurea-based drugs is that these drugs fail to keep blood glucose level under control [90].

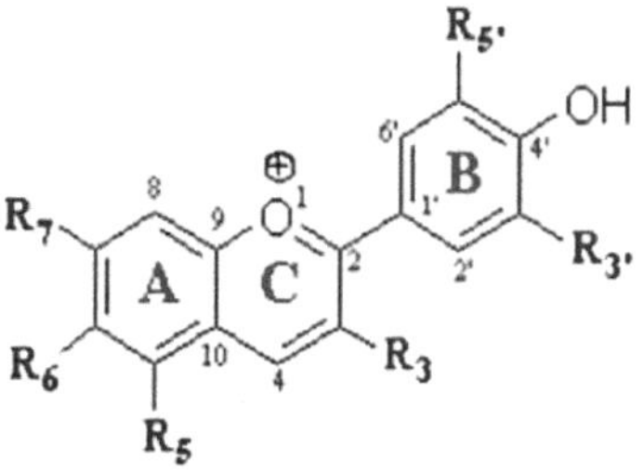

Anthocyanidin

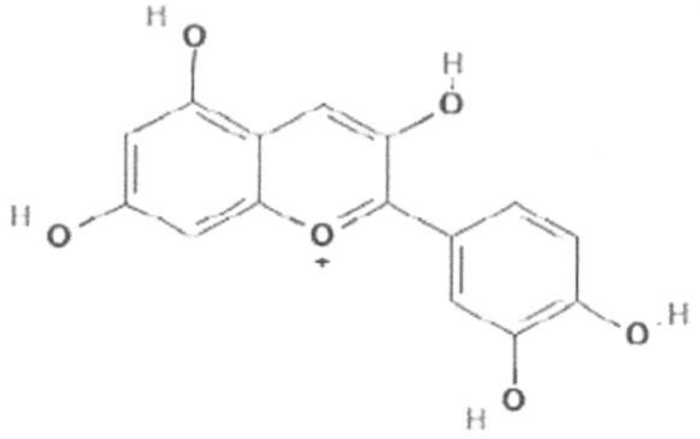

Cyanidin (Cy: YGM-3)

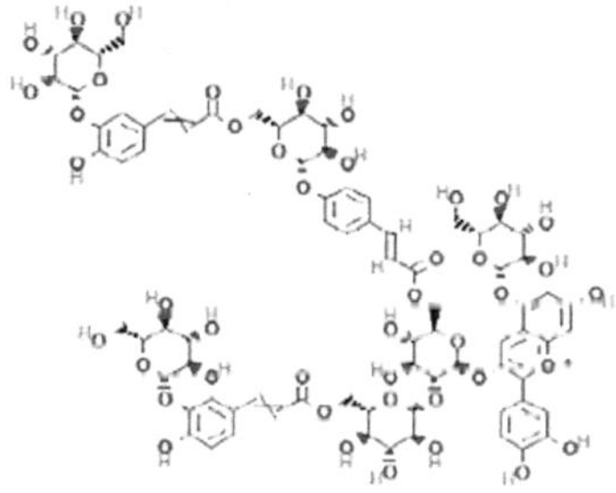

Anthocyanin

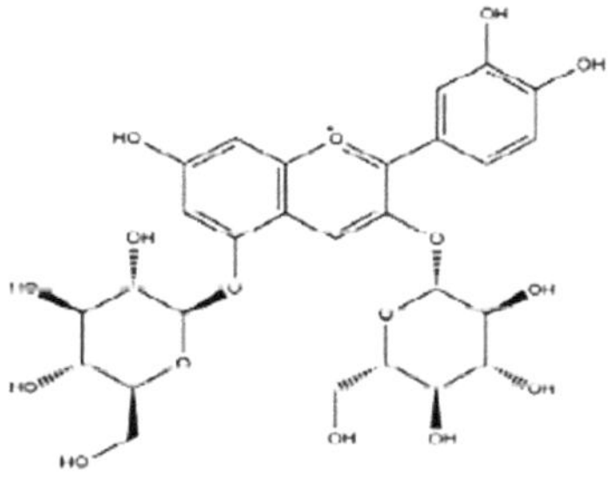

Cyanidin-3,5-glucoside

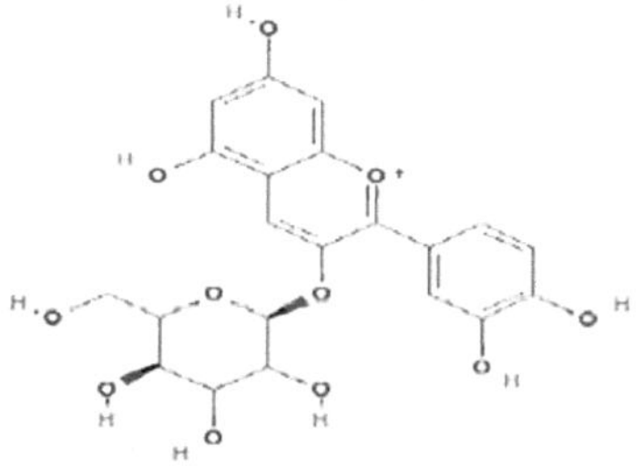

Cyanidin-3-galactoside

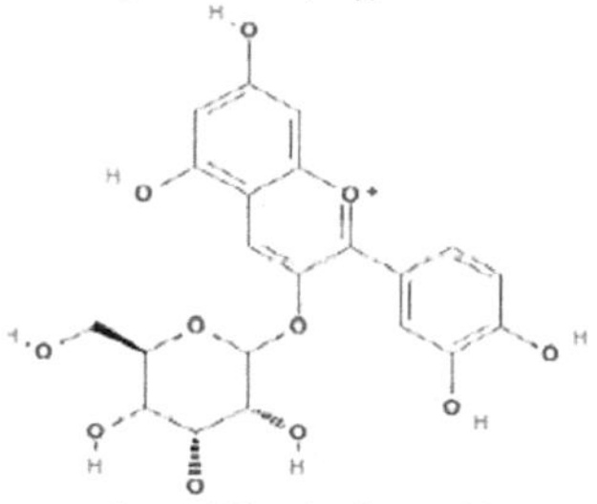

Cyanidin-3-glucoside

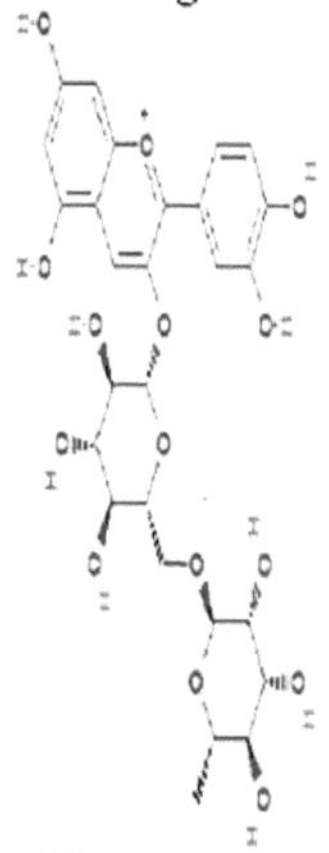

Cyaniding-3-rutinoside

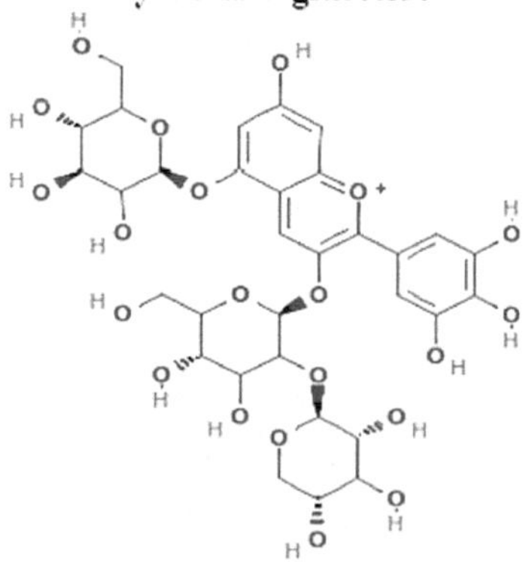

Delphinidin 3-sambubioside-
5-glucoside (D3S5G)

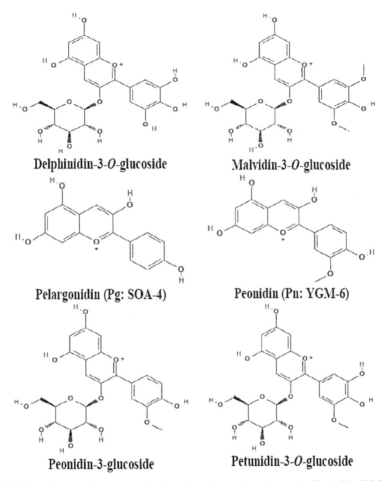

FIGURE 5.2   Chemical structure of selected anthocyanins that are indicated in Table 5.1.

5.4   POTENTIAL OF ANTHOCYANIN MOLECULES FOR DIABETES THERAPY

The increase in diabetes has become a fiscal expenditure for medical care [26]. An optimal dietary therapy is a wise approach for the management and prevention of insulin resistance and type-2 diabetes [50]. Phytochemical-rich vegetables and fruits are beneficial in the management and prevention of type-2 diabetes [22, 76]. The consumption of an antioxidant-rich diet may improve the insulin sensitivity [8, 45, 61].

Further, sulfonyl-urea-based drugs also negatively affect the ability of β-cells of islets of Langerhans to secrete consistent insulin level and cause weight-gain [90]. Therefore, it would be beneficial if blood glucose is regulated through dietary constituents. Emerging evidences have revealed potential of anthocyanins in the management of diabetes [77, 80, 102] due to its ability to decrease insulin sensitivity [62, 77] or inhibit carbohydrate hydrolyzing ability [80]. Epidemiological studies have shown that high consumption of anthocyanin-rich berry fruits decreases the risk of type-2 diabetes [15, 50].

The antidiabetic mechanism of anthocyanins is classified into two distinct groups (Figure 5.3):

- **Insulin-Dependent:** This involves the improvement of the function of β-cells of islets of Langerhans. It involves promoting β-cell proliferation, reducing oxidative stress, reducing β-cell apoptosis and increasing insulin sensitivity [84].

- **Insulin-Independent:** This includes the inhibition of enzymes meant for hydrolyzing starch and a decrease in the assimilation of glucose. It further includes the inhibition of α-amylase, α-glucosidase and glucose transporters, sodium-dependent glucose co-transporters (SGLT1: sodium glucose linked transporter 1) and GLUT4, and changes in the status of energy metabolism (AMP-activated protein kinase) [15, 84].

Anthocyanin-rich black bean coat extracts may decrease the generation of reactive oxygen species (ROS) up to 80% due to its antioxidant activity [84]. This reduction in ROS may protect and improve the ability of β-cells of islets of Langerhans, thereby suggesting an indirect role of anthocyanins in mitigation of diabetes. Dipeptidyl peptidase 4, an essential enzyme, is metabolically related to insulin secretion [84]. Anthocyanin-rich extracts obtained from the coats of black beans have been reported to inhibit Dipeptidyl peptidase 4 [84], thereby suggesting the role of anthocyanins in the prevention and management of diabetes.

Synthetic anti-diabetic drugs act by inhibiting α-amylase and α-glucosidase enzyme activity leading to reduction in the breakdown of polysaccharides. However, these synthetic drugs pose several side effects including diarrhea, flatulence, bloating, cramping etc. due to incomplete carbohydrate digestion [44, 115].

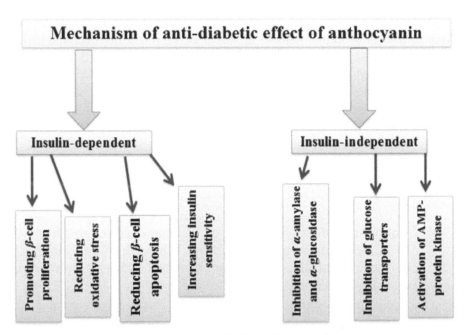

FIGURE 5.3   Mechanism of action for antidiabetic effect of anthocyanins.

Anthocyanin inhibits α-amylase through competitive inhibition [98], in which an inhibitor competes directly with the substrate for free enzyme [23]. For example, cyanidin-3-glucoside has shown inhibitory action against α-amylase through competitive inhibition. The inhibitory action, however, could be increased by either replacing 3-glucoside by 3-rutinoside or by introducing another moiety of glucose at C_5 of cyanidin-3-glucoside [98]. It can thus be concluded that the sugar (glucose) moiety at the 3-position of anthocyanin plays a vital role in the inhibitory action against enzymes.

Anthocyanin-rich extract was found to exhibit an inhibitory action against starch hydrolyzing enzymes (α-amylase and α-glucosidase) when it was used as a natural colorant in a beverage at a concentration of 1 mg/mL [84]; and it also decreased the glucose uptake at the GI level [84]. Molecular docking studies revealed a good interaction between the delphinidin (delphinidin-3-O-glucoside) from the anthocyanin-rich extract from black beans and catalytic activity of starch hydrolyzing enzymes (α-amylase and α-glucosidase) (Figure 5.4).

As indicated in this section, one of the therapeutic approaches to treat diabetes is the inhibition of α-glucosidase that acts through exo-mechanism [65] and catalyzes the release of dietary glucose from the substrate produced due to α-amylase digestion, via non-reducing end [68]. Inhibition of this enzyme prevents the upsurge of blood glucose after a carbohydrate meal [75]. The α-amylase consists of three distinguishable domains (A, B, and C) [98]. The largest domain is Domain A (containing ~330 residues) followed by Domain B (101–168 residues) and Domain C (408–496 residues) [12].

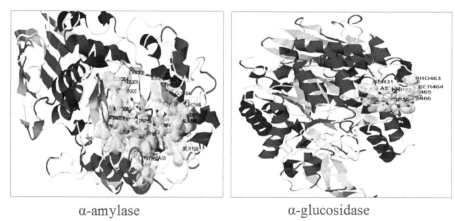

α-amylase   α-glucosidase

FIGURE 5.4   *In-silico* molecular docking showing an interaction of dolphin-3-O-glucoside with α-amylase and α-glucosidase.

Source: Reprinted with permission from Ref. [84]. © 2009 Elsevier.

The active site of α-amylase is being characterized by number non-polar residues (TRP58, TYR62, TRP59, HIS101, PRO163, IIE235, TYR258, HIS305, HIS299, and ALA307) and several side chains (ARG61, ARG165, ARG197, LYS200, ASP236, GLU233, ASP300) [12, 74]. The active site is also characterized by residues (ASP300, GLU233, and ASP197) and lies deep in a depression at the center [74]. Anthocyanin binds to α-amylase through the formation of various hydrogen bonds as can be understood by taking an example of cyanidin-3-O-glucoside (anthocyanin) as shown in Figure 5.5, wherein this molecule is encompassed by GLU240, LYS200, ILE235, GLU233, HIS305, and GLY306.

Further, the molecular docking studies revealed that the inhibitory action of anthocyanins against α-amylase is due to the binding to the side chain of GLU233 [98]. The number and position of hydroxyl groups (-OH) in flavylium cation determines the enzyme instability action of anthocyanins [4]. It is reported that the inhibitory action increases considerably with the increase in the quantum of -OH groups on B-ring of flavylium cation (Figure 5.1) [101]. Among all the anthocyanins, cyaniding-3-glucoside has the highest inhibitory action against α-amylase, obtained from porcine pancreas, with inhibition constant of Ki 0.014 mM trailed by cyaniding-3-rutinoside (Ki 0.019), cyanidin-3,5-glucoside (Ki 0.020) and peonidin-3-glucoside (Ki 0.045) [98]. Molecular docking studies have revealed that anthocyanins occupy the binding cleft of pancreatic α-amylase by way of forming hydrogen bonds. The GLU233, a common key side chain, is believed to impart enzyme inhibitory action to anthocyanin (Figure 5.5) [73, 98].

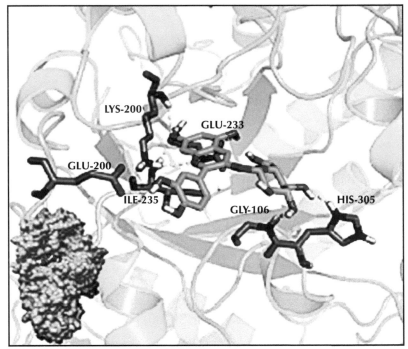

FIGURE 5.5 Binding of anthocyanin (cyaniding-3-glucoside) to α-amylase through the formation of various hydrogen bonds. The α-amylase can be viewed in surface view, while as anthocyanins are represented as sticks.

Source: Reprinted with permission from Ref. [98]. © 2009 Elsevier.

The hydroxyl (-OH) groups at C5, C6, and C7 of the A-ring of flavylium ion (Figure 5.1) impart enzyme inhibitory action to anthocyanins [88]. Double bond at 2′, 3′, hydroxyl group at C-5′, the linkage of the B-ring at 3rd position and the substitution of hydroxyl group (-OH) on the B-ring of the flavylium ion enhances inhibitory potential of anthocyanin, while the 3-OH reduces it [101].

The α-glucosidase is also called "glycomimetic" and possesses a positive charge near the glycoside bond, which acts as a competitive inhibitor. Glyco-mimetics bind to the active center of the enzyme through charge interactions with the catalytic group in addition to hydrogen bonds [107].

AMPK, the most critical factor for the balance of cell energy, is being perceived as a potential remedial focus in the counteractive action and treatment of T2D [51, 64, 96]. It is an energy-sensing enzyme and conserved kinase, the activation of which elicits the insulin-sensitizing target for T2D. AMPK gets activated when the energy levels of cell are low, which in turn stimulates the uptake of glucose in the fatty acid oxidation, skeletal muscles, and reduces the production of hepatic glucose. The dysregulation of AMPK in T2D patients by way of activating AMPK (either pharmacologically or physiologically) can improve the insulin sensitivity and metabolic health [24].

Some drugs (thiazolidinediones and metformin) and several compo-nents of food may conceivably have antidiabetic effects [21, 97]. Dietary anthocyanin-rich bilberry (European blue berries) extract significantly activates AMPK in skeletal muscle and white adipose tissue. This activation thus increases the protein expression of GLUT4, improving the take-up of glucose into these cells or tissues through an insulin-dependent mechanism [102] thus prompting the decrease in hyperglycemia. Activated AMPK addi-tionally results in the hindrance of glucose production, formation of fat and stimulation of fatty acid oxidation [105].

Activated AMPK also suppresses gluconeogenesis and thus abolishes abnormal upsurge of blood glucose in diabetic *ob/ob* mice [42]. Activated AMPK devitalizes the program of gluconeogenesis by promoting the transducer of regulated cAMP (cyclic adenosine monophosphate) response element-binding (CR E-binding) protein activity 2 (TORC2) phosphorylation and blocking its nuclear accumulation [42]. Several other studies to investigate the potential role of anthocyanins against diabetes are summarized in Table 5.1.

5.5 ANTHOCYANINS: METHODS TO INCREASE BIOAVAILABILITY, PROCESS STABILITY

Although there is an increased interest in anthocyanins as therapeutic agents against diabetes, yet their use is limited by their low chemical stability and bioavailability [52, 63]. Further, the antidiabetic potential of anthocyanins from the food supply is questionable, owing to their low bioavailability after taken orally [25, 93, 111].

TABLE 5.1 Summary of Research Studies on Role of Anthocyanins as Antidiabetic Agents

Type of Anthocyanin	Plant Sources	Anti-Diabetic Effect	Mechanism of Antidiabetic Effect	References
Anthocyanidin	—	In-vitro	Slight inhibition of α-glucosidase.	[110]
Anthocyanin	Bilberry (*Vaccinium myrtillus*); blackcurrant (*Ribesnigrum*)	In-vivo	Increase insulin sensitivity; lowers fasting plasma glucose; activates AMP; increases GLUT4 in T2D mice.	[77, 102]
	Blueberry (*Vaccinium corombosum*)		Inhibits α-amylase and α-glucosidase; more effect against α-glucosidase; up-regulates GLUT4 via activation of AMP.	[81, 103]
Cyanidin (Cy: YGM-3)	Grapes; morning glory; Cherry.	In-vitro	Inhibits α-glucosidase	[80]
Cyanidin-3,5-glucoside	—	In-vitro and in-silico	Inhibits α-amylase	[98]
Cyanidin-3-galactoside	—	In-vitro	Inhibits α-glucosidase	[99]
Cyanidin-3-glucoside	—	In-vitro and in-silico	Inhibits α-amylase	[98]
Cyaniding-3-rutinoside	Litchi; sweet cherry; black currant; capulin.	In-vitro	Inhibits α-glucosidase	[2]
Delphinidin 3-sambubioside-5-glucoside (D3S5G)	Maqui Berry (*Aristotelia chilensis*)	In-vivo	Increases down-regulation (insulin dependent) of glucose-6-phosphatase	[93]

Delphinidin-3-O-glucoside	Black beans, common pea	*In-vitro*	Inhibits α-amylase and α-glucosidase	[84]
Malvidin-3-O-glucoside	Black beans (*Vaccinum Arcto-Staphylos*)	*In-vitro*	Inhibits α-amylase and α-glucosidase	[84, 89]
Pelargonidin (Pg: SOA-4)	Blue pimpernel; blackberries; pomegranate.	*In-vitro*	Inhibits α-glucosidase	[80]
Peonidin (Pn: YGM-6)	Raw cranberries; morning glory.	*In-vitro*	Inhibits α-glucosidase	[80]
Peonidin-3-glucoside	—	*In-vitro* and *in-silico*	Inhibits α-amylase	[98]
Petunidin-3-O-glucoside	Black beans.	*In-vitro*	Inhibits α-amylase and α-glucosidase	[84]

Note: Chemical structure of the phytochemical in column 1 is shown in Figure 5.2.

Bioavailability defines the kinetics and the degree to which a bioactive agent is assimilated from any ingested food and its availability at the target site-of-action [35]. The therapeutic effects of the natural colorant (anthocyanin) are dependent on sufficient bioavailability possibly due to their complex biochemistry [82].

Anthocyanins can be absorbed directly by the epithelial cells of the small intestines [78] but it is difficult to transfer them to the GI tract due to low chemical stability in the adverse GI conditions [46, 63]. The lower bioavailability of anthocyanin was confirmed by low anthocyanin concentration in plasma [35, 59]. Further human studies have shown that <0.1% of intact anthocyanins were found in urine, affirming its low bioavailability [36].

Instead of their poor absorption from GIT, a considerable pre-systematic metabolism of some anthocyanins attributes to their low bioavailability [37]. Given their health benefits, the chemical stability of natural anthocyanins is of absolute importance [14]. The stability of anthocyanin varies widely [35] in response to pH and temperature [43]. In aqueous solutions, anthocyanin undergoes structural rearrangements in response to pH changes into four distinct molecular structures (Figure 5.6): flavylium cation (red), quinoidal base (blue), carbinol pseudo-base (colorless), and *cis*-chalcone (yellowish) forms [9–11, 13, 35, 95].

Anthocyanins possess characteristic physicochemical properties conferring these natural colorants unique stability and color [38]. They are highly reactive and sensitive to degradation reactions. The variables that influence the chemistry and subsequently the stability and color of anthocyanins are temperature [39], light [18], oxygen [40], enzymes [16], ascorbic acid [61], and pH [60].

In-vitro studies have shown that the stability of anthocyanins is decreased with the increase in the hydroxylation of B-ring [112]. The pH and temperature of the environment affects the stability and thus bioavailability of anthocyanins. The low stability of anthocyanin would directly affect their health benefits; therefore, the challenge is to protect the anthocyanin from the deterioration and increase its bioavailability. Several other studies to improve the process stability and bioavailability of anthocyanins are summarized in Table 5.2.

Using various food-processing techniques, the anti-diabetic effects of anthocyanins, administered orally, could be increased. For instance, self-micro emulsifiers (Labrasol) can significantly increase the anti-diabetic potential of anthocyanins from blueberry [47].

Microencapsulation provides a favorable and stabilization environment that protects anthocyanins from harsh environmental factors (temperature, pH, light, enzymes, etc.) [7, 52, 92].

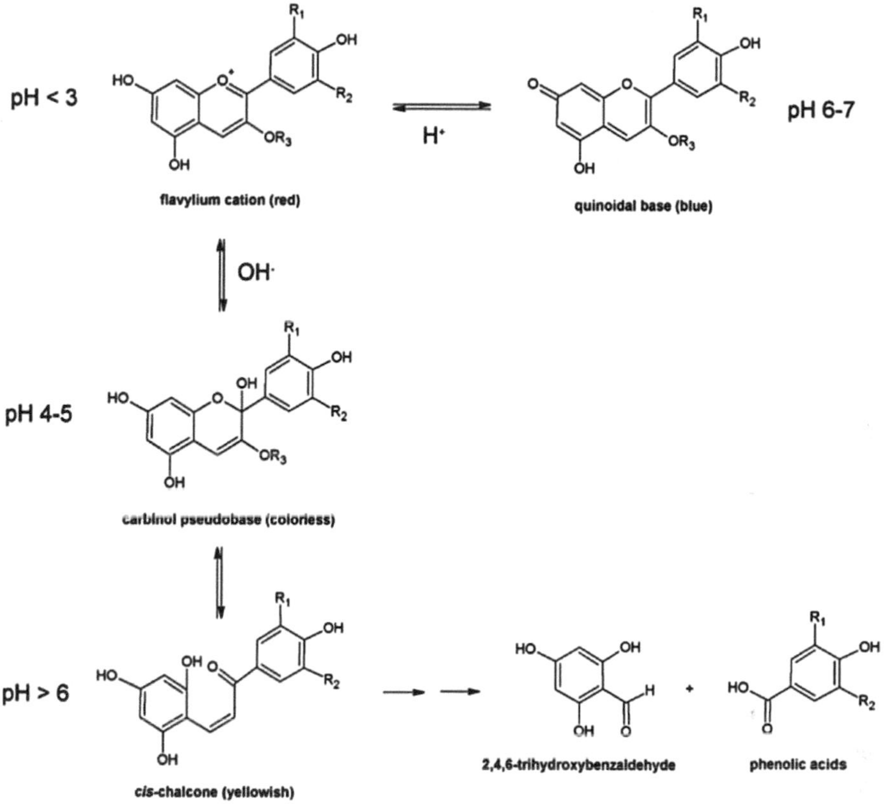

FIGURE 5.6 Effects of pH on various chemical forms and degradation reactions of anthocyanin.

Note: At different pH values, more than one chemical form can exist, and therefore, it decides the overall color of a solution.

5.6 SUMMARY

Hyperglycemia, a predisposing factor for type-2 diabetes, due to improper β-cell function, improved the insulin resistance. Controlling diabetes through the inhibition of starch hydrolyzing enzymes and increasing insulin sensitivity is an apt strategy to repress T2D. *In-vitro* and *in-vivo* studies have confirmed the bariatric therapeutic potential of anthocyanins against diabetes. The use of anthocyanin as an antidiabetic agent provides a safer and natural alternative to otherwise deleterious synthetic antidiabetic drugs. Anthocyanin shows a pleiotropic effect against diabetes through inhibition

TABLE 5.2 Summary of Techniques to Improve Process Stability and Bioavailability of Anthocyanins

Type of Anthocyanin	Source	Technique Employed	Results	References
Anthocyanins	Purple carrot	Anthocyanin-amino acid-peptide (L-phenylalanine; L-tyrosine; L-tryptophan; and e-poly-L-lysine) system.	Amino acid or peptide prolongs the stability of anthocyanin in beverages.	[19]
	Blueberry	Freeze-drying followed by pulverization. Acetone, ethanol, and methanol extracts of pulverized powder was prepared.	Inhibitory action against α-amylase and α-glucosidase was retained to a maximum by acetone extract of rehydrated powder.	[41]
	Purple potato peel	Co-pigmentation with ascorbic acid, glucose, and citric-acid monohydrate in both liquid and solid states.	In the liquid state, co-pigmentation of ascorbic acid decreased anthocyanin stability, while glucose and citric-acid monohydrate increased the anthocyanin stability. In solid-state, co-pigmentation with ascorbic acid, glucose, and citric-acid monohydrate increased anthocyanin stability.	[114]
	Choke-berry juice	Addition of β-cyclodextrin and refrigeration.	Addition of β-cyclodextrin at concentration of 3% to the choke berry juice (pH 3.6) protected anthocyanins, with 81% and 95% retention after 8 months of storage at 25°C and 4°C, respectively.	[71]
	Hibiscus sabdariffa	Addition of β-cyclodextrin	Thermal stability of anthocyanins was increased in the presence of β-cyclodextrin	[86]

Source	Method	Effect	Reference
Skin of grapes	Co-pigmentation with isoquercitrin modified enzymatically.	Increase in enzymatically modified isoquercetin significantly extended the half-life of anthocyanins. Color stability was also increased.	[113]
Black bean coats	Co-pigmentation with β-cyclodextrin	Co-pigmented anthocyanin-β-cyclodextrin showed extended half-life up to 13 months.	[1]
— Chrysanthemum (Cyanidin-3-O-glucoside); Myrtillin (Delphynidin-3-glucoside); Callistephin (Pelargonidin-3-glucoside)	Addition of Fe^{3+} and polysaccharide alginate	The stability of anthocyanin with added Fe^{3+} and polysaccharide alginate improved on exposure to deleterious conditions (60°C, 80 minutes).	[100]
Purple carrot Cyanidin-3-O-glucoside	Co-pigmentation with pectins (citrus pectin, beet pectin) and whey (native or denatured)	Denatured whey gave the best results and improved the stability of Cyanidin-3-O-glucoside by reducing the perverse effect of vitamin C on relative absorbance by half.	[20]

of starch hydrolyzing enzymes, inhibition of glucose transporters, and improvement in the function of β-cells. However, its therapeutic effects are limited by its instability and low bioavailability, which can be improved by microencapsulation. Although anthocyanins have shown promising results against diabetes, yet clinical evidence for the use of anthocyanins or anthocyanin-rich extracts in diabetes is not convincing.

KEYWORDS

- anthocyanidin
- anthocyanin
- diabetes
- glucose transporters
- α-amylase
- α-glucosidase

REFERENCES

1.   Aguilera, Y., Mojica, L., & Rebollo-Hernanz, M., (2016). Black bean coats: New source of anthocyanins stabilized by β-cyclodextrin co-pigmentation in a sport beverage. *Food Chemistry, 212*, 561–570.
2.   Akkarachiyasit, S., Yibchok, S., Wacharasindhu, S., & Adisakwattana, S., (2011). *In vitro* inhibitory effects of cyandin-3-rutinoside on pancreatic α-amylase and its combined effect with acarbose. *Molecules, 16*, 2075–2083.
3.   Alonso-Castro, A. J., & Miranda-Torres, A. C., (2008). *Cecropia obtusifolia* bertol and its active compound, chlorogenic acid stimulate 2-NBD glucose uptake in both insulin-sensitive and insulin-resistant 3T3 adipocytes. *Journal of Ethnopharmacology, 120*, 458–464.
4.   Anderson, A. R., Broadhurst, C. L., & Polansky, M. M., (2004). Isolation and characterization of polyphenol type-A polymers from cinnamon with insulin-like biological activity. *Journal of Agricultural and Food Chemistry, 52*, 65–70.
5.   Anderson, R., & Polansky, M., (2002). Tea enhances insulin activity. *Journal of Agricultural and Food Chemistry, 50*, 7182–7186.
6.   Beers, E. H. V., Buller, H. A., Grand, R. J., Einerhand, A. W., & Dekker, J., (1995). Intestinal brush-border glycohydrolases: Structure, function, and development. *Critical Reviews in Biochemistry and Molecular Biology, 30*, 197–262.
7.   Betz, M., Steiner, B., Schantz, M., Oidtmann, J., & Mader, K., (2012). Antioxidant capacity of bilberry extract microencapsulated in whey protein hydrogels. *Food Research International, 47*(1), 51–57.

8. Blakely, S., Herbert, A., Collins, M., & Jenkins, M., (2003). Lutein interacts with ascorbic acid more frequently than with alpha-tocopherol to alter biomarkers of oxidative stress in female Zucker obese rats. *Journal of Nutrition, 133*, 2838–2844.

9. Brouillard, R., & Dubois, J. E., (1977). Mechanism of structural transformations of anthocyanins in acidic media. *Journal of the American Chemical Society, 99*, 1359–1364.

10. Brouillard, R., & Delaporte, B., (1977). Chemistry of anthocyanin pigments, II: Kinetic and thermodynamic study of proton transfer, hydration, and tautomeric reactions of malvidin 3-glucoside. *Journal of the American Chemical Society, 99*, 8461–8468.

11. Brouillard, R., & Lang, J., (1990). The hemiacetal-cis-chalcone equilibrium of malvidin, a natural anthocyanin. *Canadian Journal of Chemistry-Revue Canadienne De Chimie, 68*, 755–761.

12. Buisson, G., Duée, E., Haser, R., & Payan, F., (1987). Three-dimensional structure of porcine pancreatic alpha-amylase at 2.9Å resolution: Role of calcium in structure and activity. *The EMBO Journal, 6*, 3909–3916.

13. Cabrita, L., Petrov, V., & Pina, F., (2014). The thermal degradation of anthocyanidins, cyanidin. *RSC Advances, 4*, 18939–18944.

14. Castaneda-Ovando, A., & De Lourdes, P. H. M., (2009). Chemical studies of anthocyanins: Review. *Food Chemistry, 113*, 859–871.

15. Castro-Acosta, M. L., Lenihan-Geels, G. N., Corpe, C. P., & Hall, W. L., (2016). Berries and anthocyanins, promising functional food ingredients with postprandial glycaemia-lowering effects. *Proceedings of the Nutrition Society, 75*, 342–343.

16. Cavalcanti, R. N., Santos, D. T., & Meireles, M. A., (2011). Non-thermal stabilization mechanisms of anthocyanins in model and food systems-an overview. *Food Research International, 44*(2), 499–509.

17. Chinese Nutrition Society (CNS), (2013). *Chinese DRIs (Dietary Requirement Intakes) Handbook* (p. 416). First edition, Beijing (China), Standards Press of China.

18. Chiste, R. C., Lopes, A. S., & Faria, L. J. D., (2010). Thermal and light degradation kinetics of anthocyanin extracts from mangosteen peel (*Garcinia mangostana* L.). *International Journal of Food Science and Technology, 45*(9), 1902–1908.

19. Chung, C., Rojanasasithara, T., Mutilangi, W., & McClements, D. J., (2017). Stability improvement of natural food colors, impact of amino acid and peptide addition on anthocyanin stability in model beverages. *Food Chemistry, 218*, 277–284.

20. Chung, C., Rojanasasithara, T., Mutilangi, W., & McClements, D. J., (2015). Enhanced stability of anthocyanin-based color in model beverage systems through whey protein isolate complexation. *Food Chemistry, 76*, 761–768.

21. Collins, Q. F., Liu, H. Y., Pi, J., Liu, Z., Quon, M. J., & Cao, W., (2007). Epigallocatechin-3-gallate (EGCG), a green tea polyphenol, suppresses hepatic gluconeogenesis through 59-AMP-activated protein kinase. *The Journal of Biological Chemistry, 282*, 30143–30149.

22. Cooper, A. J., Forouhi, N. G., Ye, Z., & Buijsse, B., (2012). Fruit and vegetable intake and incidence of type-2 diabetes, EPIC-InterAct prospective study, and meta-analysis. *European Journal of Clinical Nutrition, 66*, 1082–1092.

23. Copeland, R., (2013). *Evaluation of Enzyme Inhibitors in Drug Discovery: A Guide for Medicinal Chemists and Pharmacologists* (2nd edn., p. 572). New Jersey: John Wiley & Sons, Inc.

24. Coughlan, K. A., Valentine, R. J., Ruderman, N. B., & Saha, A. K., (2014). AMPK activation, a therapeutic target for type-2diabetes. *Diabetes, Metabolic Syndrome and Obesity, Targets and Therapy, 7*, 241–253.

25. Crozier, A., Jaganath, I. B., & Clifford, M. N., (2009). Dietary phenolics, chemistry, bioavailability and effects on health. *Natural Product Reports*, *26*(8), 1001–1043.

26. Daar, A. S., Singer, P. A., Persad, D. L., & Pramming, S. K., (2007). Grand challenges in chronic non-communicable diseases. *Nature*, *450*, 494–496.

27. Dahlqvist, A., (1962). Specificity of human intestinal disaccharides and implications for hereditary disaccharide intolerance. *Journal of Clinical Investigation*, *41*, 463–469.

28. Dahlqvist, A., Auricchio, S., Prader, A., & Semenza, G., (1963). Human intestinal disaccharides and hereditary disaccharide intolerance - hydrolysis of sucrose, isomaltose, palatinose (isomalto-malto-oligosaccharide) preparation. *Journal of Clinical Investigation*, *42*, 556–562.

29. Dahlqvist, A., & Telenius, U., (1969). Column chromatography of human small intestinal maltase isomaltase and invertase activities. *Biochemistry Journal*, *111*, 139–146.

30. DeFronzo, R. A., (1999). Pharmacologic therapy for type-2diabetes mellitus. *Annals of Internal Medicine*, *131*(4), 281–303.

31. Dhital, S., Lin, A. M., Hamaker, B. R., Gidley, M. J., & Muniandy, A., (2013). Mammalian Mucosal α-glucosidases coordinate with a-amylase in the initial starch hydrolysis stage to have a role in starch digestion beyond glucogenesis. *PLoS One*, *8*(4), e-article ID: 62546. doi: 10.1371/journal.pone.0062546.

32. Dubois, S., (2014). *The Ability of Berry Extracts to Inhibit Alpha-Glucosidase in Vitro* (p. 216). Maine, USA: University of Maine Press.

33. Edirisinghe, I., & Burton-Freeman, B., (2016). Anti-diabetic actions of berry polyphenols: Review on proposed mechanisms of action. *Journal of Berry Research*, *6*, 237–250.

34. Eid, H. M., Martineau, L. C., Saleem, A., & Muhammad, A., (2010). Stimulation of AMP-activated protein kinase and enhancement of basal glucose uptake in muscle cells by quercetin and quercetin glycosides, active principles of the antidiabetic medicinal plant *Vaccinium vitis-idaea*. *Molecular Nutrition and Food Research*, *54*, 991–1003.

35. Fang, J., (2014). Bioavailability of anthocyanins. *Drug Metabolism Reviews*, *46*(4), 508–520.

36. Fang, J., (2015). Classification of fruits based on types of anthocyanin and relevance to their health. *Nutrition*, *31*(11–12), 1301–1306.

37. Fang, J., (2014). Some anthocyanins could be efficiently absorbed across the gastrointestinal mucosa, extensive presystemic metabolism reduces apparent bioavailability. *Journal of Agricultural and Food Chemistry*, *62*, 3904–3911.

38. Fernandes, I., Faria, A., Calhau, C., De Freitas, V., & Mateus, N., (2014). Bioavailability of anthocyanins and derivatives. *Journal of Functional Foods*, *7*, 54–66.

39. Fischer, U. A., Carle, R., & Kammerer, D. R., (2013). Thermal stability of anthocyanins and colorless phenolics in pomegranate (*Punica granatum* L.) juices and model solutions. *Food Chemistry*, *138*(2–3), 1800–1809.

40. Fleschhut, J., Kratzer, F., Rechkemmer, G., & Kulling, S. E., (2006). Stability and biotransformation of various dietary anthocyanins *in vitro*. *European Journal of Nutrition*, *45*(1), 7–18.

41. Flores, F. P., Singh, R. K., Kerr, W. L., Pegg, R. B., & Kong, F., (2013). Antioxidant and enzyme inhibitory activities of blueberry anthocyanins prepared using different solvents. *Journal of Agricultural and Food Chemistry*, *61*, 4441–4447.

42. Foretz, M., Ancellin, N., Andreelli, F., & Saintillan, Y., (2005). Short-term over expression of a constitutively active form of AMP-activated protein kinase in the liver leads to mild hypoglycemia and fatty liver. *Diabetes, 54,* 1331–1339.

43. Gaede, P., Vedel, P., Larsen, N., & Jensen, G. V., (2003). Multi factorial intervention and cardiovascular disease in patients with type-2 diabetes. *The New England Journal of Medicine, 348,* 383–393.

44. Ghadyale, V., Takalikar, S., Haldavnekar, V., & Arvindekar, A., (2012). Effective control of postprandial glucose level through inhibition of intestinal α-glucosidase by *Cymbopogon martinii* (Roxb.). *Evidence-Based Complementary and Alternative Medicine,* p. 6. Open access e-article: 372909. http://dx.doi.org/10.1155/2012/372909 (accessed on 18 May 2020).

45. Ghosh, D., & Konishi, T., (2007). Anthocyanins and anthocyanin-rich extracts, role in diabetes and eye function. *Asia Pacific Journal of Clinical Nutrition, 16*(2), 200–208.

46. Gonzalez-Barrio, R., Borges, G., Mullen, W., & Crozier, A., (2010). Bioavailability of anthocyanins and ellagitannins following consumption of raspberries by healthy humans and subjects with an ileostomy. *Journal of Agricultural and Food Chemistry, 58,* 3933–3939.

47. Grace, M. H., Ribnicky, D. M., & Kuhn, P., (2009). Hypoglycemic activity of a novel anthocyanin-rich formulation from low-bush blueberry, *Vaccinium angustifolium Aiton. Phytomedicine, 16*(5), 406–415.

48. Gray, A., Clarke, P., Farmer, A., & Holman, R., (2002). Implementing intensive control of blood glucose concentration and blood pressure in type-2 diabetes in England, cost analysis: United Kingdom Prospective Diabetes Study Group (UKPDS63). *British Medical Journal, 325,* 860–866.

49. Gray, A., Raikou, M., McGuire, A., & Fenn, P., (2000). Cost effectiveness of an intensive blood glucose control policy in patients with type-2 diabetes, economic analysis alongside randomized trial: United Kingdom prospective diabetes study (UKPDS 41) Group. *British Medical Journal, 320,* 1373–1378.

50. Guo, X., Yang, B., Tan, J., Jiang, J., & Li, D., (2016). Associations of dietary intakes of anthocyanins and berry fruits with risk of type-2 diabetes mellitus: A systematic review and meta-analysis of prospective cohort studies. *European Journal of Clinical Nutrition,* 1–8.

51. Hardie, D., (2008). Role of AMP-activated protein kinase in the metabolic syndrome and in heart disease. *FEBS Letters, 582,* 81–89.

52. He, B., Ge, J., Yue, P., Yue, X., Fu, R., Liang, J., & Gao, X., (2017). Loading of anthocyanins onchitosan nanoparticles influences anthocyanin degradation in gastrointestinal fluids and stability in a beverage. *Food Chemistry, 221,* 1671–1677.

53. Hernandez-Herrero, J. A., & Frutos, M. J., (2015). Influence of rutin and ascorbic acid in color, plum anthocyanins, and antioxidant capacity stability in model juices. *Food Chemistry, 173,* 495–500.

54. Heymann, H., & Gunther, S., (1994). Calculation of subsite affinities of human small intestinal glucoamylase-maltase. *Biological Chemistry Hoppe-Seyler, 375,* 451–455.

55. Hizukuri, S., Abe, J., & Hanashiro, I., (1996). Starch and analytical aspects. In: *Carbohydrates in Food* (pp. 347–429). New York: Marcel Dekker.

56. Horii, S., Fukase, H., Matsuo, T., Kameda, Y., Asano, N., & Matsui, K., (1986). Synthesis and α-D-glucosidase inhibitory activity of N-substituted valiolamine derivatives as potential oral antidiabetic agents. *Journal of Medicinal Chemistry, 29*(6), 1038–1046.

57. Hossain, M. K., Dayem, A. A., & Han, J., (2016). Molecular mechanisms of the anti-obesity and anti-diabetic properties of flavonoids. *International Journal of Molecular Sciences, 17*(4), 569–577.

58. Howard, L. R., Brownmiller, C., Prior, R. L., & Mauromoustakos, A., (2013). Improved stability of chokeberry juice anthocyanins by β-cyclodextrin addition and refrigeration. *Journal of Agricultural and Food Chemistry, 61*, 693–699.

59. Hribar, U., & Poklar, U. N., (2014). The metabolism of anthocyanins. *Current Drug Metabolism, 15*, 3–13.

60. Hurtado, N. H., Morales, A. L., & Gonzalez-Miret, M. L., (2009). Color, pH stability, and antioxidant activity of anthocyanin rutinosides isolated from tamarillo fruit (*Solanum betaceum* Cav.). *Food Chemistry, 117*(1), 88–93.

61. Ide, T., Ashakumary, L., Takahashi, Y., & Kushiro, M., (2001). Sesamin (a sesame lignan) decreases fatty acid synthesis in rat liver accompanying the downregulation of sterol regulatory element binding protein-1. *Biochimica et Biophysica Acta, 1534*, 1–13.

62. Jayaprakasam, B., Vareed, S. K., Olson, L. K., & Nair, M. G., (2005). Insulin secretion by bioactive anthocyanins and anthocyanidins present in fruits. *Journal of Agricultural and Food Chemistry, 53*, 28–31.

63. Kamiloglu, S., Capanoglu, E., Bilen, F. D., & Gonzales, G. B., (2015). Bioacessibility of polyphenols from plant-processing by-products of black carrot (*Daucus carota* L.). *Journal of Agriculture and Food Chemistry, 64*, 2450–2458.

64. Kim, E. J., Jung, S. N., Son, K. H., & Kim, S. R., (2007). Antidiabetes and antiobesity effects of cryptotanshinone via activation of AMP-activated protein kinase. *Molecular Pharmacology, 72*, 62–72.

65. Kimura, K., Lee, J. H., Lee, I. S., & Lee, H. S., (2004). Two potent competitive inhibitors discriminating α-glucosidase family I from family II. *Carbohydrate Research, 339*, 1035–1040.

66. Klonoff, D. C., & Schwartz, D. M., (2002). An economic analysis of interventions for diabetes. *Diabetes Care, 23*, 390–404.

67. Kreft, S., Knapp, M., & Kreft, I., (1999). Extraction of rutin from buckwheat (*Fagopyrum esculentum moench*) seeds and determination by capillary electrophoresis. *Journal of Agricultural and Food Chemistry, 47*, 4649–4652.

68. Kumar, S., Narwal, S., Kumar, V., & Prakash, O., (2011). α-glucosidase inhibitors from plants: A natural approach to treat diabetes. *Pharmacognosy Research, 5*(9), 19–29.

69. Laar, F. A., (2008). α-glucosidase inhibitors in the early treatment of type-2 diabetes. *Vascular Health and Risk Management, 4*(6), 1189–1195.

70. Laar, F. A., Lucassen, P. L., & Akkermansetal, R. P., (2005). Alpha-glucosidase inhibitors for type-2 diabetes mellitus. *The Cochrane Database of Systematic Reviews, 2*(2), 8. e-article ID: 003639.

71. Landrault, N., Poucheret, P., & Azay, J., (2003). Effect of a polyphenols enriched chardonnay white wine in diabetic rats. *Journal of Agricultural and Food Chemistry, 51*, 311–318.

72. Laokuldilok, T., Shoemaker, C. F., Jongkaewwattana, S., & Tulyathan, V., (2011). Antioxidants and antioxidant activity of several pigmented rice brans. *Journal of Agricultural and Food Chemistry, 59*, 193–199.

73. Laoufi, H., Benariba, N., Adjdir, S., & Djaziri, R., (2017). *In vitro* α-amylase and α-glucosidase inhibitory activity of *Ononis angustissima* extracts. *Journal of Applied Pharmaceutical Science, 7*(02), 191–198.

74. Larson, S. B., Day, J. S., & McPherson, A., (2010). X-ray crystallographic analyses of pig pancreatic α-amylase with limit dextrin, oligosaccharide, and α-cyclodextrin. *Biochemistry, 49*, 3101–3115.
75. Lebovitz, H., (1997). α-Glucosidase inhibitors. *Endocrinology Metabolism Clinics of North America, 26*, 539–551.
76. Ley, S. H., Hamdy, O., Mohan, V., & Hu, F. B., (2014). Prevention and management of type-2 diabetes, dietary components, and nutritional strategy. *Lancet, 383*, 1999–2007.
77. Li, D., Zhang, Y., Liu, Y., Sun, R., & Xia, M., (2015). Purified anthocyanin supplementation reduces dyslipidemia, enhances antioxidant capacity, and prevents insulin resistance in diabetic patients. *The Journal of Nutrition*, pp. 742–748.
78. Liu, Y., Zhang, D., Wu, Y., Wang, D., Wei, Y., Wu, J., & Ji, B., (2014). Stability and absorption of anthocyanins from blueberries subjected to a simulated digestion process. *International Journal of Food Sciences and Nutrition, 65*(4), 440–448.
79. Lohachoompol, V., Mulholland, M., Srzednicki, G., & Craske, J., (2008). Determination of anthocyanins in various cultivars of high-bush blueberries. *Food Chemistry, 111*, 249–254.
80. Matsui, T., Ueda, T., Oki, T., Sugita, K., & Terahara, N., (2001). α glucosidase inhibitory action of natural acylated anthocyanins, II: α-glucosidase inhibition by isolated acylated anthocyanins. *Journal of Agricultural and Food Chemistry, 49*, 1952–1956.
81. McDougall, G. J., & Stewart, D., (2005). The inhibitory effects of berry polyphenols on digestive enzymes. *Biofactors, 23*, 189–195.
82. McGhie, T. K., & Walton, M. C., (2007). The bioavailability and absorption of anthocyanins, towards a better understanding. *Journal of Agricultural and Food Chemistry, 51*, 702–713.
83. Min, S. W., Ryu, S. N., & Kim, D. H., (2010). Anti-inflammatory effects of black rice: Cyanidin-3-O-β-D-glycoside, glycoside, and its metabolites, cyanidin and protocatechuic acid. *International Immunopharmacology, 10*, 959–966.
84. Mojica, L., Berhow, M., & Mejia, E. G., (2017). Black bean anthocyanin-rich extracts as food colorant, physicochemical stability, and anti-diabetes potential. *Food Chemistry, 229*, 628–639.
85. Monami, M., Adalsteinsson, J. E., & Desideri, C. M., (2013). Fasting and postprandial glucose and diabetic complication: A meta-analysis. *Nutrition, Metabolism and Cardiovascular Diseases, 23*, 591–598.
86. Mourtzinos, I., Makris, D. P., & Yannakopoulou, K., (2008). Thermal stability of anthocyanin extract of *Hibiscus sabdariffa* L. in the presence of β-cyclodextrin. *Journal of Agricultural and Food Chemistry, 56*, 10303–10310.
87. Murase, T., Mizuno, T., Omachi, T., & Onikawa, K., (2001). Dietary diacylglycerol suppresses high fat and high sucrose diet-induced body fat accumulation in C57BL/6J mice. *Journal of Lipid Research, 42*, 372–378.
88. Ng, K., Gu, C., Zhang, H., & Patri, C., (2015). Evaluation of alpha-amylase and alpha-glucosidase inhibitory activity of flavonoids. *International Journal of Nutritional Sciences, 2*(6), 1–6.
89. Nickavar, B., & Amin, G., (2010). Bioassay-guided separation of an α-amylase inhibitor anthocyanin from *Vaccinium arctostaphylos*berries. *Verlag der Zeitschriftfür Naturforschung, Tübingen (Publisher of the Journal of Natural Science, Tübingen), 65*, 567–570.

90. Pfeiffer, A., (2003). Oral hypoglycemic agents, sulfonylureas, and meglitinides. In: Goldstein, B., & Muller-Wieland, D., (eds.), *Text Book of Type-2 Diabetes* (pp. 77–85). London: Martin Dunitz Ltd.

91. Prestesa, M., Gayarreb, M. A., & Elgarta, J. F., (2017). Multistrategic approach to improve quality of care of people with diabetes at the primary care level, study design and baseline data. *Primary Care Diabetes, 11*(2), 193–200.

92. Robert, P., Gorena, T., & Romero, N., (2010). Encapsulation of polyphenols and anthocyanins from pomegranate (*Punica granatum*) by spray drying. *International Journal of Food Science and Technology, 45*(7), 1386–1394.

93. Rojo, L. E., Ribnicky, D., & Logendra, S., (2012). *In vitro* and *in vivo* anti-diabetic effects of anthocyanins from Maqui Berry (*Aristotelia chilensis*). *Food Chemistry, 131*, 387–396.

94. Sandborn, W. J., & Faubion, W. A., (2000). Clinical pharmacology of inflammatory bowel disease therapies. *Current Gastroenterology Reports, 2*, 444–445.

95. Semaana, D. G., Igoli, J. O., Young, L., Marrero, E., Gray, A. I., & Rowan, E. G., (2017). *In vitro* anti-diabetic activity of flavonoids and pheophytins from *Allophylus cominia* Sw. on PTP1B, DPPIV, α-glucosidase and α-amylase enzymes. *Journal of Ethnopharmacology, 203*, 39–46.

96. Shen, Q. W., Zhu, M. J., Tong, J., Ren, J., & Du, M., (2007). Ca^{2+}/calmodulin-dependent protein kinase is involved in AMP-activated protein kinase activation by α-lipoic acid in C2C12 myotubes. *American Journal of Physiology-Cell Physiology, 293*, 1395–1403.

97. Sriwijitkamol, A., Coletta, D. K., Wajcberg, E., & Balbontin, G. B., (2007). Effect of acute exercise on AMPK signaling in skeletal muscle of subjects with type-2diabetes, a time-course and dose-response study. *Diabetes, 56*, 836–848.

98. Sui, X., Zhang, Y., & Zhou, W., (2016). *In vitro* and *in silico* studies of the inhibition activity of anthocyanins against porcine pancreatic α-amylase. *Journal of Functional Foods, 21*, 50–57.

99. Tabopda, T. K., Ngoupayo, J., & Awoussong, P. K., (2008). Triprenylated flavonoids from *Dorsteniap silurus* and their α-glucosidase inhibition properties. *Journal of Natural Products Research, 71*, 2068–2072.

100. Tachibana, N., Kimura, Y., & Ohno, T., (2014). Examination of molecular mechanism for the enhanced thermal stability of anthocyanins by metal-cations and polysaccharides. *Food Chemistry, 143*, 452–458.

101. Tadera, K., Minami, Y., Takanatsu, K., & Matsuoka, T., (2006). Inhibition of α-glucosidase and α-amylase by flavonoids. *Journal of Nutritional Science and Vitamology, 52*, 99–103.

102. Takikawa, M., Inoue, S., Horio, F., & Tsuda, T., (2010). Dietary anthocyanin-rich bilberry extract ameliorates hyperglycemia and insulin sensitivity via activation of AMP-activated protein kinase in diabetic mice. *The Journal of Nutrition, 140*, 527–533.

103. Tundis, R., Loizzo, M. R., & Menichini, F., (2010). Natural products as α-amylase and α-glucosidase inhibitors and their hypoglycemic potential in the treatment of diabetes: An update. *Mini-Reviews in Medicinal Chemistry, 10*(4), 315–331.

104. UK Prospective Diabetes Study (UKPDS) Group, (1998). Tight blood pressurecontrol and risk of macrovascular and microvascular complications in type-2 diabetes. *British Medical Journal, 317*, 703–713.

105. Viollet, B., Foretz, M., Guigas, B., & Horman, S., (2006). Activation of AMP-activated protein kinase in the liver, a new strategy for the management of metabolic hepatic disorders. *The Journal of Physiology, 574*, 41–53.

106. Wallace, T. C., & Giusti, M. M., (2015). Anthocyanins. *Advances in Nutrition*, *6*(5), 620–622.
107. Wang, H., Liu, T., & Huang, D., (2013). Starch hydrolase inhibitors from edible plants. *Advances in Food and Nutritional Research*, *70*, 103–137.
108. Wang, L., & Stoner, G. D., (2008). Anthocyanins and their role in cancer prevention. *Cancer Letters*, *269*, 281–290.
109. Watanabe, J., Kawabata, J., Kurihara, H., & Niki, R., (1997). Isolation and identification of alpha-glucosidase inhibitors from tochu-cha (*Eucommia ulmoides*). *Bioscience Biotechnology Biochemistry*, *61*(1), 177–178.
110. Wenzel, U., (2013). Flavonoids as drugs at the small intestinal level. *Current Opinion in Pharmacology*, *13*(6), 864–868.
111. Wilson, T., Meyers, S. L., Singh, A. P., Limburg, P. J., & Vorsa, N., (2008). Favorable glycemic response of type-2 diabetics to lowcalorie cranberry juice. *Journal of Food Science*, *73*(9), 241–245.
112. Woodward, G., Kroon, P., Cassidyetal, A., & Kay, C., (2009). Anthocyanin stability and recovery, implications for the analysis of clinical and experimental samples. *Journal of Agricultural and Food Chemistry*, *57*, 5271–5278.
113. Yan, Q., Zhang, L., Zhang, X., Liu, X., Yuan, F., Hou, Z., & Gao, Y., (2013). Stabilization of grape skin anthocyanins by co-pigmentation with enzymatically modified isoquercitrin (EMIQ) as a co-pigment. *Food Research International*, *50*, 603–609.
114. Zhang, C., Ma, Y., Zhao, X., & Mu, J., (2009). Influence of co-pigmentation on stability of anthocyanins from purple potato peel in both liquid state and solid state. *Journal of Agricultural and Food Chemistry*, *57*, 9503–9508.
115. Zhang, H., Wang, G., Beta, T., & Dong, J., (2015). Inhibitory properties of aqueous ethanol extracts of propolis on α-glucosidase. *Evidence-Based Complementary and Alternative Medicine* (p. 7). Open access, Article ID 587383: http://dx.doi.org/10.1155/2015/587383 (accessed on 18 May 2020).
116. Zhu, W., Jia, Q., Wang, Y., Zhang, Y., & Xia, M., (2012). The anthocyanin cyanidin-3-O increases hepatic glutathione synthesis and protects hepatocytes against reactive oxygen species during hyperglycemia, involvement of a camp-pka-dependent signaling pathway. *Free Radical Biology and Medicine*, *52*, 314–327.
117. Zimmet, P. Z., (2011). The growing pandemic of type-2 diabetes a crucial need for prevention and improved detection. *Medicographia*, *33*, 15–21.

Part II:
Secondary Phytocompounds From Marine Sources

CHAPTER 6

THERAPEUTIC POTENTIAL OF MARINE FOODS: A REVIEW

ZOHAIB HASSAN, MUHAMMAD KAMRAN KHAN, FARHAN SAEED, SADIA HASSAN, and HAFIZ ANSAR RASUL SULERIA

ABSTRACT

Marine species particularly seaweeds and fish have been used to cure several diseases and have shown numerous health benefits. Marine-based remedies are safe, potent, and cheaper contributing to consumer's health longevity and beauty. This chapter emphasizes on the consumption of seaweeds (mainly marine algae and fish) as a rich source of bioactive compounds with antioxidant, antitumor, antiviral, anti-obesity, cardio-protective, antiosteoporosis, and skin whitening properties.

6.1 INTRODUCTION

Natural food stimulates optimum strength and diminishes the uncertainty of disease in which minerals, pigments, vitamins trace elements contribute to a beneficial role [53]. More than 70% of the earth's surface is occupied by oceans while different marine species cover almost one half of the total biodiversity in the world [155]. Marine organisms live in a very insistent, fierce, and offensive environment, which is totally different from the terrestrial ecosystem. Additionally from marine organisms, different kind of phytochemicals has been derived [161].

Marine foods are considered as a powerhouse for production of bioactive, healthy food substances, such as bioactive peptides, enzymes, natural pigments, essential minerals (e.g., iron, zinc, selenium, and calcium), vitamins (A, B3, B6, B12, E, and D), polysaccharides, and polyunsaturated fatty acids (FAs). Due to the presence of these substances, fish consumption results in different nutritional benefits [168]. Crustaceans, mollusks, marine algae, and marine fish among marine organisms though have long been used in different food diets, yet are still known as under-exploited food sources [75].

Marine species show numerous health benefits based on different epidemiological studies (Table 6.1). The modern age is full of challenges in the field of medicine; therefore, there is an urgent need to find solutions, which are safe, potent, and cheaper against the chronic diseases. Therefore, scientists emphasize on the consumption of seaweeds, mainly marine algae that are rich sources of bioactive components and act as an antioxidant, antitumor, anticoagulant, antiviral, antimicrobial, hypo-cholesterolemic, and anti-inflammatory agents [41, 86]. In recent years, scientists have focused on marine algae that are being studied for their health benefits, nutritional, and pharmaceutical profiles.

Consumption of marine fish is for the world economy. Common species include anchovy, cod, flounder, herring, mullet, salmon, and tuna; thus, fulfilling almost 16% protein requirements in the world [72]. Various research studies have demonstrated that peptides produced from fermented fish through enzymatic applications have therapeutic potential against different common acute and chronic diseases including Alzheimer's disease, cancer, hypertension, and viral infections, etc. [31, 174].

As a substitute for the collagen of mammals, fish collagen can be used immunogenic agent [57]. Fish and fish products are major sources of different essential nutrients in our diet due to having good quality protein, vitamin D, and minerals like iodine and selenium, and a rich source of omega-3 FAs. On the other side, seafood also comprises of unwanted compounds, such as heavy metals and organic pollutants. The main unacceptable components are dioxins, dioxin-like polychlorinated biphenyls, methyl mercury, and brominated flame retardants [167].

The positive influence of fish consumption on cardiovascular health is due to the presence of long-chain omega-3 polyunsaturated FAs and fish protein [112]. Within a healthy eating pattern, an association of fish consumption with decreased body weight has been approved through different anthropometric studies [143]. However, intervention dietary research studies are limited to the addition of fish in a weight-loss diet [161]. Other fish nutritional ingredients like vitamin B and iron, fluorine, and iodine are only concisely debated; however, in portions of the Western diet, these substances have suboptimal consumption levels.

TABLE 6.1   Biocomponents with Health-Care Benefits from Fish

Compounds	Components	Level in 100 g of Fish	Daily Requirement		References
			For Beneficial Effect	For Cure Deficiency	
		A	**B**	**C**	
Nitrogenous compounds	Taurine	6–176 mg	NA	NA	[A: 48]
	Proteins	15–20 g	NA	0.66–0.87g/ kg body weight	[A: 179] [C: 102]
Fats/Lipids	Choline (from lecithin)	74–77 mg	NA	125–550 mg	[A: 36] [C: 40]
	Omega-3 FAs	118–2000 mg	650–800 mg	200–400 mg	[A: 49] [B & C: 54]
Vitamins	Vitamin A	2–90 µg RE	NA	210–625 µg RE	[A: 62] [C: 102]
	Vitamin B$_2$	0.01–0.14 mg	NA	0.3–1.6 mg	[A: 179] [C: 173]
	Vitamin B$_6$	0.06–0.83 Mg	NA	0.1–2 mg	[A: 179] [C: 173]
	Vitamin D	50–1000 IU	1000 IU	200–800 IU	[A: 85] [B: 10] [C: 173]
Dietary minerals	Iron	0.36–3.3 mg	NA	3–8.1 mg	[A: 75, 158] [C: 102]
	Fluorine	0.004–0.4 mg [a])	NA	1.5–4.0 mg	[A: 62, 99] [C: 172]
	Calcium	250–380 mg in canned bony fish	1000 mg	400–1000 mg	[A: 161] [B: 2] [C: 173]
	Phosphorus	200–400 mg	NA	100–1250 mg	[A: 62] [C: 40]
	Selenium	13.6–90 µg	200–300 µg	55–90 µg	[A: 39, 70] [B: 85] [C: 161]
	Iodine	30–1270 µg/ 100 g fillet	NA	120–150 µg	[A: 99] [C: 173]

This chapter focuses on health benefits from marine foods due to the presence of secondary metabolites.

6.2 POTENTIAL HEALTH BENEFITS FROM ALGAE: WOMEN HEALTH, LONGEVITY AND BEAUTY

6.2.1 ANTIOXIDANT ACTIVITY

Antioxidants produce positive effects on health care by acting as a shield to our bodies. They stabilize free radicals (reactive oxygen species), which react with large molecules such as lipids, proteins, DNA that cause diseases, e.g., mellitus, diabetes, aging, cancer, and neurodegenerative disorders [105]. Due to the number of health benefits produced by the antioxidants, they are commercially synthesized and deliberately added to the food and pharmaceutical items. TBHQ (tert-butylhydroquinone), BHA (butylated hydroxyl anisole), BHT (butylated hydroxyl toluene), and PG (propylene glycol) are a few examples of commercially available synthetic antioxidants. These synthetic antioxidants are also used to overcome the oxidation process and different peroxidation reactions.

They have the potential to cause different health risks, therefore, proper knowledge and strict regulatory limits regarding their best usage should be followed [117]. However, natural antioxidants are safe, cheaper, and fittest alternatives in food production and processing [120]. Currently, pharmaceutical industries are conducting extensive research to develop natural antioxidants based on plants and marine sources. Among marine flora, marine algae are abundant sources having safe and natural antioxidants [91, 135].

Sulfated polysaccharides (SPs) produced by marine algae are considered a good source of dietary fiber (Table 6.3), at the same time different research works also explored the great antioxidant potential of SPs. Fucoidan, laminaran, and alginic acid are a few main classes of SPs which showed antioxidant activity [132, 170]. Antioxidant potential of SPs can be analyzed through different analyses like DPPH (2-diphenyl-1-picrylhydrazyl), FRAP (ferric reducing antioxidant power), NO (nitric oxide) scavenging, ABTS assays [2,2'-azino-bis (3-ethylbenzothiazoline-6-sulphonic acid)], hydroxyl radical scavenging, and lipid peroxide inhibition methods. Kim et al. [70] demonstrated as compared to commercially available antioxidants (BHA, α-tocophorol), *Sargassum fulvellum* having SPs, is a comparatively more effective scavenger. Different structural properties of SPs are directly related to their antioxidant activities like sulfating degree, molecular mass, categories of the main sugar and glycosidic bonding [123, 188].

SPs having lower molecular mass have revealed as more effective antioxidants as compared to SPs of higher molecular mass [152]. Various types of ulvans (*Ulva pertusua*) having different molecular mass, varied in their ability of H_2O_2 degradation and antioxidant activities [123]. Qi et al. [123] also stated high antioxidant potential of lower molecular mass ulvans. It was explained that lower molecular weight ulvans may easily and efficiently integrate into the cells; therefore, they are more effective in proton donation. Among fucoidan, alginate, and laminaran from *Turbinaria conoides*, fucoidan revealed maximum antioxidant activity [20].

TABLE 6.2   Antioxidant Compounds Derived from Marine Algae

Source	Biocompound	Health Benefits	References
Actinobacteria	Glycosides	Antioxidant, Antimicrobial	[141]
Actinomycetes	Carotenoids	Antioxidant	[30]
Eisenia bicyclis	Pyropheophytin A	Antioxidant	[15]
Enteromorpha prolifera	Pheophorbide A	Antioxidant	[22]
Enteromorpha prolifera; *Fucus*	Chlorophyll A	Antioxidant	[79]
Padina tetrastromatic; *Hijikia fusiformis*	Fucoxanthin	Antioxidant	[107, 139, 182]
Porphyra sp.	Phycoery-Throbilin	Antioxidant	[180]

Marine algae-based natural pigments (fucoxanthin, phycocrythrobilin, chlorophyll) can also contribute to antioxidative functions (Tables 6.2–6.4). Yan et al. [182] investigated that main antioxidant compound (fucoxanthin) from *Hizikia fusiformis* showed maximum radical scavenging activity due to double allenic bonds at C-70 position. Sachindra et al. [136] isolated fucoxanthin from *Undaria pinnatifida* and further synthesized two metabolites known as fucoxanthinol and halocynthiaxanthin. Antioxidant potential of these formulated metabolites was assessed through DPPH assay, scavenging of hydroxyl radicals, and quenching of singlet oxygen. Fucoxanthin showed the highest scavenging activity followed by fucoxanthinol and halocynthiaxanthin [136].

Phycobiliproteins in marine algae are extensively applied as natural dye in foods and cosmetic industry. Extensively used compounds as dyes in food and cosmetic industry are C-phycocyanin, C-phycocyanin, and

R-phycoerythrin [144]. Compared to ardenia and indigo to give bright blue color in food items, phycobili proteins (despite their decreased stability under heat and light) are approved for multipurpose use in foods. Antioxidant functions of phycoerythrobilin isolated from *Porphyra* sp. were also explored [180]. Based on research studies on biocompounds from marine organisms, natural pigments are more effective and non-poisonous having the highest antioxidant activity.

TABLE 6.3 Marine Algae-Based Soluble, Insoluble, and Total Dietary Fiber

Type of Algae	Species	Soluble Fiber	Insoluble Fiber	Total Fiber	References
		% of Dry Weight			
Rhodophyceae:	*Chondrus crispus*	22.25	12.04	34.29	[135]
	Porphyra tenera	14.56	19.22	33.78	[135]
Phaeophycea:	*Fucus vesiculosus*	9.80	40.29	50.09	[135]
	Laminaria digitate	9.15	26.98	36.12	[135]
	Durvillaea antarctica	27.7	43.7	71.4	[113]
	Eisenia bicyclis	59.7	14.9	74.6	[78]
	Hijiki fusiformis	16.3	32.9	49.2	[78]
Chlorophyceae:	*Enteromorpha sp.*	17.2	16.2	33.4	[14]
	Ulva lactuca	21.3	16.8	38.1	[14]

6.2.2 ANTICANCER ACTIVITY

External factors that are responsible for cancer are tobacco, contagious organisms, malnutrition, chemical substances, and radiation along with inherited mutations, which may result from any reason. Globally, breast cancer has been recognized as a major type of cancer that causes the death of females [45]. Each year, >1.1 million female victims of breast cancer are identified, representing approximately 10% of all cancer cases [5, 32]. This figure is projected to increase with an increasing population [64]. It is assessed that 13.1 million deaths in 2030 could occur due to breast, colon, liver, lungs, and stomach cancer [177].

In the etiology of breast cancer, genetics has insignificant part in comparison to environment-based aspects, e.g., food items and diets. Among leading causes of deaths in females, estrogen-based cancerous infections are the major reason (causing breast, endometrial, and ovarian cancer) [126]. Various epidemiological studies revealed that number of cases of estrogen-based cancerous infections is higher in Western countries compared to Asian countries [118]. East Asian population consumed more fish and marine algae compared to Western inhabitants. Large number of health benefits is associated with the consumption of marine algae, fish, and their products.

In a research-based study, lower rate of estrogen-dependent cancer in Asian countries was associated with consumption of diets rich in soy-protein and marine algae. In healthy postmenopausal females, metabolism of estrogen and phytoestrogen were modified with the diet rich with *Alaria esculenta* and soy protein [157]. The second most common type of cancer is cervical cancer [35, 67].

In developing countries, 80% of cases of cervical cancer were identified [186]. In- *in vitro* studies, extracts from different species of marine algae (like *Udotea flabellum, Nereocystis leutkeana, Laminaria setchellii, Udotea conglutinate, Palmaria palmate* and *Macrocystis integrifolia*) showed their capacity to control the proliferation of cervical cancers in females [97, 186]. In recent cancer therapies, basic goal is to reduce the number of tumors. Therefore, the anti-proliferation ability of marine algae has proved beneficial in the case of cervical cancer.

Free radicals can induce the formation of cancerous cells in the human body. Therefore, to control their activity, anti-cancer drugs are used in the form of chemo-preventive; these drugs scavenge the free radicals rendering them ineffective to cause cancer. Marine algae are rich sources of phloro-tannins, sulfated polysaccharides, carotenoids, carbamol derivatives; and these compounds can be effective to stabilize the carcinogenic agents in females. Marine algae with its secondary metabolites showed remarkable anti-cancerous activities and potential of improving the health status of females by reducing cancer risks and chances. They can be used in the form of nutraceuticals. However, extensive research is needed for their synergistic effects, defining algae species, dose, time of intake, and most importantly preparation methods.

Marine algae rich in phytochemical compounds have long been used as medicine since the 17th century. The anti-cancerous components of the marine algae kill cancers cells through apoptosis or by affecting mechanisms of cell signaling by activating the components of protein kinase-c family (signaling enzymes) [148]. Number of anti-cancerous compounds is produced by the marine algae, and few are listed in Table 6.4.

TABLE 6.4   Anti-Carcinogenic Compounds Isolated from Marine Algae

Type of Algae	Species	Health Benefits	Compounds	Type of Studies	References
Phaeophyceae	*U. pinnatifida*; *Saccharina japonica*	Prevent proliferation and colony formation in breast cancer.	Fucoidans	*In vitro*	[110]
	Laminaria japonica; *Euchema cottonii*; *Gelidiumb amansii.*	colon cancer cells and mammary tumors.	Crude extracts	*In vivo* and *in vitro*	[43, 104]
	Undaria pinnatifida; *Saccharina japonica.*	Antitumor activity against human breast cancer and melanoma cell lines.	Sulfated polysaccharides	*In vitro*	[109]
Chlorophyceae	*Bryopsis sp.*	Actions against breast and prostate tumors.	Kahalaides	*In vitro*	[74]
	Avrainvillea nigricans	Antimitotic activity against human breast cancer	Nigriccanoides A and B	*In vitro*	[175]
Cyanobacteria	*Aphanizomenon flosaquae*	Anticancer	Aphanorphne, Siatoxin	*In vitro*	[97]
Dinophyceae	*Amphidinium sp.*	Cytotoxic effect against lung cancer cell lines	Lingshioils A	*In vitro*	[171]

6.2.3 ANTIVIRAL ACTIVITY

Cervical cancer also arises due to the infection caused by the HPV virus (human papillomavirus) that is present in the female genital. Therefore, the prevention and control of this virus has attained major consideration through different scientific studies [83]. Vaccine for the first generation of HPV is now commercially available to control its infections [115]. Though, in developing countries, people cannot afford its cost as it is too expensive to purchase. Therefore, scientists are in a quest to search the anti-HPV medicine, which would be less expensive and easily available. Various studies had proved that natural bioactive compounds and their derivatives have high potential in this regard. These compounds can be used in the form of functional food as a new generation anti-HPV. These products are relatively less expensive, more effective with fewer side effects.

Marine algae are excellent sources of sulfated polysaccharides (SPs) (Table 6.5) that show effective results against HPV [17]. Gerber et al. [44] observed the antiviral role of polysaccharides from marine algae, such as *Gelidium cartilagenium*. Such highly sulfated polysaccharides showed protection of embryonic eggs against mump virus and influenza-B [58]. SPs from carrageenan are composed of D-galactose and 3,6-anhydro-D-galactose and are usually obtained from the *Rhodophyceae*; and these have been used extensively in various food items. *In vitro* study proved that carrageenan has activities against the HPV virus [17].

TABLE 6.5 Marine-Based SPs Against Viral (Herpes Simplex Virus, HSV) Activity

Major Sugar of SPs	Source	Virus Strain	References
Chlorophyceae:			
Galactose	*Caulerpa racemose*	HSV-2	[46]
Galactose	*Codium fragile*	HSV-2	[108]
Phaeophyceae:			
Fucose	*Undaria pinnatifida*	HSV-2	[160]
Fucose	*Sargassum horneri*	HSV-1	[55]
Fucose	*Sargassum patens*	HSV-1	[189]
Rhodophyceae:			
Mannose	*Nemalion helminthoides*	HSV-1	[127]
Galactose	*Cryptonemia crenulata*	HSV-1	[156]

Buck et al. [13] investigated that carrageenan can inhibit HPV activity three-times more affectively compared to heparin, which is an extremely efficient model to inhibit HPV. Carrageenan mainly ceases the attachment of the HPV virus to cells and stops HPV proliferation. This reaction mechanism is highly constant with the structure of the carrageenan resembling with heparin sulfate (HPV-cell attachment factor). Besides these, some milk products were also identified having inherent carrageenan, whose activity to block HPV was proved through *in vitro* studies [13]. In other research studies on female mouse model of cervicovaginal, carrageenan inhibited the genital transmission of HPV [131].

The SPs isolated from seaweeds showed great potential as a vaginal antiviral agent in different formulations, without interrupting important functions of the vaginal epithelial cells and normal bacterial flora [9]. Although a wide range of SPs compounds having anti-viral activity are available, yet it is a challenge to choose the most promising one. These compounds have number of advantages over the anti-viral drugs, e.g., reduced production charges, a wide range of antiviral functions, decreased toxicity, safe to use, extensive suitability, and novel action approach. Therefore, marine algal SPs will be most effective drug component.

Further research studies are needed along with clinical trials. These findings show that carrageenan is an efficient substitute as novel therapeutic ingredient against HPV and it can easily be added in food in the form of additives. Similarly, carrageenan also contains numerous advantages against various other types of antiviral compounds, such as reduced production cost, a wide range of antiviral activities, less cytotoxic, safe to use, widely acceptable, and novel mode of action. Yet extensive research work along with clinical trials is required for its validation.

6.2.4 ANTIOSTEOPOROSIS ACTIVITY

Osteoporosis (skeletal disorder) is mainly recognized by a reduction in bone mineral density (BMD) that causes higher risks of fractures in bones [8]. Cases of osteoporosis are more frequent in females compared to males [142]. For example, the male has higher (30–50%) bone mass compared to female at their skeletal maturity [106]; and with the passage of age, reduction in BMD is more prominent in female compared to male [130].

In the lab-scale experiment, extract (rich in minerals) from marine red algae (*Lithothamnion calcareum*) has successfully improved the bone strength and increased the mineral content in female mice on a Western-diet [6, 42]. However, extensive research is needed on marine algae to identify the specific mineral to improve bone structure and function. The extract (Aquamin (GRAS 000028)) obtained from marine algae is now available in the form of a food supplement. This is used in many products made for human use in developed countries.

It was reported that treatment of osteoclastogenesis with 2.5M fucoxanthin also produced apoptosis along with caspase-3 activation in osteoclast-like cells in aged females. *In vitro* studies showed that consumption of fucoxanthin reduces the chances of osteoclastogenesis through suppressing of differentiation process of osteoclast and its intake also induces the apoptosis process in osteoclasts [26]. Therefore, the ingestion of fucoxanthin in the diet may inhibit osteoclastogenesis. To prevent bone-related disorders, marine algae can be the efficient source for the development of pharmaceutical products and functional foods. Furthermore, it is extremely vital to assess other marine algae species for the control and inhibition of osteoporosis in females.

6.2.5 ANTI-OBESITY ACTIVITY

Excess of body weight in the form of fat is known as obesity [73, 128]. Numerous studies have been conducted to investigate the reasons of obesity [60]. It may arise due to *diabetes mellitus* (type-II), heart-related maladies, cancer, and breathing disorders [82]. In developed countries, obesity is continuously increasing especially in teens and senior-females [68]. Many studies revealed that obesity has a detrimental effect on the female reproductive system [121].

Social norms and mass media influence continuously pressurizing young females to be thin. This dilemma has resulted in disappointment, ingestion disturbances, downheartedness, and anxiety among females [151]. For a reduction in weight (desire to be thin), numerous groups of natural and artificial drugs as anti-obesity agents are frequently being used by females to decrease body fat. Numerous side effects have been identified associated with synthetic anti-obesity agents. They may cause unacceptable tachycardia, hypertension, may improve lipid blood levels/ glucose metabolism/ female reproductive system disturbances, etc. [7].

It is evident from research studies that oxidative stress is associated with obesity, which can reduce life expectancy and increase health issues [95,

176]. Various tools have been developed for the identification of symptoms of obesity, such as waist to hip ratio (WHR) and body mass index (BMI) (Table 6.6).

TABLE 6.6   International Classification of Adult Underweight, Overweight, and Obesity According to Body Mass Index

Categories	BMI, kg/m²	Categories	BMI, kg/m²
Underweight	< 18.5	Obesity type-I	30.0–34.9
Normal weight	18.5–24.9	Obesity type-II	35.0–39.9
Overweight	25.0–29.0	Obesity type-III	≥ 40
Obesity	≥ 30		

Many investigators found that obesity can be decreased with the consumption of soluble dietary fibers. Marine algae are rich in soluble and insoluble dietary fibers [78]. *Eisenia bicyclis* (Arame) is a rich source of dietary fiber, having almost 50% soluble dietary fiber compared to 40% of insoluble fiber in brown algae species (*F. vesiculosus*) [135]. These two types of dietary fibers have a different fate in the human digestive system thus behaving differently. Soluble dietary fiber (carrageenan agar) obtained from marine algae cause hypocholesterolemic and hypoglycemic effects in the human body [116]. In another study, it was shown that alginates dietary fiber improves appetite and energy-intake by using an acute feeding model. Alginate reacts with gastric acid (acid-soluble calcium source) after consumption, which resultantly forms ionic gelation and ultimately alginate help to sluggish stomach emptying process.

Stimulation of gastric stretch receptors, intestinal nutrient uptake reduction, and glycaemic response stimulation are associated with the intake of alginate [28]. In overweight females, the intake of calcium-gelled alginate-pectin two times daily caused reduction in spontaneous food intake [119]. On the other hand, insoluble dietary fiber (such as cellulose, xylans, mannans) are strongly associated with effects, such as increase in fecal bulk, excretion of bile acids, and reduced intestinal transit time [14].

White adipose tissues (WAT) can be reduced through the consumption of fucoxanthin. Uncoupling protein-1 (UCP1) in obese KKAy mice was also reported by Maeda et al. [87] by dietary intake of fucoxanthin. In fact, C-70 is the position of unusual double bonds in fucoxanthin that is the main reason for weight reduction from WAT [96]. Mammals contain a significant amount of WAT (called "fat" in mammals) [163]. Higher release of adipokines associated with pro-inflammatory features and chemokines has been associated with WAT [25]. These substances play a significant role in the onset of weight gain [163]. Therefore, weight reduction by reducing fat in WAT through consumption of fucoxanthin in female mice was a useful and efficient anti-obesity candidate.

1-(30,50-dihydroxyphenoxy)-7-(200,400,6-trihydroxyphenoxy)-2,4,9-trihydroxydibenzo-1,4-dioxin, and Dioxinodehydroeckol (two phloroglucinol derivatives isolated from *Ecklonia cava*) have remarkably attenuated adipocyte differentiation in 3T3-L1 cells, which suggest that these compounds have great potential as functional ingredient in obesity-related issues.

Moreover, clinical trials evidently revealed the anti-obesity effect of xanthigen, which is an anti-obesity supplement containing fucoxanthin in addition, pomegranate seed oil. In obese non-diabetic women, xanthigen endorsed weight loss, attenuated body, and liver fat content, and enhanced liver function tests [1].

Intake of fucoxanthin together with NPs (natural products) may reduce or prevent the obesity in females. They also stimulate lipid metabolism in adipose tissues. Number of advantages is associated with fucoxanthin derived from marine algae due to relatively low production cost, low cytotoxicity, safety, and wide acceptability. Moreover, NPs (natural products, e.g., fucoxanthin) from marine algae has more potential and it is a favorable food supplement, slimming agent, and supplement in the inhibition and managing the obesity in the female.

In view of these results, it is concluded that marine algae are efficient sources of functional foods with health benefits, particularly for obesity. Consequently, extracts obtained from marine algae would be natural sources and it can give rise to functional foods and medicinal products to reduce the one set of obesity.

6.2.6 ANTIDIABETIC ACTIVITY

The ratio of prevalence of *Diabetes mellitus* is collectively increased with an increase in obesity and age. There are >150 million persons in the world suffering with it and this will expand to 300 million in 2025 [69]. Worldwide, the number of patients with type-II diabetes (T2D) surpasses 90% of all patients with diabetic conditions [159]. Research for novel anti-diabetic drugs to complement those under clinical trials has intensified over the years [3].

The α-Glucosidase activity is prohibited through callyspongynic acid (also known as polyacetylenic acid) that is derived from sponge *Callyspongia truncate* [103]. To prevent the glycogen hydrolysis, α-Glucosidase is interfaced within the mechanism because it is responsible to keep the concentration of glucose in the blood at a desirable low level [181]. Another polybromodiphenyl ether compound isolated from Indonesian marine sponge (*Lamellodysidea herbacea*) inhibits protein tyrosine phosphatase 1B (an important target for diabetes treatment) [166]. Marine omega-3 FAs show beneficial effects on the prevention of T2D in the Asian population, which suggests an association between consumption of fish or fish oil and the development of Type-2 diabetes on geographical location [183].

6.2.7   SKIN WHITENING ACTIVITY

Skin whitening products mostly hold tyrosinase inhibition [169]; and this enzyme catalyzes the rate limiting step in skin pigmentation process, therefore it facilitates to achieve the skin hypo-pigmentation. Several tyrosinase inhibitors have been investigated via *in vitro* studies to reduce skin whitening but only few showed promising effects in clinical trials. The potential marine organisms with skin whitening effects are particularly based on tyrosinase inhibitor enzyme.

Tyrosinase inhibitors from natural and safe sources have drawn attention among the scientific community [150]. Cha et al. [19] studied forty-three native marine algae for their tyrosinase restricting effects. They showed that extracts of *Ecklonia cava* and *Sargassum silquastrum* show exceptional inhibitory effects on the pigmentation process of zebrafish. In a study on effects of Fucoxanthin (from *Laminaria japonica*) on melanogenesis in UVB-irradiated mice, the tyrosinase activity was reduced. Dietary intake of fucoxanthin predominantly decreased the skin mRNA expression, which is directly associated with melanogenesis. Therefore, it was suggested that the melanogenesis factor was negatively regulated by fucoxanthin at the transcriptional level [146].

Fucoxanthin and astaxanthin together have photoprotective characteristics in fibroblast cells of humans through inhibition of DNA damage along with the enhancement of antioxidant activity [52]. Additionally, diphlorethohydroxycarmalol isolated from *Ishigeokamurae* showed potential skin whitening effects [51, 52].

Commonly available secondary metabolites (Phloroglucinol derivatives), which can be extracted from brown algae have tyrosinase inhibitory potential. This secondary metabolite chelates the copper, which inactivates this enzyme [65]. Few phlorotannins (e.g., 7-phloroeckol and dioxinodehydroeckol) also showed the potential to inhibit tyrosinase activity. Their inhibition is stronger compared with inhibition of arbutin and kojic acids [185].

Also, biologically active metabolites (such as flavonoid glycoside obtained from *Hizikia fusiformis*) obtained from marine-based algae have the potential to inhibit tyrosinase enzyme [124]. They can efficiently be used as skin faire products. Utilization of marine algae has a beneficial edge due to having less toxic potential, wide acceptability, easy to use along with the decreased cost of production. These characteristics of marine algae can efficiently enhance female beauty. Nevertheless, extensive studies along with clinical trials are required for their whitening properties.

6.2.8 POTENTIAL BENEFITS FROM FISH: WOMEN HEALTH AND LONGEVITY

Fish is a source of numerous phytochemicals, which are significant for human health and proper body functions. These substances are comprised of several proteins required for cancer prevention and building of muscle mass, e.g.:

- Carnosine to enhance skeletal muscle working [27] and for antiaging and antioxidative effects [168].
- Choline to reduce cardiovascular diseases (CVD); and to improve memory [111].
- Creatinine to improve athletic activity.
- Micro-substances, such as copper to maintain cardiovascular and brain system properly [47].
- Peptides to regulate blood pressure [155].
- Phosphorus to improve heart and kidney functioning and to speed up healing of broken bones [102].
- Taurine to lower high blood pressure and protect from neurodegenerative disorders [12].
- Ubiquinone coenzyme to protect from suspected cardio effects and to slow down Alzheimer's and Parkinson's disease [29].
- Vitamin A for protection from night blindness, cancer, and for fresh skin [48].

As compared to our current knowledge, future research may disclose other indications about these fish ingredients regarding their contribution to more health benefits. A possible component of a fish protein to reduce risks of T2D is an example in this direction [114].

6.2.9   FISH: BIOACTIVE COMPOUNDS

Among all bioconstituents of fish, omega-3 FAs are best known for their health benefits. Seafoods like fish and algae among food sources are basic sources of long-chain FAs including docosahexaenoic acid (DHA) and eicosapentaenoic acid (EPA). One-third of fat in the fish muscle is composed of these long-chain FAs [50]. All known beneficial health effects of fish fat are due to both DHA and EPA. The term "fish oil" relates with intake of DHA and EPA either through fish consumption or through intake of isolated fish-oil products having DHA and EPA. For human health, both FAs are very crucial: (i) DHA is major fatty acid during development for human brain [88] and nerve endings [59]; (ii) human muscle and liver tissues has the highest level of EPA [125]. Human body requires alpha-linolenic acid (ALA) to initiate the synthesis process of these FAs because both of these FAs cannot be synthesized by our body. Due to having the first double bond at position of third carbon from the tail starts, all of three DHA, EPA, and ALA belong to omega-3 fatty acid group.

Omega-3 FAs decrease the level of arachidonic acid (AA) in the phospholipid membrane of endothelial cells, inflammatory cells, and platelets, which result in the decreased production of AA-derived pro-inflammatory mediators including hydroxyl-eicosatetraenoic acid, leukotriene, prostaglandin, and thromboxane [183]. Consumption of EPA for a long duration prevents the activity of Rho-kinase that is responsible to down-regulate endothelial nitric oxide synthase and to up-regulate pro-inflammatory molecules [21].

In inflammatory response, the downregulation of (NF)-kB activity has the main contribution in the regulation of gene expression. The omega-3 FAs decrease the transcription of inflammatory cytokines and activate the peroxisome proliferator-activated receptor to inhibit the NF-kB activity [93]. All these effects are responsible for the potential of anti-inflammatory activity [11] of omega-3 FAs that also stabilize the unstable plaques [184]. Both EPA and DHA change the concentration of phospholipid in mitochondrial membrane leading to improved mitochondrial function and efficient generation of ATP, and all these effects result in cardioprotective action of omega-3 FAs [33].

EPA decreases thromboxane A2 synthesis (strong platelet agonist) and enhances the thromboxane A3 synthesis (relatively inactive) and conducts these reactions by competing with AA for lipo-oxygenase and cyclo-oxygenase enzymes. EPA also enhances three series PGs (propylene glycols), TXs (thromboxanes), and different groups of eicosanoids). Therefore, an increased dose of omega-3 FAs result in:(i) inhibitory effect of platelets [81]; (ii) reduction in the level of triglycerides by reducing the assembly of very-low-density lipoproteins and their secretion; (iii) inhibition of lipogenesis by reducing the activity of sterol receptor element-binding protein-1c; (iv) reduction in the fatty acid substrate to synthesize triglyceride; (v) enhancement of β-oxidation in peroxisomes and mitochondria. All these activities of omega-3 FAs are conducted through the activation of peroxisome PPAR-α (peroxisome proliferator-activated receptors).

6.3   OMEGA-3 POLYUNSATURATED FATTY ACIDS (PUFAS) VERSUS TARGET DISEASES

6.3.1   CARDIOVASCULAR DISEASES (CVDS)

The two most common diseases of the cerebrovascular and coronary heart cause more than 38% and 42% deaths, respectively [18]. Numerous risk factors are linked to the progress of CVD, including diabetes, dyslipidemia, hypertension, obesity, and smoking [92], and oxidative stress. Imbalance of antioxidants (e.g., glutathione) with pro-oxidants reactive oxygen species leads to the development of oxidative stress in our body. Main CVD-related disorders are atherosclerotic, hypertensive diseases, ischemic heart disease, and stroke along with heart failure [11, 187].

Women with diabetes consuming fish and omega-3 FAs in higher quantities were at a significantly lower rate of CHD incidence. Women consuming fish had a risk factor of CHD of 0.70 for <one serving per month; 0.64 for once in a week; 0.60 for 2–4 times per week; 0.36 of 5 servings per week. The consumption of fish of ≥5 times per week significantly lowered the incidence of CHD in diabetic women [56].

The plasma lipid modulatory effect was showed by EPA and DHA in CVD and diabetic women [93]. Omega-3 FAs lowered the triglyceride level by decreasing the VLDL assembly and secretion. The reduced activity of sterol receptor element-binding protein-1c by ω-3 FAs leads to inhibition of lipogenesis. By activating the peroxisome PPAR-α and ω-3 FAs favor the β-oxidation in mitochondria and peroxisomes; and also reduces the fatty acid substrate for triglyceride synthesis [33].

The main cause of all CVD issues (claudication, heart failure, myocardial infarction, and stroke) is the atherosclerosis. Atherosclerosis mostly occurred in the innermost layers of middle and large sized arteries, mainly where vessels become separated [16]. Atherosclerosis disease develops in a long-time period; therefore, former lipid management can prevent the atherosclerotic vascular diseases [134].

In the prevention of CVD, the focus is to decrease low-density lipoprotein-cholesterol. It is assessed that every 1% decrease in LDL-cholesterol results in 1% to 2% reduction of CVD risks [39]. Even though in overall risk determination, a considerable percentage of CVD issues occur due to series of genetic factors and aging, which can be regulated, and these modifiable factors include lifestyle factors, such as diet and physical exercise, hyperlipidemia, hypertension, obesity, diabetes, and insulin resistance [23].

6.3.2   CARDIOVASCULAR RISK FACTORS VERSUS OMEGA-3 POLYUNSATURATED FATTY ACIDS (FAS)

Several data are available highlighting the ability of omega-3 PUFAs to affect arrhythmia [66], endothelial function [100], inflammation, and blood pressure, lipid profile [133], and platelet activity [80, 154] as shown Figure 6.1. Risk factors affecting CVDs are levels of [154]:

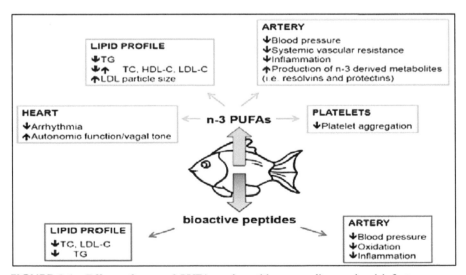

FIGURE 6.1   Effects of omega-3 PUFAs and peptides on cardiovascular risk factors.

- HDL-C (high density lipoprotein-cholesterol);
- LDL-C (low density lipoprotein-cholesterol);
- O-3 PUFAs (omega-3 polyunsaturated FAs);
- TC (total cholesterol);
- TG (triglycerides).

6.3.3 ARRHYTHMIAS

Ventricular arrhythmias are associated with muscle contraction controlling electrophysiological mechanisms [129]. Data from animal and *in vitro* studies exhibited that *n*-3 PUFAs influence cardiac ionic channels [80]. The *n*-3 PUFAs have noticeable preventive influences on sodium, potassium, sodium-calcium exchanger, and L-type calcium channels thus lower excitability based on data obtained from cardiomyocytes [84].

Furthermore, *n*-3 PUFAs may change the fluidity of membrane that in turn can affect ionic transport [76]. Reduction in ischemia-induced ventricular fibrillation in dogs and pigs are done by treating them with *n*-3 PUFAs through action on potassium channels [164]. An action of *n*-3 PUFAs on autonomic control may mediate its anti-arrhythmic effects, especially through an enhanced vagal tone [24]. Overall all, these mechanisms are constantly linked with anti-arrhythmic reactions and reduced levels of abrupt cardiac deaths detected in some human studies [77]. Table 6.7 indicates organizations recommending fish consumption for human health benefits.

TABLE 6.7 Different Organizations Recommending Fish Consumption for Human Health Benefits

Organizations	Recommendation	Region	Year	References
Scientific Advisory Committee for Nutrition (SACN)	450 mg DHA + EPA daily	UK	2004	[137]
International Society for the Study of Fatty Acids and Lipids (ISSFAL)	500 mg DHA + EPA daily	Europe	2004	[63]
European Food Safety Association (EFSA)	250 mg DHA + EPA daily	UK	2010	[34]
World Health Organization (WHO)/ Food and Agricultural Organization of the United Nations (FAO)	At least 1–2 100 g servings of fatty fish per week	World wide	2011	[38]
The Norwegian Directorate of Health/ VKM	Must eat fish 2–3 times per week in dinner	Norway	2014	[149]
The American Heart Association	Eat fish two times in a week	USA	2015	[4]

6.3.4   BLOOD PRESSURE

Different research studies on normotensive and hypertensive subjects have specified that the blood pressure is lowered by taking a high dose of n-3 PUFAs, though this influence was more distinct in hypertensive subjects [101]. Minihane et al. [94] have revealed that ingesting of 2–3 servings of oily fish per week or 2 capsules of fish oil per day to provide EPA + DHA was able to decrease the systolic blood pressure by 5 mmHg in systolic hypertensive subjects [94, 138] due to (i) the capability of n-3 PUFAs to reduce the synthesis of thromboxane A2; (ii) to increase the production of nitric oxide; and (iii) to influence the autonomic nerve function.

6.4   SUMMARY

Biologically active metabolites from marine foods have the potential to inhibit tyrosinase enzyme, which can efficiently be used in as skin fair products. A number of advantages are associated with fish, algae, and their products, such as low production costs, a broad spectrum of skin whitening characteristics, less cytotoxic, safe to use, widely acceptable, and novel modes of action. Therefore, nutritious foods from marine sources can efficiently be used to enhance health and beauty. Nevertheless, extensive studies along with clinical trials are required for their whitening properties.

KEYWORDS

- **flavonoids**
- **fucoxanthin**
- **marine medicinal activity**
- **nutraceuticals**
- **osteoporosis**
- **prostate cancer**
- **skin whitening**
- **tyrosinase**
- **women health**

REFERENCES

1. Abidov, M., Ramazanov, Z., Seifulla, R., & Grachev, S., (2010). The effects of Xanthigen in the weight management of obese premenopausal women with non-alcoholic fatty liver disease and normal liver fat. Diabetes, *Obesity and Metabolism, 12*, 72–81.

2. Abrahamsen, B., Masud, T., Avenell, A., Anderson, F., & Meyer, H. E., (2010). Patient level pooled analysis of 68,500 patients from seven major vitamin D fracture trials in US and Europe. *British Medical Journal, 340*(b5463), 1–8.

3. Agyemang, K., Han, L., Liu, E., Zhang, Y., Wang, T., & Gao, X., (2013). Recent advances in *Astragalus membranaceus* antidiabetic research: Pharmacological effects of its phytochemical constituents. *Evidence-Based Complementary and Alternative Medicine, 9*, 654643.

4. *AHA Recommendation.*, (2020). Online, The American Heart Association, Dallas, TX, USA. http://www.heart.org/HEARTORG/GettingHealthy/NutritionCenter/Fish-101 UCM_305986_Article.jsp#aha_recommendation (accessed on 18 May 2020).

5. Anderson, B. O., Shyyan, R., Eniu, A., Smith, R. A., Yip, C. H., Bese, N. S., Chow, L. W. C., et al., (2006). Breast cancer in limited-resource countries: An overview of the breast health global initiative 2005 guidelines. *The Breast Journal, 12*, S3–S15.

6. Aslam, M., Kreider, J., & Paruchuri, T., (2010). Mineral-rich extract from the red marine algae *Lithothamnion calcareum* preserves bone structure and function in female mice on a western-style diet. *Calcified Tissue International, 86*, 313–324.

7. Bays, H., (2004). Current and investigational anti-obesity agents and obesity therapeutic treatment targets. *Obesity Research, 12*, 1197–1211.

8. Beikler, T., & Flemmig, T., (2003). Implants in the medically compromised patient. *Critical Reviews in Oral Biology and Medicine, 14*, 305.

9. Beress, A., Wassermannv, O., Bruhn, T., & Beress, L., (1993). A new procedure for the isolation of anti-HIV compounds (polysaccharides and polyphenols) from *Fucus vesiculosus. Journal of Natural Products, 56*, 478–488.

10. Bischoff-Ferrari, H. A., Giovannucci, E., Willet, W. C., & Dietrich, T., (2006). Estimation of optimal serum concentrations of vitamin D for multiple health outcomes. *American Journal of Clinical Nutrition, 84*(1), 18–28.

11. Block, R. C., Harris, W. S., Reid, K. J., Sands, S. A., & Spertus, J. A., (2008). EPA and DHA in blood cell membranes from acute coronary syndrome patients and controls. *Journal of Atherosclerosis and Thrombosis, 197*(2), 821–828.

12. Bouckenooghe, T., Remacle, C., & Reusens, B., (2006). Is taurine a functional nutrient? *Current Opinion in Clinical Nutrition and Metabolic Care, 9*(6), 728–733.

13. Buck, C. B., Thompson, C. D., Roberts, J. N., & Muller, M., (2006). Carrageen is a potent inhibitor of papillomavirus infection. *PLoS Pathog., 2*, E69–E74.

14. Burtin, P., (2003). Nutritional value of seaweeds. *Electronic Journal of Environmental, Agricultural and Food Chemistry, 2*, 6–10.

15. Cahyana, A. H., Shuto, Y., & Kinoshita, Y., (1992). Pyropheophytin a as an antioxidative substance from the marine alga, arame (*Eisenia bicyclis*). *Bioscience, Biotechnology and Biochemistry, 56*, 1533–1535.

16. Campbell, K. A., Lipinski, M. J., & Doran, A. C., (2012). Lymphocytes and the adventitial immune response in atherosclerosis. *Circulation Research, 110*, 889–900.

17.   Campo, V. L., Kawano, D. F., Da Silva, D. B., & Carvalho, I., (2009). Carrageenan: Biological properties, chemical modifications, and structural analysis-review. *Carbohydrate Polymers, 77*, 167–180.

18.   *Cardiovascular Diseases (CVDs): Fact Sheet no. 317.*, (2020). World Health Organization: Geneva, Switzerland. Available online: http://www.who.int/mediacentre/factsheets/fs317/en/index.html (accessed on 18 May 2020).

19.   Cha, S. H., Ko, S. C., Kim, D., & Jeon, Y. J., (2010). Screening of marine algae for potential tyrosinase inhibitor: Those inhibitors reduced tyrosinase activity and melanin synthesis in zebra fish. *The Journal of Dermatology, 37*, 1–10.

20.   Chattopadhyay, N., Ghosh, T., Sinha, S., Chattopadhyay, K., Karmakar, P., & Ray, B., (2010). Polysaccharides from *Turbinaria conoides*: Structural features and antioxidant capacity. *Food Chemistry, 118*, 823–829.

21.   Chen, C., Yu, X., & Shao, S., (2015). Effects of Omega-3 fatty acid supplementation on glucose control and lipid levels in type-2 diabetes: Meta-analysis. *PLoS One, 2*(10), 7, e-article ID: 0139585.

22.   Cho, M. L., Lee, H. S., Kang, I. J., Won, M. H., & You, S. G., (2011). Antioxidant properties of extract and fractions from *Enteromorpha prolifera*, a type of green seaweed. *Food Chemistry, 127*, 999–1006.

23.   Chomistek, A. K., Manson, J. E., Stefanick, M. L., & Lu, B., (2013). Relationship of sedentary behavior and physical activity to cardiovascular disease: Results from Women's Health Initiative (WHI). *Journal of the American College of Cardiology, 61*, 2346–2354.

24.   Christensen, J. H., & Schmidt, E. B., (2007). Autonomic nervous system, heart rate variability and n-3 fatty acids. *Journal of Cardiovascular Medicine, 8*, S19–S22.

25.   Curat, C., Wegner, V., Sengenes, C., & Miranville, A., (2006). Macrophages in human visceral adipose tissue: Increased accumulation in obesity and a source of resistin and visfatin. *Diabetologia, 49*, 744–747.

26.   Das, S. K., Ren, R., Hashimoto, T., & Kanazawa, K., (2010). Fucoxanthin induces apoptosis in osteoclast-like cells differentiated from RAW264.7 cells. *Journal of Agricultural and Food Chemistry, 58*, 6090–6095.

27.   Derave, W., Everaert, I., Beeckman, S., & Baguet, A., (2010). Muscle carnosine metabolism and beta-alanine supplementation in relation to exercise and training. *Sports Medicine, 40*(3), 247–263.

28.   Dettmar, P. W., Strugala, V., & Craig, R. J., (2011). The key role alginates play in health. *Food Hydrocoll., 25*, 263–266.

29.   Dhanasekaran, M., & Ren, J., (2005). The emerging role of coenzyme Q-10 in aging, neurodegeneration, cardiovascular disease, cancer, and diabetes mellitus. *Current Neurovascular Research, 2*(5), 447–459.

30.   Dharmaraj, S., Ashokkumar, B., & Dhevendaran, K., (2009). Fermentative production of carotenoids from marine actinomycetes. *Iranian Journal of Microbiology, 1*, 36–41.

31.   Di-Bernardini, R., Harnedy, P., & Bolton, D., (2011). Antioxidant and antimicrobial peptidic hydrolysates from muscle protein sources and by-products. *Food Chemistry, 124*, 1296–1307.

32.   Dikshit, R. G. P. C., & Ramasundarahettige, C., (2012). Cancer mortality in India: A nationally representative survey. *Lancet*, 1–10.

33.   Duda, M. K., O'Shea, K. M., & Stanley, W. C., (2009). Omega-3 polyunsaturated fatty acid supplementation for the treatment of heart failure: Mechanisms and clinical potential. *Cardiovascular Research, 84*, 33–41.

34. *EFSA Sets European Dietary Reference Values for Nutrient Intakes*, (2010). European Food Safety Authority (EFSA), Parma, Italy. Available online: http://www.efsa.europa. eu/en/press/news/nda100326.htm (accessed on 18 May 2020).

35. El-Hage, A., (2005). *Hazardous to Teen Health* (pp. 1–4). North Carolina Family Policy Council, North Carolina, USA.

36. Engel, R. W., (1942). The choline content of animal and plant products. *Journal Nutrition, 25*(5), 441–446.

37. *Eurostat*, (2016). News Releases, Product Code: 3-24052016-AP. Available online: http://ec.europa.eu/eurostat/documents/2995521/7335847/3-24052016-APEN. pdf/4dd0a8ad-5950-4425-9364-197a492d3648 (accessed on 18 May 2020).

38. FAO/WHO, (2011). *Report of the Joint FAO/WHO Expert Consultation on the Risks and Benefits of Fish Consumption* (p. 52). Food and Agriculture Organization of the United Nations, World Health Organization: Rome, Italy and Geneva, Switzerland.

39. Ference, B. A., Yoo, W., Alesh, I., & Mahajan, N., (2012). Effect of long-term exposure to lower low-density lipoprotein cholesterol beginning early in life on the risk of coronary heart disease: Mendelian randomization analysis. *Journal of American College of Cardiology, 60*, 2631–2639.

40. Food and Nutrition Board, Institute of Medicine (IOM), (1997). *Dietary Reference Intakes (DRIs) for Calcium, Magnesium, Phosphorus, Vitamin D, and Fluoride* (p. 102). Washington, DC: National Academy Press.

41. Frestedt, J. L., Kuskowski, M. A., & Zenk, J. L., (2009). Natural seaweed derived mineral supplement (Aquamin F) for knee osteoarthritis: A randomized, placebo-controlled pilot study. *Journal of Nutrition, 2*, 7–11.

42. Fujita, T., Ohue, T., Fujii, Y., Miyauchi, A., & Takagi, Y., (1996). Heated oyster shell-seaweed calcium (AAA Ca) on osteoporosis. *Calcified Tissue International, 58*, 226–230.

43. Funahashi, H., Imai, T., & Mase, T., (2001). Seaweed prevents breast cancer. *Japanese Journal of Cancer Research, 92*, 483–487.

44. Gerber, P., Dutcher, J., Adams, E., & Sherman, J., (1958). Protective effect of seaweed extracts for chicken embryos infected with influenza virus B or mumps virus. *Proceedings of the Society for Experimental Biology and Medicine, 99*, 590–593.

45. Geyer, C., Forster, J., Lindquist, D., & Chan, S., (2006). Lapatinib plus capecitabine for HER2-positive advanced breast cancer. *The New England Journal of Medicine, 355*, 2733–2739.

46. Ghosh, P., Adhikari, U., Ghosal, P. K., & Pujol, C. A., (2004). *In vitro* anti-herpetic activity of sulfated polysaccharide fractions from *Caulerpa racemosa*. *Phytochemistry, 65*, 3151–3157.

47. Goodman, B. P., Bosch, E. P., & Ross, M. A., (2009). Clinical and electrodiagnostic findings in copper deficiency myeloneuropathy. *Journal of Neurology Neurosurgery Psychiatry, 80*(5), 524–527.

48. Gormley, T. R., Neumann, T., Fagan, J. D., & Brunton, N. P., (2007). Taurine content of raw and processed fish fillets/portions. *European Food Research Technology, 225*(5–6), 837–842.

49. Harris, W. S., (2004). Fish oil supplementation: Evidence for health benefits. *Cleveland Clinic Journal of Medicine, 71*(3), 208–221.

50. He, K., (2009). Fish, long-chain omega-3 polyunsaturated fatty acids and prevention of cardiovascular disease-eat fish or take fish oil supplement? *Progress in Cardiovascular Diseases, 52*(2), 95–114.

51. Heo, S. J., Ko, S. C., Kang, S. M., & Cha, S. H., (2010). Inhibitory effect of diphloretho-hydroxy-carmalol on melanogenesis and its protective effect against UV-B radiation-induced cell damage. *Food and Chemical Toxicology, 48*, 1355–1361.

52. Heo, S., Ko, S., Cha, S., Kang, D., Park, H., & Choi, Y., (2009). Effect of phlorotannins isolated from *Ecklonia cava* on melanogenesis and their protective effect against photo-oxidative stress induced by UV-B radiation. *Toxicology In Vitro, 23*, 1123–1130.

53. Holdt, S. L., & Kraan, S., (2011). Bioactive compounds in seaweed, functional food applications and legislation. *Journal of Applied Phycology, 23*, 543–597.

54. Holub, D. J., & Holub, B. J., (2004). Omega-3 fatty acids from fish oils and cardiovascular disease. *Molecular Cell Biochemistry, 263*(1–2), 217–225.

55. Hoshino, T., Hayashi, T., Hayashi, K., Hamada, J., Lee, J. B., & Sankawa, U., (1998). An antiviral active sulfated polysaccharide from *Sargassum horneri* (Turner) Cagardh. *Biological Pharmaceutical Bulletin, 21*, 730–734.

56. Hu, F. B., Cho, E., Rexrode, K. M., Albert, C. M., & Manson, J. E., (2015). Fish and long-chain omega-3 fatty acid intake and risk of coronary heart disease and total mortality in diabetic women. *Circulation Research, 107*(14), 1852–1857.

57. Hu, G. P., Yuan, J., Sun, L., & She, Z. G., (2011). Statistical research on marine natural products based on data obtained between 1985 and (2008). *Marine Drugs, 9*, 514–525.

58. Huheihel, M., Ishanu, V., Tal, J., & Arad, S. M., (2002). Activity of *Porphyridium sp.* polysaccharide against herpes simplex viruses *in vitro* and *in vivo. Journal of Biochemical and Biophysical Methods, 50*, 189–200.

59. Hutcheson, R., & Rocic, P., (2012). The metabolic syndrome, oxidative stress, environment, and cardiovascular disease: The great exploration. *Experimental Diabetes Research*, p. 9, e-article ID: 271028.

60. Inoue, S., Zimmet, P., & Caterson, I., (2000). The Asia-Pacific perspective by health communications Australia Pvt. Limited on behalf of the steering committee. *Redefining Obesity and its Treatment*, 1–55.

61. Institute of Medicine (IOM), (1998). *Dietary Reference Intakes for Thiamin, Riboflavin, Niacin, Vitamin B6, Folate, Vitamin B12, Pantothenic Acid, Biotin, and Choline* (p. 56). Washington (DC): National Academies Press.

62. ISGEM., (2020). *Food Database of Icelandic Food and Biotech R&D.* https://english.matis.is/isgem/ (accessed on 19 August 2020).

63. ISSFAL, (2004). Recommendations for intake of polyunsaturated fatty acids in healthy adults (Online). *Proceedings of the International Society for the Study of Fatty Acids and Lipids*. Brighton, UK. Online: http://www.issfal.org/statements/pufa recommendations/statement-3 (accessed on 18 May 2020).

64. Jha, P., (2009). Avoidable global cancer deaths and total deaths from smoking. *Nature Reviews Cancer, 9*, 655–664.

65. Kang, H., Kim, H., Byun, D., Son, B., Nam, T., & Choi, J., (2004). Tyrosinase inhibitors isolated from the edible brown alga *Ecklonia stolonifera. Archives of Pharmacal. Research, 27*, 1226–1232.

66. Kang, J. X., & Leaf, A., (1996). Antiarrhythmic effects of polyunsaturated fatty acids. *Recent Studies Circulation, 94*, 1774–1780.

67. Kaplan, N., & Dollin, J., (2007). Cervical cancer awareness and HPV prevention in Canada. *Canadian Family Physician, 53*(693–696), 697–708.

68. Kelishadi, R., (2007). Childhood overweight, obesity, and the metabolic syndrome in developing countries. *Epidemiologic Reviews, 29*, 62–76.

69. Kim, K. Y., Nam, K. A., Kurihara, H., & Kim, S. M., (2008). Potent α-glucosidase inhibitors purified from the red alga *Grateloupia elliptica*. *Phytochemistry, 69*(16), 2820–2825.

70. Kim, S. H., Choi, D. S., & Athukorala, Y., (2007). Antioxidant activity of sulfated polysaccharides isolated from *Sargassum fulvellum*. *Journal of Food Science and Nutrition, 12*, 65–73.

71. Kim, S. K., & Wijesekara, I., (2010). Development and biological activities of marine derived bioactive peptides-review. *Journal of Functional Foods, 2*, 1–9.

72. Kim, S., & Pallela, R., (2012). Medicinal foods from marine animals: Current status and prospects. *Advance Food and Nutrition Research, 65*, 1–9.

73. Kong, C. S., Kim, J. A., & Kim, S. K., (2009). Anti-obesity effect of sulfated glucosamine by AMPK signal pathway in 3T3-L1 adipocytes. *Food and Chemical Toxicology, 47*, 2401–2406.

74. Konitza, I., Stavri, M., Zloh, M., Vagias, C., Gibbons, S., & Roussis, V., (2008). New metabolites with antibacterial activity from the marine angiosperm *Cymodocea nodosa*. *Tetrahedron, 64*, 1696–1702.

75. Krelowskakulas, M., (1995). Content of some metals in mean tissue of salt-water and fresh-water fish and their products *Nahrung (Food), 39*(2), 166 172.

76. Kromhout, D., Yasuda, S., Geleijnse, J. M., & Shimokawa, H., (2012). Fish oil and omega-3 fatty acids in cardiovascular disease: Do they really work? *European Heart Journal, 33*, 436–443.

77. Kumar, S., Sutherland, F., & Rosso, R., (2011). Effects of chronic omega-3 polyunsaturated fatty acid supplementation on human atrial electrophysiology. *Heart Rhythm, 8*, 562–568.

78. Lahaye, M., (1991). Marine algae as sources of fibers: Determination of soluble and insoluble dietary fiber contents in some Lsquo sea-vegetable srsquo. *The Journal of the Science of Food and Agriculture, 54*, 587–594.

79. Le Tutour, B., Benslimane, F., & Gouleau, M., (1998). Antioxidant and pro-oxidant activities of the brown algae, *Laminaria digitata, Himanthalia elongata, Fucus vesiculosus, Fucus serratus* and *Ascophyllum nodosum*. *Journal of Applied Phycology, 10*, 121–129.

80. Leaf, A., & Weber, P. C., (1988). Cardiovascular effects of n-3 fatty acids. *The New England Journal of Medicine, 318*, 549–557.

81. Lee, J. H., O'Keefe, J. H., Lavie, C. J., Marchioli, R., & Harris, W. S., (2008). Omega-3 for cardio-protection. *Mayo Clinic Proceedings, 83*, 324–332.

82. Lee, W. J., Koh, E. H., Won, J. C., Kim, M. S., Park, J. Y., & Lee, K. U., (2005). Obesity: The role of hypothalamic AMP-activated protein kinase in body weight regulation. *The International Journal of Biochemistry and Cell Biology, 37*, 2254–2259.

83. Lehtinen, M., & Dillner, J., (2002). Preventive human papillomavirus vaccination. *Sexually Transmitted Infections, 78*, 4–6.

84. London, B., Albert, C., & Anderson, M. E., (2007). Omega-3 fatty acids and cardiac arrhythmias: Prior studies and recommendations for future research. Omega-3 fatty acids and their role in cardiac arrhythmogenesis workshop, national heart, lung, and blood institute and office of dietary supplements. *Circulation Research, 116*, e320–e335.

85. Lu, Z., Chen, T. C., & Zhang, A., (2007). An evaluation of the vitamin D3 content in fish: Is the vitamin D content adequate to satisfy the dietary requirement for vitamin D? *Journal of Steroid Biochemistry and Molecular Biology, 103*(3–5), 642–644.

86.   Macartain, P., Gill, C. I., Brooks, M., Campbell, R., & Rowland, I. R., (2007). Nutritional value of edible seaweeds. *Nutrition Reviews, 65*, 535–543.

87.   Maeda, H., Hosokawa, M., Sashima, T., & Miyashita, K., (2007). Dietary combination of fucoxanthin and fish oil attenuates the weight gain of white adipose tissue and decreases blood glucose in obese/diabetic KK-Ay mice. *Journal of Agricultural and Food Chemistry, 55*, 7701–7706.

88.   Marik, P. E., & Varon, J., (2009). Omega-3 dietary supplements and the risk of cardiovascular events: Systematic review. *Clinical Cardiology, 32*(7), 365–372.

89.   Mathers, C. D., Lopez, A. D., & Murray, C. J. L., (2006). The burden of disease and mortality by condition: Data, methods, and results for 2001. In: Lopez, A. D., Mathers, C. D., Ezzati, M., Jamison, D. T., & Murray, C. J. L., (eds.), *Global Burden of Disease and Risk Factors* (pp. 45–240). Washington, DC: International Bank for Reconstruction and Development/World Bank.

90.   Matis., (2020). *Internal Data of Icelandic Food and Biotech R&D.* www.matis.is (accessed on 18 May 2020).

91.   Mayer, A. M. S., & Hamann, M. T., (2002). Marine pharmacology: Compounds with antibacterial, anticoagulant, antifungal, anthelmintic, anti-inflammatory, anti platelet, antiprotozoal and antiviral activities affecting the cardiovascular, endocrine, immune and nervous systems, and other miscellaneous mechanisms of action. *Comparative Biochemistry and Physiology Part C, 132*, 315–339.

92.   Mendis, S., Puska, P., & Norrving, B., (2011). *Global Atlas on Cardiovascular Disease Prevention and Control* (p. 261). World Health Organization: Geneva, Switzerland.

93.   Minihane, A. M., Armah, C. K., & Miles, E. A., (2016). Consumption of fish oil providing amounts of eicosapentaenoic acid and docosahexaenoic acid that can be obtained from the diet reduces blood pressure in adults with systolic hypertension: Retrospective analysis. *Journal Nutrition, 146*, 516–523.

94.   Minihane, A. M., (2016). Metabolic Syndrome, diabetes mellitus, cardiovascular and neurodegenerative disease: Fish oil omega fatty acids and cardio-metabolic health, alone or with statins. *European Journal of Clinical Nutrition, 67*, 536–540.

95.   Miyashita, K., (2014). Anti-obesity therapy by food component: Unique activity of marine carotenoid, fucoxanthin. *International Journal of Obesity and Control Therapies, 1*, 1–4.

96.   Miyashita, K., & Hosokawa, M., (2009). Anti-obesity effect of allenic carotenoid, fucoxanthin. *Nutrigenomics and Proteomics in Health and Disease: Impact of Food Factors-Gene Interactions* (pp. 145–160). New York: Wiley-Blackwell Publisher.

97.   Moo-Puc, R., Robledo, D., & Freile-Pelegrin, Y., (2009). *In vitro* cytotoxic and antiproliferative activities of marine macroalgae from Yucatan Mexico. *Ciencias Marinas (Marine Science), 35*, 345–358.

98.   Moore, R. E., (1978). Algal non-isoprenoids. In: Scheuer, P. J., (ed.), *Marine Natural Products, Chemical and Biological Perspective* (pp. 44–171) New York: Academic Press.

99.   Moren, M., Malde, M. K., & Olsen, R. E., (2007). Fluorine accumulation in Atlantic salmon (*Salmo salar*), Atlantic cod (*Gadusmorhua*), rainbow trout (*Onchorhyncus mykiss*) and Atlantic halibut (*Hippoglossus hippoglossus*) fed diets with krill or amphipod meals and fish meal-based diets with sodium fluoride. *Aquaculture, 269*(1–4), 525–531.

100. Morgan, D. R., Dixon, L. J., & Hanratty, C. G., (2006). Effects of dietary omega-3 fatty acid supplementation on endothelium-dependent vasodilation in patients with chronic heart failure. *American Journal of Cardiology, 97*, 547–551.

101. Mori, T. A., (2010). Omega-3 fatty acids and blood pressure. *Cell Molecular Biology*, *56*, 83–92.
102. Moshfegh, A., Goldman, J., & Cleveland, L., (2005). *What do We Eat in America, Usual Nutrient Intakes from Food Compared to Dietary Reference Intakes?* (p. 89) Washington, DC: U.S. Department of Agriculture - Agricultural Research Service.
103. Nakao, Y., Uehara, T., & Matunaga, S., (2002). Callyspongynic acid, a polyacetylenic acid which inhibits α-glucosidase, from the marine sponge *Callyspongia truncate. Journal of Natural Products*, *65*(6), 922–924.
104. Namvar, F., Mohamed, S., & Fard, S. G., (2012). Polyphenol-rich seaweed (*Eucheumacottonii*) extract suppresses breast tumor via hormone modulation and apoptosis induction. *Food Chemistry*, *130*, 376–382.
105. Ngo, D. H., Wijesekara, I., & Vo, T. S., (2010). Marine food-derived functional ingredients as potential antioxidants in the food industry: An overview. *Food Research International*, *44*, 523–529.
106. Nieves, J. W., Formica, C., Ruffing, J., & Zion, M., (2005). Males have larger skeletal size and bone mass than females, despite comparable body size. *Journal of Bone and Mineral Research*, *20*, 529–535.
107. Nomura, T., Kikuchi, M., Kubodera, A., & Kawakami, Y., (1997). Proton donative antioxidant activity of fucoxanthin with 1,1-diphenyl-2-picrylhydrazyl (DPPH). *IUBMB* (*Int. Union of Biochem. and Molecular Biology*) *Life*, *42*, 361–370.
108. Ohta, Y., Lee, J. B., Hayashi, K., & Hayashi, T., (2009). Isolation of sulfated galactan from Codium fragile and its antiviral effect. *Biological Pharmaceutical Bulletin*, *32*, 892–898.
109. Olesya, S. V., Dariya, V. T., Svetlana, P. E., & Tatyana, N. Z., (2012). Structural characteristics and biological activity of fucoidans from the brown algae (*Saccharina japonica*) of different reproductive status. *Chemistry and Biochemistry*, *9*, 817–828.
110. Olesya, V. S., Olga, V. S., & Daria, M. S., (2011). Modern fluorescent proteins: From chromophore formation to novel intracellular applications. *Bio Techniques*, *51*, 313–327.
111. Olthof, M. R., Brink, E. J., Katan, M. B., & Verhoef, P., (2005). Choline supplemented as phosphatidylcholine decreases fasting and post-methionine loading plasma homocysteine concentrations in healthy men. *American Journal of Clinical Nutrition*, *82*(1), 111–117.
112. Ortega, R. M., (2006). Importance of functional foods in the Mediterranean diet. *Public Health Nutrition*, *9*, 1136–1140.
113. Ortiz, J., Romero, N., Robert, P., & Araya, J., (2006). Dietary fiber, amino acid, fatty acid and tocopherol contents of the edible seaweeds *Ulva lactuca* and *Durvillea antarctica. Food Chemistry*, *99*, 98–104.
114. Ouellet, V., Weisnagel, S. J., Marois, J., & Bergeron, J., (2008). Dietary cod protein reduces plasma c-reactive protein in insulin-resistant men and women. *Journal of Nutrition*, *138*(12), 2386–2391.
115. Paczos, T., Liu, W., & Chen, F., (2010). Human papillomavirus vaccine: A comprehensive review. *North American Journal of Medical Sciences*, *3*, 201–204.
116. Panlasigui, L., Baello, O., Dimatangal, J., & Dumelod, B., (2003). Blood cholesterol and lipid-lowering effects of carrageenan on human volunteers. *Asia Pacific Journal of Clinical Nutrition*, *12*, 209–214.
117. Park, P., Jung, W., Nam, K., Shahidi, F., & Kim, S., (2001). Purification and characterization of antioxidative peptides from protein hydrolysate of lecithin-free egg yolk. *Journal of the American Oil Chemists Society*, *78*, 651–656.

118. Parkin, D., Bray, F., Ferlay, J., & Pisani, P., (2002). Global cancer statistics. *Cancer Journal for Clinicians*, *55*, 74–82.

119. Pelkman, C., Navia, J., Miller, A., & Pohle, R., (2007). Novel calcium-gelled, alginate-pectin beverage reduced energy intake in nondieting overweight and obese women: Interactions with dietary restraint status. *The American Journal of Clinical Nutrition*, *86*, 1595–1692.

120. Penta-Ramos, E. A., & Xiong, Y. L., (2001). Antioxidative activity of whey protein hydrolysates in a liposomal system. *Journal of Dairy Science*, *84*, 2577–2583.

121. Pettigrew, R., & Hamilton-Fairley, D., (1997). Obesity and female reproductive function. *British Medical Bulletin*, *53*, 341–358.

122. Pilarczyk, B., Tomza-Marciniak, A., & Mituniewicz-Malek, A., (2010). Selenium content in selected products of animal origin and estimation of the degree of cover daily Se requirement in Poland. *International Journal of Food Science and Technology*, *45*(1), 186–191.

123. Qi, H., Zhang, Q., Zhao, T., Chen, R., Zhang, H., & Niu, X., (2005). Antioxidant activity of different sulfate content derivatives of polysaccharide extracted from *Ulva pertusain vitro*. *International Journal of Biological Macromolecules*, *37*, 195–199.

124. Ranathunga, S., Rajapakse, N., & Kim, S. K., (2006). Purification and characterization of antioxidative peptide from conger eel. *Eur. Food Res. Technol.*, *222*, 31–315.

125. Rangel-Zúñiga, O. A., Camargo, A., Marín, C., & Peña-Orihuela, P., (2015). Proteome from patients with metabolic syndrome is regulated by quantity and quality of dietary lipids. *BMC Genomics*, *16*, 509.

126. Raposo, M. F. J., Morais, R. M. S. C., & Morais, A. M. M. B., (2013). Bioactivity and applications of sulfated polysaccharides from marine microalgae (review). *Mar Drugs*, *11*(1), 233–252.

127. Recalde, M. P., Noseda, M. D., & Pujol, C. A., (2009). Sulfated mannans from the red seaweed nemalion helminthoides of the South Atlantic. *Phytochemistry*, *70*, 1062–1068.

128. Rennie, K., & Jebb, S., (2005). Prevalence of obesity in Great Britain. *Obesity Reviews*, *6*, 11–12.

129. Richardson, E. S., Iaizzo, P. A., & Xiao, Y. F., (2011). Electrophysiological mechanisms of the anti-arrhythmic effects of omega-3 fatty acids. *Journal of Cardiovascular Translational Research*, *4*, 42–52.

130. Riggs, B. L., Melton, L., Robb, III. R. A., & Camp, J. J., (2008). Population-based assessment of rates of bone loss at multiple skeletal sites: Evidence for substantial trabecular bone loss in young adult women and men. *Journal of Bone and Mineral Research*, *23*, 205–214.

131. Roberts, J. N., Buck, C. B., & Thompson, C. D., (2007). Genital transmission of HPV in a mouse model is potentiated by nonoxynol-9 and inhibited by carrageenan. *Nature Medicine*, *13*, 857–861.

132. Rocha, D. S. M. C., Marques, C. T., & Dore, C. M. G., (2007). Antioxidant activities of sulfated polysaccharides from brown and red seaweeds. *Journal of Applied Phycology*, *19*, 153–160.

133. Roche, H. M., & Gibney, M. J., (2000). Effect of long-chain n-3 polyunsaturated fatty acids on fasting and postprandial triacylglycerol metabolism. *American Journal of Clinical Nutrition*, *71*, 232S–237S.

134. Rosin, S., Ojansivu, I., & Kopu, A., (2015). Optimal use of plant stanol ester in the management of hypercholesterolemia. *Cholesterol*, e-article ID: 706970. Online: doi: 10.1155/2015/706970.

135. Ruperez, P., & Saura-Calixto, F., (2001). Dietary fiber and physicochemical properties of edible Spanish seaweeds. *The Journal of European Food Research and Technology*, *212*, 349–354.

136. Sachindra, N., Sato, E., Maeda, H., Hosokawa, M., Niwano, Y., & Kohno, M., (2007). Radical scavenging and singlet oxygen quenching activity of marine carotenoid fucoxanthin and its metabolites. *Journal of Agricultural and Food Chemistry*, *55*, 8516–8522.

137. SACN, (2004). Advice on fish consumption: Benefits and risks. In: *Proceedings EDID Collection* (p. 23). Scientific Advisory Committee on Nutrition (SACN), Committee on Toxicity, Norwich, UK.

138. Saravanan, P., Davidson, N. C., Schmidt, E. B., & Calder, P. C., (2010). Cardiovascular effects of marine omega-3 fatty acids. *Lancet*, *376*, 540–550.

139. Sasaki, K., Ishihara, K., Oyamada, C., Sato, A., Fukushi, A., & Arakane, T., (2008). Effects of fucoxanthin addition to ground chicken breast meat on lipid and color stability during chilled storage, before and after cooking. *Asian-Australasian Journal of Animal Sciences*, *21*, 1067–1072.

140. Schaafsma, G., & Martinez, J. A., (2007). Randomized trial of weight-loss-diets for young adults varying in fish and fish oil content. *International Journal of Obesity (London)*, *31*, 1560–1566.

141. Schneemann, I., Nagel, K., Kajahn, I., Labes, A., Wiese, J., & Imhof, J. F., (2010). Comprehensive investigation of marine actinobacteria associated with the sponge *Halichondria panacea*. *Applied and Environmental Microbiology*, *76*, 3702–3714.

142. Schuit, S. C. E., Van, D. K. M., & Weel, A. E. A. M., (2004). Fracture incidence and association with bone mineral density in elderly men and women: *The Rotterdam Study: Bone*, *34*, 195–202.

143. Schulze, M. B., Fung, T. T., Manson, J. E., Willett, W. C., & Hu, F. B., (2006). Dietary patterns and changes in body weight in women. *Obesity (Silver Spring)*, *14*, 1444–1453.

144. Sekar, S., & Chandramohan, M., (2008). Phycobiliproteins as a commodity: Trends in applied research, patents, and commercialization. *Journal of Applied Phycology*, *20*, 113–136.

145. Shahidi, F., & Alasalvar, C., (2011). Marine oils and other marine nutraceuticals. In: *Handbook of Seafood Quality, Safety, and Health Applications* (pp. 444–454). New York: Wiley Online Library.

146. Shimoda, H., Tanaka, J., Shan, S., & Maoka, T., (2010). Anti-pigmentary activity of fucoxanthin and its influence on skin mRNA expression of melanogenic molecules. *Journal of Pharmacy and Pharmacology*, *62*, 1137–1145.

147. Sibilla, S., Martin, G., Sarah, B., Anil, B. R., & Licia, G., (2015). An overview of the beneficial effects of hydrolyzed collagen as a nutraceutical on skin properties: Scientific background and clinical studies. *Open Nutraceuticals Journal*, *8*, 29–42.

148. Sithranga-Boopathy, N., & Kathiresan, K., (2010). Anticancer drugs from marine flora: An overview. *Journal of Oncology*, 1–18.

149. Skåre, J. U., Brantsæter, A. L., & Frøyland, L., (2015). Benefit-risk assessment of fish and fish products in the Norwegian diet: An update. *European Journal of Nutrition and Food Safety*, *5*, 260–266.

150. Solano, F., Briganti, S., Picardo, M., & Ghanem, G., (2006). Hypo-pigmenting agents: An updated review on biological, chemical, and clinical aspects. *Pigment Cell Research*, *19*, 550–571.
151. Stice, E., Maxfield, J., & Wells, T., (2003). Adverse effects of social pressure to be thin on young women: An experimental investigation of the effects of "fat talk". *International Journal of Eating Disorders*, *34*, 108–117.
152. Sun, L., Wang, C., Shi, Q., & Ma, C., (2009). Preparation of different molecular weight polysaccharides from *Porphyridium cruentum* and their antioxidant activities. *International Journal of Biological Macromolecules*, *45*, 42–47.
153. Synytsya, A., Kim, W. J., Kim, S. M., & Pohl, R., (2010). Structure and antitumor activity of fucoidan isolated from sporophyll of Korean seaweed *Undaria pinnatifida*. *Carbohydrate Polymers*, *81*, 41–48.
154. Tagawa, H., Shimokawa, H., & Tagawa, T., (1999). Long-term treatment with eicosapentaenoic acid augments both nitric oxide-mediated and non-nitric oxide-mediated endothelium-dependent forearm vasodilatation in patients with coronary artery disease. *Journal of Cardiovascular Pharmacology and Therapeutics*, *33*, 633–640.
155. Takahashi, K., (2004). Fish-derived peptides: From fish to human physiology and diseases. *Editorial Peptides*, *25*, 1575–1576.
156. Talarico, L. B., Zibetti, R. G. M., & Faria, P. C. S., (2004). Anti-herpes simplex virus activity of sulfated galactans from the red seaweeds *Gymnogongrus griffithsiae* and *Cryptonemia crenulata*. *International Journal of Biological Macromolecules*, *34*, 63–71.
157. Teas, J., Hurley, T., Hebert, J., Franke, A., Sepkovic, D., & Kurzer, M., (2009). Dietary seaweed modifies estrogen and phytoestrogen metabolism in healthy postmenopausal women. *Journal of Nutrition*, *139*, 939.
158. Tepe, Y., (2009). Metal concentrations in eight fish species from Aegean and Mediterranean Seas. *Environ. Monit. Assess, 159*(1–4), 501–509.
159. Tewari, N., Tiwari, V. K., & Mishra, R. C., (2003). Synthesis and bioevaluation of glycosyl urea as α-glucosidase inhibitors and their effect on mycobacterium. *Bioorganic and Medicinal Chemistry*, *11*(13), 2911–2922.
160. Thompson, K. D., & Dragar, C., (2004). Antiviral activity of *Undaria pinnatifida* against herpes simplex virus. *Phytotherapy Research*, *18*, 551–555.
161. Thorsdottir, I., Tomasson, H., & Gunnarsdottir, I., (2007). System to assess the quality of food sources of calcium. *Journal of Food Composition and Analysis*, *20*(8), 717–724.
162. Torstensen, B., (2015). Benefit-risk assessment of fish and fish products in the Norwegian diet: An update. *European Journal of Nutrition and Food Safety*, *5*, 260–266.
163. Trayhurn, P., & Wood, I. S., (2005). Signaling role of adipose tissue: Adipokines and inflammation in obesity. *Biochemical Society Transactions*, *33*, 1078–1081.
164. Tsuburaya, R., Yasuda, S., Ito, Y., & Shiroto, T., (2011). Eicosapentaenoic acid reduces ischemic ventricular fibrillation via altering monophasic action potential in pigs. *Journal of Molecular and Cellular Cardiology*, *51*, 329–336.
165. Tufail, T., Saeed, F., Abbas, M., & Arshad, M. U., (2018). Marine bioactives: Potential to reduce the incidence of cardiovascular disorders. *Current Topics in Nutraceutical Research*, *16*(1), 47–56.
166. Venugopal, V., (2009). *Marine Products for Healthcare: Functional and Bioactive Nutraceutical Compounds from the Ocean* (p. 528). Boca Raton: CRC Press.

167. VKM (Norwegian Scientific Committee for Food Safety), (2006). *Fish and Seafood Consumption in Norway-Benefits and Risks* (p. 135). Report by Norwegian Scientific Committee for Food Safety, Norway.

168. Wang, A. M., Ma, C., Xie, Z. H., & Shen, F., (2000). Use of carnosine as a natural anti-senescence drug for human beings. *Biochemistry (Moscow), 65*(7), 869–871.

169. Wang, C., Huang, C., Hu, C., & Chan, H., (1997). The effect of glycolic acid on the treatment of acne in Asian skin. *Dermatologic Surgery, 23*, 23–29.

170. Wang, J., Zhang, Q., Zhang, Z., & Li, Z., (2008). Antioxidant activity of sulfated polysaccharide fractions extracted from *Laminaria japonica*. *International Journal of Biological Macromolecules, 42*, 127–132.

171. Washida, K., Koyama, T., Yamada, K., Kita, M., & Uemura, D., (2006). Karatungiols A and B, two novel antimicrobial polyol compounds from the symbiotic marine *dinoflagellate Amphidinium* sp. *Tetrahedron Letters, 47*, 2521–2525.

172. Whitney, E. R., Hamilton, E. M. N., & Rolfes, S. R., (1990). *Understanding Nutrition* (5th edn.) West Publishing Company, MN, USA, E-book online, ISBN: 0314578315.

173. WHO/FAO, (2004). Vitamin and mineral requirements in human nutrition (2nd edn., p. 89) Report Food and Agricultural Organization (FAO), Rome.

174. Wijesekara, I., & Kim, S. K., (2010). Angiotensin-i-converting enzyme (ACE) inhibitors from marine resources: Prospects in the pharmaceutical industry. *Marine Drugs, 8*, 1080–1093.

175. Williams, D. E., Sturgeon, C. M., Roberge, M., & Andersen, R. J., (2007). Nigricanosides A and B, antimitotic glycolipids isolated from the green alga *Avrainvillea nigricans* collected in Dominica. *Journal of the American Chemical Society, 129*, 5822–5823.

176. World Health Organization (WHO), (2018). *Global Database on Body Mass Index: An Interactive Surveillance Tool for Monitoring Nutrition Transition: BMI Classification.* Online: http://www.who.int/news-room/fact-sheets/detail/obesity-and-overweight (accessed on 18 May 2020).

177. World Health Organization (WHO), (2012). *February Cancer Fact Sheet* (Vol. 297, p. 5). Factsheet.

178. World Health Organization. *The Top 10 Causes of Death.*, (2018). Online, https://www.who.int/news-room/fact-sheets/detail/the-top-10-causes-of-death (accessed on 19 August 2020).

179. www.nifes.no (accessed on 18 May 2020).

180. Yabuta, Y., Fujimura, H., Kwak, C. S., Enomoto, T., & Watanabe, F., (2010). Antioxidant activity of the phycoerythrobilin compound formed from a dried Korean purple laver *(Porphyra sp.)* during *in vitro* digestion. *Food Science and Technology Research, 16*, 347–352.

181. Yamazaki, H., Sumilat, D. A., & Kanno, S., (2013). Poly-bromodiphenyl ether from an Indonesian marine sponge *Lamellodysidea herbacea* and its chemical derivatives inhibit protein tyrosine phosphatase 1B, an important target for diabetes treatment. *Journal of Natural Medicines, 67*(4), 730–735.

182. Yan, X., Chuda, Y., Suzuki, M., & Nagata, T., (1999). Fucoxanthin as the major antioxidant in *Hijikia fusiformis*: A common edible seaweed. *Bioscience, Biotechnology, and Biochemistry, 63*, 605–607.

183. Yanani, H., Hamasahim, H., & Sako, A., (2015). Effects of intake of fish or fish oils on the development of diabetes. *Journal of Clinical Medicine Research, 7*(1), 8–12.

184. Yasuda, S., & Shimokawa, H., (2010). Potential usefulness of fish oil in the primary prevention of acute coronary syndrome. *European Heart Journal, 31,* 15–16.
185. Yoon, N., Eom, T., Kim, M., & Kim, S., (2009). Inhibitory effect of phloro-tannins isolated from *Ecklonia cava* on mushroom tyrosinase activity and melanin formation in mouse B16F10 melanoma cells. *Journal of Agricultural and Food Chemistry, 57,* 4124–4129.
186. Yuan, Y. V., & Walsh, N. A., (2006). Antioxidant and antiproliferative activities of extracts from a variety of edible seaweeds. *Food and Chemical Toxicology, 44,* 1144–1150.
187. Yusuf, H. R., Giles, W. H., Croft, J. B., Anda, R. F., & Casper, M. L., (1998). Impact of multiple risk factor profiles on determining cardiovascular disease risk. *Preventive Medicine, 27,* 1–9.
188. Zhang, Q., Li, N., Zhou, G., Lu, X., Xu, Z., & Li, Z., (2003). *In vivo* antioxidant activity of polysaccharide fraction from *Porphyra haitanensis* (Rhodephyta) in aging mice. *Pharmacological Research, 48,* 151–155.
189. Zhu, W., Chiu, L. C. M., Ooi, V. E. C., Chan, P. K. S., & Ang, P. O., (2006). Antiviral property and mechanisms of a sulfated polysaccharide from the brown alga *Sargassum patens* against herpes simplex virus type 1. *Phytomedicine, 13,* 695–701.

CHAPTER 7

MARINE NATURAL PRODUCTS FOR HUMAN HEALTH CARE

K. L. SREEJAMOLE

ABSTRACT

The marine environment is an untapped resource compared to the terrestrial ecosystem, which has been utilized extensively for natural products. Marine pharmacology is relatively a recent field that started its journey in 1950s, and since then around 14,000 phytocompounds have been discovered with a plethora of bioactivities. Currently, there is a growing interest in this field and more researchers are involved in this novel field of pharmacology. This chapter focuses on the marine natural products; the processes involved in marine natural product research (from the collection, extraction, fractionation, bioassay-screening to the structural elucidation of purified products). An outline on the evolutionary and ecological implications of marine secondary metabolites is also discussed. This chapter also summarizes major secondary metabolites isolated from marine organisms against inflammation, cancer, hypertension, pain, cardiovascular diseases, HIV (human immunodeficiency virus), other viruses, and bacteria. A brief outline on the limitations and challenges faced by the marine drug industry and future prospects has also been presented.

7.1 INTRODUCTION

The marine ecosystem covers more than 70% of the earth's surface representing 95% of the biosphere and contains more than 300,000 known species of plants and animals representing 34 out of the 36 phyla of life. Some of them exclusively belong to the marine ecosystem, such as Ctenophores,

Echinodermates, Sponges, Phoronids, Brachiopods, and Chaetognaths and invertebrates comprising of approximately 60% of all marine life diversity [15]. Considering the evolution and biodiversity, the marine ecosystem seems to be superior to the terrestrial ecosystem with approximately a half of the total biodiversity, and hence contributes to an enormous source of useful drugs and bioactive compounds [71]. The greatest biodiversity is found in ecosystems, such as rocky coasts, kelp beds, and coral reefs, where species diversity and population density are exceedingly high [106].

The ocean contains a broad spectrum of organisms one of the diversified environments offered by different oceanic zones and contains more than 80% of diverse plant and animal species in the world [84]. Though this field of research is relatively recent, yet the bioprospecting of new marine natural products has already produced thousands of novel compounds [51, 86]. The discovery of novel marine natural products is expected to increase in the years to come, yielding better novel drugs for human illnesses, along with other advanced products [194, 218, 314].

Marine organisms especially invertebrates such as sponges, tunicates, fishes, soft corals, nudibranchs, sea hares, opisthobranch mollusks, echinoderms, bryozoans, prawns, shells, and sea slugs including marine microorganisms are rich sources of bioactive compounds and nutraceuticals thus form potential candidates for the treatment of several human diseases. The first biologically active marine natural product was formally reported in late 1950 by Bergmann [22]. Arabino and ribo-pentosyl nucleosides extracted from *Cryptotethia crypta* sponge were the first demonstrations that naturally occurring nucleosides could contain sugars other than ribose and deoxyribose [20, 21, 23]. However, interest in marine natural product research has increased [309]. In the late 1970s, it was established that marine plants and animals are genetically and biochemically unique. It was also observed that molecules of marine origin can be accepted by humans with minimal manipulation.

7.2 MARINE NATURAL PRODUCTS

Pharmacognosy was earlier dealt exclusively with the study of drugs derived from terrestrial plants and animals. However, it was in the 1950s that marine organisms were identified as an excellent source of new biologically active compounds. Later in 1960s, the effective exploration for tapping the marine organisms was studied as a source of bioactive metabolites that may be directly utilized as drugs or serve as lead structures for drug development

[104]. Since then,>14,000 unique natural compounds have been described [182] with hundreds of additional compounds being discovered annually [85]. Out of these compounds reported till now, 30% of these have been isolated from sponges. Despite four decades of intensive research, marine pharmacognosy is still considered a relatively younger field with only a fraction of organisms has so far been investigated [71] compared to terrestrial pharmacognosy. However, recent developments in this field show that marine environment can potentially provide us with an even more structurally diverse array of biocompounds in future.

The natural product (NP) is any biological molecule, but the term is usually reserved for the secondary metabolites, which are small molecules (with molecular weight of<1500 amu ca) produced by an organism [254]. Secondary metabolites have evolved from primary metabolites [40] that offer evolutionary advantages to the host organisms [90]. It is believed that secondary metabolites with adaptive characteristics may contribute to the survival of new strains [87]. While secondary metabolism has different functions than primary metabolism, they cannot be sharply distinguished from one another. The genes-coding for enzymes present in secondary metabolisms are different, but the precursors are the same [43].

Marine natural products are generally secondary metabolites. Marine secondary metabolites include compounds, which can differ fundamentally from terrestrial secondary metabolites and incorporation of halogen is a very characteristic feature [85, 244].

Natural products derived from marine organisms have been reported for the last 20 years, and most of them are from invertebrates and identified as steroids, terpenoids, isoprenoids, and sesquiterpenes. These have shown potential of becoming therapeutic agents. Unlike terrestrial environments, where plants are greatly richer in secondary metabolites, marine invertebrates and microorganisms have yielded comparatively more bioactive natural products than from seaweeds [85, 231]. In the last 30 years, the interest in marine bioprospecting has increased among researchers throughout the world [155, 204].

7.3   MARINE NATURAL PRODUCTS: EVOLUTIONARY AND ECOLOGICAL ROLES

The evolutionary process of many marine organisms has prepared themselves with proper defense mechanisms to survive in a rather unpredictable environment like extreme temperatures, change in salinity and pressure, bacterial, and viral pathogens along with the effects of mutation [141].

The most popular compounds isolated from marine habitats strengthen this assumption. Marine secondary metabolites involved in the host defense are often active at minute concentrations owing to the dilution factor, hence it must be extremely potent if they are to be active.

For evaluating the biomedical potential of marine organisms, it is important to know their history of evolution and chemical ecology. The secondary metabolites are assumed to be randomly evolved from primary metabolites. Although it is not clear whether any precise evolutionary pressures have led to chemically rich organisms, yet the surviving organism needs to perfect its chemical defenses during the long periods of evolution. Having a long evolutionary history, the sessile marine invertebrates had plenty of opportunities. Chemical defense mechanisms and potential biomedical activity cannot be directly associated, but in reality, the two correlate well [84].

The best sources of pharmacologically active biocompounds among the many phyla found in the oceans are algae, dinoflagellates, and invertebrates (like sponges, echinoderms, soft corals, gorgonians, sea hares, nudibranchs, bryozoans, and tunicates). The reason behind the production of these potent chemicals lies in the fact that they are susceptible to very high predation. Therefore, these compounds function as chemical weapons and have evolved into highly potent inhibitors of physiological processes of the prey, predators, or competitors of the concerned organisms [106].

The soft-bodied organisms like invertebrates usually have sedentary lifestyle, necessitating chemical means of defense-system. It reflects the ecology of the respective invertebrates and is therefore the prime candidates to possess bioactive metabolites. The physical and chemical conditions prevailing in the marine environment, force aquatic organisms to produce molecules that differ substantially in structural terms from substances of terrestrial origin. Almost every class of marine organism produces a variety of molecules with unique structural features. The organisms have evolved biochemical and physiological mechanisms to produce bioactive compounds, which fulfill the purposes of reproduction, communication, and defense against predation, infection, and competition [113, 115].

7.4 CHEMO DIVERSITY OF MARINE NATURAL PRODUCTS: ECOLOGICAL PERSPECTIVE

Marine chemo-diversity can largely be attributed to the composition of seawater itself. Concentrations of halides in seawater (19000 mg/l Cl^-, 65 mg/l Br^- and 5×10^{-4} mg/l/IO_3^-) are reflected in the number of compounds incorporating these elements and presence of sulfated compounds can be attributed to the relatively high concentration of sulfur (2700 mg/l) in seawater [89]. The incorporation of halides in the marine environment is exemplified by the comparison of the monoterpene aldehyde isolated from the boll weevil and the halogenated marine equivalent isolated from the red alga *Ochtodes crockeri* [89]. Both compounds are biosynthesized from geraniol; however, the abundance of Br⁻ and Cl⁻ ions in seawater has allowed these halides to be incorporated into the chemical structure.

The ecology of the habitat of an organism plays a major role in determining the type of metabolites they produce. It is evident from the fact that majority of the drug candidates from marine invertebrates so far isolated claim its origin from tropical and subtropical seas, where the grazing pressure by predator such as fishes is higher than any other ecosystems of the world. The intense competition and predation in these regions have led to the evolution of a wider range of secondary metabolites. Secondary metabolites (i.e., exceptional compounds not involved in primary metabolism) seem to be plentiful among tropical benthic organisms that are exposed to high rates of predation by consumers on coral reefs [116, 217, 220]. However, in temperate [281] and Antarctic benthic communities [186], and probably in pelagic communities worldwide [129, 215], secondary metabolites have played significant roles.

Secondary metabolites in marine organisms, especially in invertebrates, are often attributed a defensive function because they are either sessile or slow-moving and usually lack physical defenses like protective shells or spines, thus necessitating chemical defense mechanisms (such as the ability to synthesize toxic and/or deterrent compounds). Chemically defended organisms often produce multiple secondary metabolites, which opens the possibility of synergistic or additive effects among various metabolites [115]. A good example of this is the Spanish dancer nudibranch (*Hexabranchus sanguineus*) [220], which feed on sponges in the genus *Halichondria* that has oxazole macrolides. The organism modifies these compounds slightly and concentrates it in dorsal mantle and egg-masses and uses it to defend against consumers.

Some compounds produced by marine invertebrates keep the predators and competitors away [80]. Moreover, the fishes can avoid the mollusk *Saccoglossus kowalevskii* due to the presence of a deterrent molecule 2,3,4-tribromopyrrole [150]. Marine organisms form an exceptional source of secondary metabolites to provide defense against competitors, consumers, and pathogens.

Although the ecological and evolutionary consequences of these compounds have only recently been studied in detail, their influence on marine biodiversity is already known. They have an important role in the

organization and structure of marine systems at the genetic, species, population, and ecosystem levels. The strong bioactivity and structural diversity of
many of these compounds render them excellent candidates for development
as agrochemicals, pharmaceuticals, molecular probes, and growth-regulating
substances, etc.

7.5   SCREENING OF MARINE ORGANISMS FOR NATURAL PRODUCTS

7.5.1   SAMPLE COLLECTION

The first step in marine bioprospecting is the collection of specimens, which
is much more difficult in the marine environment compared to the collection
of terrestrial organisms. Collecting samples from the marine environment
poses so many difficulties, such as lack of sufficient biological material
and problems associated with taxonomic identification. The occurrence of
symbionts (such as fungi, bacteria, microalgae living on or inside the macro-
organisms being studied) poses ambiguity about the exact metabolic origin
of bioactive compounds isolated. Moreover, it is further complicated by the
difficulties in standardizing methodologies for ambient growth and cultiva-
tion of marine organisms, mainly invertebrates (i.e., sponges, bryozoans)
and microbes [89].

There are several factors to be considered before preparing for the collection of marine organisms: (i) the selection of the environment is of prime
importance as it determines the success of the discovery of a new compound.
The areas-where the animals are abundantly available for future works and
which are less screened for biological studies-must be preferred; (ii) prior to
collection, intrinsic, and extrinsic factors, method of collection and sampling,
sampling details (day/night, season, preservation, and life cycle) should be
considered; (iii) the major risks that can ensue while sampling is contamina-
tion due to epiphytes or other attached organisms hence care should be taken
to remove them prior to processing.

The oceanic environment is less explored for natural products in earlier
times [102, 142]. The amount of the organism to be collected is based on
its availability [122]. Voucher specimens must be ready through preserving
and keeping an entire organism for later use and must be a representative of
the whole population. Algal specimens are generally preserved in the 5%
formalin in seawater. In the case of animals, a small section of tissue of
about 2–5 cm can be preserved. Both exosome and endosome are crucial for

the precise identification of sponges. In the case of soft corals and tunicates, usually a part of the organism or the entire organism must be collected, including the "root" [122].

The number of samples screened determines the probability of finding useful active metabolites; hence, the primary selection tests should be cost-effective, rapid, and reproducible. During this stage, only minute quantity of biological samples is spent, but once the active organism is found, large amount not less than ≤ 1 kg of biomaterial will be needed just to get the minimal amount of pure biocompound to detect its bioactivity and elucidate the chemical structure.

7.5.2   HANDLING AND PROCESSING

Handling of marine organisms during collection is a very serious step, since many of the invertebrates contain highly irritating components. Standard precautions should be taken while working with compounds of unknown biological properties. More often compounds from marine organisms especially from invertebrates are highly toxic hence prior precautions should be taken not to get exposed to the compounds in anyway, wearing proper protective outfits (like, facial masks, gloves, etc.). This is applicable to both the raw samples and the extracts derived from them.

Marine organisms are so fragile that many of them die instantly on exposure to air and quickly get degraded by enzymatic, oxidative, or polymerization process. The entire organism can be frozen at -20°C immediately after collection until further workup can be carried out. Another method is to keep the organisms in alcohol (such as methanol, ethanol, or isopropanol) and then preserved and stored.

7.5.3   EXTRACTION OF MARINE ORGANISMS: OUTLINE ON SAMPLE PREPARATION AND EXTRACTION TECHNIQUES

The type of solvent used is either non-polar, medium-polar or polar that determines the efficiency with which they get extracted according to Cannel and Durate et al. [40, 74]. Following objectives must be considered before starting any screening process [118]:

- Efforts should be taken to reduce material losses;
- Minimal costs for sample preparation.

After collection from their natural habitats or after freeze-drying, the marine organisms is extracted by means of methanol or ethanol [267].

The solid-solid extraction offers some measure of preliminary separation of compounds. Initially, the samples are crushed into fine particles to attain maximum solvent penetration and first extracted thoroughly with non-polar solvents followed by water or water-alcohol mixtures with occasional stirring or sonication to increase the diffusion rate.

The tissue residue after centrifugation is again extracted with the second portion of the solvent and this process is continued until no further color is observed in the solvent. In cases when the extracts are colorless, the successive extracts can be concentrated separately. The filtration process can be speeded up with filter-aids and vacuum. Table 7.1 shows examples of solvents frequently used for the extraction of marine bioactive compounds.

TABLE 7.1 Solvents Commonly Used for Extraction of Bioactive Compounds

Type of Bioactive Compound-Based on Polarity	Bioactive Compounds	Commonly Used Solvents for Extraction
Polar organic compounds	Alkaloids	n-butanol
	Anthocyanins	Methanol
	Amino acids	Ethanol
	Polyketides	Ethyl acetate
	Saponins	Water
	Glycoside	Chloroform
	Quinones	Acetone
Medium polar compounds	Peptides	Dichloromethane
		Methanol
		Carbon tetrachloride
Non-polar compounds	Fatty acids	Ethyl acetate
	Hydrocarbons	Carbon tetrachloride
	Sterols	Hexane
	Terpenes	Petroleum ether

The main advantage of this extraction procedure over others is low processing cost, ease of operation and does not need any sophisticated equipment. The main disadvantage is the requirement of more energy for their evaporation. Kupchan's extraction method [161] is probably the most popular liquid-liquid fractionation procedure, based on the principle of separation of compounds. After centrifugation, the aqueous extract was removed and then lyophilized [122].

Another advanced technique involves the use of supercritical fluids (SCFs), which is increasingly replacing organic solvents in natural product extraction and isolation. Supercritical fluid extraction (SFE), especially that employing supercritical CO_2, has become the method of choice being an eco-friendly alternative for conventional extraction techniques. Supercritical fluid is any substance at a temperature and pressure above its critical point, where distinct liquid and gas phases do not exist. In the case of water, the critical temperature and pressure values are 374°C and 220 atmospheres.

SCFs have the advantage of low viscosity, good solvation power, and superior mass transfer properties. They easily penetrate microporous materials, making them appropriate for marine natural product extraction. It has other benefits, such as chemical inertness, nontoxicity, non-flammability, non-corrosiveness, and cost effectiveness. Supercritical carbon dioxide easily evaporates into the atmosphere after extraction, hence is the most preferred solvent and can be used at low temperatures [14].

7.5.4   FRACTIONATION OF MARINE EXTRACTS

Marine extracts are exceptionally complex, comprising of mixture of neutral, acidic, basic, amphiphilic, and lipophilic compounds [96]. There is no general fractionation rule for extraction, which can serve all purposes. Experience has a vital role in isolation of marine natural products irrespective of the recent advances in separation technology [122]. Proper measures should be taken by referring to the available comprehensive databases for known compounds to avoid isolation of already-known compounds (Table 7.2).

7.5.4.1   PRE-FRACTIONATION OF CRUDE EXTRACT

Crude extracts are complex mixtures and are often composed of hundreds of different constituents. Fractionation of crude extracts prior to further investigation is known as pre-fractionation and may result in any number of fractions ranging from few to several hundreds [39]. Pre-fractionation increases the chances of detecting bioactivity of the actual secondary metabolite in the extracts via removal of salts, sugars, and lipids. Fractionation will reduce the complexity of crude extracts and hence can decrease risk of masking the activity of secondary metabolite by other interfering compounds.

TABLE 7.2　Marine Natural Products and Their Databases

Database	URL	Description
ChemSpider	http://www.chemspider.com/	ChemSpider is an aggregated database of organic molecules containing more than 20 million compounds from many different providers.
Database of marine natural products	https://www.bio.org/articles/database-marine-natural-products	This database provides a place for those in academia and the private sector to share information regarding their available collections with other biotech companies and researchers.
Dictionary of marine natural products	http://dmnp.chemnetbase.com/intro/	The "Dictionary of Marine Natural Products" is a comprehensive database containing over 30,000 compounds, based on 25 years of literature. It is a subset of the dictionary of natural products (DNP) database.
Marine bacteria metabolite database	http://mbmsearcher.ust.hk/	Marine bacteria metabolite database is a chemical database based on secondary metabolite profiles of marine bacteria with the support from the China Mineral Resources Research and Development Association.
Marine drug discovery database	http://www.bdu.ac.in/MDDD/	Marine Drug Discovery Database is supported and maintained by Bharathidasan University, Tiruchirappalli, India.
MarinLit	http://pubs.rsc.org/marinlit/	MarinLit is a database dedicated to marine natural products research. The database was established in 1970s
NAPRALERT (Natural Product Alert) database	http://www.cas.org/ONLINE/DBSS/napralertss.html	NAPRALERT® is a relational database of natural products, including ethno-medical information, pharmacological/biochemical information on extracts of organisms *in vitro, in situ, in vivo,* in human (case reports, non-clinical trials) and clinical studies.
NCI data search	https://cactus.nci.nih.gov/ncidb2.2/	Database contains 250250 open structures ready for searching.
Seaweed metabolite database	http://www.swmd.co.in/	Seaweed metabolite database is an open access database of secondary metabolites from seaweeds, providing text and structure search access of chemical structures.

The first step of pre-fractionation includes solvent partitioning, which allow us to identify extracts in terms of their polarity and content and to narrow down them before elaborate fractionation steps are carried out. HPLC a high-resolution separation technique is usually used as the third step with the aim of purifying target compounds. The fourth stage enables the subsequent structural elucidation with NMR, Mass spectrophotometry, FT-IR, etc. These four procedures are discussed in detail in this section.

7.5.4.2   SOLVENT PARTITIONING

Solvent partitions of the active extract at early stages eliminate most of the bulky inactive material, although the active fractions of these fractions are still much chemically complex. The extensive fractions, thus obtained, are further separated with several types of column chromatography. A modified form of solvent partition procedure developed by Kupchan has been described by Houssan and Jaspers [122]. It can be used for defatting and desalting. In this technique, n-hexane fraction takes up most of the fats, while the aqueous fraction takes up all inorganic salts.

7.5.4.3   DESALTING

Large amount of inorganic salts interferes with the chromatographic separations including gel-filtration by giving false results in bioassays. West and Northcote, and others suggested the most effective method for desalting of marine extracts [311, 312]. This process is repeated to make all the hydrophobic compounds get adsorbed onto the resin. The resin is desalted by washing with plenty of water. Methanol or acetone in water in different proportions is used to attain a certain extent of fractionation and to elute the adsorbed compounds.

If the size of the inorganic salts and the low molecular weight compounds are identical, then absolute methanol can be used as a suitable method for desalting of the freeze-dried aqueous extract of marine organisms. Most of the salt present is removed by extracting the residue with absolute methanol, followed by removal of solvent and the procedure is repeated three to four times to make further desalting easier [24].

7.5.5   ROLE OF BIOASSAYS IN SCREENING OF MARINE NATURAL PRODUCTS

According to Suffness and Pezzuto [287], four major roles of bioassays are: Pre-screening, screening, monitoring, and secondary testing.

In pre-screening, large numbers of initial samples are subjected to bioassay to determine whether or not they have any desired bioactivity. Such bioassays must have high capacity, low cost, and must give rapid answers and need not be quantitative. Appropriate biological screening methodologies for marine sources are of great significance because of the great biodiversity existing in marine environment. Basically, the screening assay is selected depending upon the disease of interest. The following factors need to be taken care of while selecting an assay for testing [273]:

- High throughput.
- Reproducibility and reliability.
- Sensitivity and competence to detect the presence of very low concentrations of potentially active substances.
- Tolerance to DMSO (commonly used solvent for dissolving extracts).
- Tolerance to several impurities present in crude extracts.

In natural product drug discovery, the bioassays are utilized at different stages of the drug discovery process starting from the initial phase of the bioprospecting pipeline with crude extracts or fractions to the isolation of active component. Usually combinations of bioassays are used. Multiple samples are screened during primary biological screening. It is estimated that the screening of 5 million extracts will generate 1000 positive hits, from which ten leads will be generated. Out of these, five compounds will enter clinical trials and, in the end, one will become a marketed drug [59]. The initial bioactivity screening is followed by a more careful examination of the bioactive crude extract or fraction, in which a positive result nominates the sample for dereplication. Finally, the bioassays can be used to elucidate the bioactivity profiles of isolated compounds.

If during the primary screening, a positive hit is observed to eliminate false positives then the follow-up assays should be performed [195]. List of common bioassays used for natural product screening is given in the Table 7.3.

7.5.5.1   TYPES OF BIOASSAYS

- **Target-Based Assays:** These assays measure the effect of compounds on a single, defined target [253]. The targets are typically proteins, with key roles in disease pathogenesis. Examples of targets are G protein coupled receptors and kinases [288].
- **Phenotypic Assays:** In these assays, cells, tissues, or whole living organisms are used to detect an activity. The aim is to discover a

desired effect on the selected system, independent of any defined target, and therefore does not require any prior knowledge to the pathophysiology of the disease.

TABLE 7.3 Bioassays Employed for Marine Natural Product Screening

Bioactivity	Name of Bioassays
Analgesic activity	Tail flick model, hot plate model, acetic acid induced writhing, formalin test.
Anti-allergic activity	Mouse and rat passive cutaneous anaphylaxis (PCA) tests
Antibacterial activity	Poison food technique; disc diffusion method; tube dilution method and microtiter technique.
Anti-cancer	Protein kinase assay, Ascites tumor in mice, solid tumor in mice
Anti-inflammatory activity	**Acute Models:**
	Carrageenan-induced edema in Mice
	Sub-acute Models:
	Cotton Pellet Test
	Granuloma Pouch Test
	Formaldehyde-induced Arthritis
	Chronic Models:
	Adjuvant-induced Arthritis
Antioxidant activity	DPPH scavenging activity
	Hydrogen peroxide scavenging (H_2O_2) assay
	Nitric oxide scavenging activity
	ABTS radical cation decolorization assay
	Total radical-trapping antioxidant parameter (TRAP) method
	Ferric reducing antioxidant power (FRAP) assay
	Superoxide radical scavenging activity (SOD)
	Hydroxyl radical scavenging activity
	Beta-carotene linoleic acid method
Anti-ulcer activity	Absolute Ethanol induced ulcer model in rats, Pyloric ligation in rats
Anti-viral activity	Plaque inhibition assay
	Plaque reduction assay.
	Inhibition of virus-induced cytopathic effect (CPE)
	Virus yield reduction assay
	End point titration technique (EPTT)

TABLE 7.3   *(Continued)*

Bioactivity	Name of Bioassays
	Assays-Based on Measurement of Specialized Functions and Viral Products:
	For example, hemagglutination, and hemadsorption tests (myxoviruses), inhibition of cell transformation (EBV), immunological tests detecting antiviral antigens in cell cultures (EBV, HIV, HSV, and CMV).
	Estimation of viral nucleic acids, uptake of radioactive isotope labeled precursors or viral genome copy numbers.
Cytotoxicity	Brine shrimp lethality, Trypan blue method, MTT assay, MTS assay, XTT assay
Hepato-protective activity	carbon tetrachloride, D-glactosamine, paracetamol, thioacetamide, monocrotaline, and aflatoxin B1 induced hepato-toxicity

Analysis of the origin of new FDA approved drugs between 1999 and 2008 suggests that phenotypic screening strategies have been more productive than target-based approaches in drug discovery. It has been shown that utilization of phenotypic assays early in the screening cascade generates hits of higher quality, as opposed to target-based screening. This is because many other factors, in addition to compound-target interactions, come into play when a compound is to be used as a drug. Examples of these are: membrane permeability, unspecific protein binding, and metabolism.

7.5.5.2   HIGH THROUGHPUT SCREENING (HTS) FOR THE DETECTION OF BIOACTIVITY IN CRUDE EXTRACTS OR FRACTIONS

High throughput screening (HTS) is the process of assaying huge numbers of crude extracts or fractions against selected targets in a relatively short span of time. To conduct bioactivity screening in a high throughput manner, validated drug targets and assays suitable for detecting the bioactivity of a compound or an extract need to be developed. In addition to this, necessary equipment, like microtiter plates and laboratory automation techniques, are needed to make HTS executable [190]. In HTS of crude extracts or fractions, the assays need to detect desired bioactivity properties of constituents of complex samples. The assays are designed to possess a high efficiency to deliver rapid results at relatively low cost. In addition, they should be convenient, reliable, and sensitive and require little material. HTS is typically performed at a single concentration, and a positive hit is followed by additional testing to confirm the potency and target/ phenotypic specificity. Additional testing ensures elimination of false positives caused by nonspecific activities of constituents of the assayed crude extracts or fractions.

Development of the HTS technology started in 1950s during the screening for bioactivity in samples from microorganisms [54]. During 1980s and 1990s, HTS analysis mainly evolved around screening for bioactivity in small molecule libraries generated from combinatorial chemistry. This failed to increase the output of new pharmaceuticals; however, later in the beginning of the 21st century, HTS of crude extracts or fractions have regained much popularity. As the number of available targets increased the possibility of testing "old" crude extracts, fractions, or isolated compounds for new activities opened-up.

In the systematic attempt to identify bioactive crude extracts or fractions, selected biochemical and/or phenotypical targets are assayed in an HTS manner as part of the bioassay-guided isolation process. The bioassays constituting an HTS program are chosen based on the research area of interest for the individual bioprospecting laboratories. An HTS program may consist of bioassays devoted to detecting bioactivity within one area of interest, for example anticancer agents. For this purpose, cell-based anticancer assays, as well as kinase and caspase inhibition assays may be used. It may also consist of bioassays for detecting a variety of activities towards a range of diseases or interest areas.

7.5.6 ISOLATION

In natural product drug discovery, isolation of the active pure compound is an essential step in the identification of a new chemical entity. The isolation of secondary metabolites from a crude extract is known to be one of the bottlenecks in natural product drug discovery. A purified compound allows for chemical characterization, confirmation, and further evaluation of its bioactivity. Isolation can be rather easy and rapid when the desired compound is present as the major metabolite in the extract. However, it is typically not the case, as the target compounds often exist in trace quantities in a matrix of dozens of other constituents.

If the chemical property of the active compound is not known, then the process of isolation becomes complicated making it impossible to design the isolation procedure for a compound in many cases. Besides, the reactive functional groups of the active principles can readily undergo reaction. The combination of classical techniques (such as thin-layer chromatography, HPLC, Column chromatography, etc.) will help in the partial purification and identification of the active components. Analysis of samples by UV-visible spectroscopy, IR spectroscopy, mass spectrometry (MS), and nuclear magnetic resonance spectroscopy (NMR) often enables the unambiguous structural elucidation of pure compounds.

7.5.6.1 THIN LAYER CHROMATOGRAPHY (TLC)

Analytical TLC is used in natural product isolation to identify the degree of polarity of various chemical components. TLC is also widely used for the detection of unknown compounds; and assists in measuring the degree of purity of isolated compounds. Spraying reagents will specifically react with certain classes of compounds on TLC plates to impart color changes, which can be detected easily. Combining TLC with bioassay (bio-autography) is an effective method to get information on the active component within the extract, especially in the case of antimicrobial agents [109]. Following are most commonly used spraying reagents for marine natural products [40]:

1. Alkaloids:

   - The developed TLC plates were sprayed with 37% formaldehyde in concentrated H_2SO_4 (1:10) immediately after removing from the chamber. No heating was required. The occurrence of various colored spots indicates the presence of alkaloids.
   - Dragendorff's reagent: 40% aqueous solution of potassium iodide (10 ml) was added to 10 ml of the solution containing 0.85 g of basic bismuth sub-nitrate in acetic acid (10 ml) and 50 ml of distilled water. The resulting solution was diluted with acetic acid and water in the ratio 1:2:10. The plates were heated if the reaction is not spontaneous, a dark orange to red coloration indicates the presence of alkaloids.

2. Phenolics:

   - The developed TLC plate was sprayed with 5% ferric chloride in 0.5 N HCl. No heat was required or gently heat.

3. Flavonoids:

   - Spray the developed TLC plate with 10% solution of antimony (III) chloride in chloroform. Fluorescing spots at 360 nm.

- Spray TLC plate with 1% aluminum chloride. Yellow fluorescence spots at 360 nm.

4. Steroids and Terpenes:

- **Vanillin / sulfuric acid**: The spray reagent was made by dissolving 4 g of vanillin in concentrated H_2SO_4 (100 ml). The plates on subsequent development were sprayed in fume cupboard and heated at 100°C until the color appears. It is a universal spray reagent for terpenes; and many of them give red and blue colors.
- **Perchloric acid:** A 20% (w/v) aqueous perchloric acid solution was prepared and sprayed it on the developed TLC plate using a spraying apparatus. The plates were heated at 100°C until the coloration appears. Violet and pink coloration indicate the presence of steroids and terpenes.

7.5.6.2 HIGH-PERFORMANCE LIQUID CHROMATOGRAPHY (HPLC)

Several chromatographic techniques can be utilized for the separation and purification of biologically active molecules from complex mixtures. When the natural product is a known compound and if the standards are available, then HPLC may be used as a method of purification to get a high degree of the outcome. HPLC is the most versatile and robust technique for secondary metabolite isolation and offers high resolving power and can be scaled up as well as automated.

Among the available chromatographic techniques, preparative HPLC has emerged as the method of choice for secondary metabolite isolation. The term "preparative" refers to a chromatographic analysis, where the objective is to collect a valuable product after it is separated from the other sample constituents. Preparative HPLC is the most convenient tool for separating mixtures of the crude extract [122]. Other chromatography detectors include: flame ionization detector (FID), flame photometric detector (FPD) and nitrogen phosphorus detector (NPD), etc.

7.5.6.2.1 UV Diode Array Detector (DAD)

HPLC and UV diode arrays have become indispensable components of the natural product separation process. Today most natural product experts analyze hundreds of purified natural products using HPLC-UV. With UV photodiode array detectors, we can collect UV-absorbance data simultaneously at different wavelengths and thus facilitate purity assessment peaks. Most modern DADs contain UV spectral libraries of previously reported molecules. The operating software assists in the generation of spectral library allowing rapid identification of known biocompounds [280].

7.5.6.2.2   Evaporative Light-Scattering Detector (ELSD)

ELSD has the advantage of detecting compounds with weak UV absorption. Here, the HPLC effluent is allowed to nebulize and then vaporize in a heated drift tube. It results in a cloud of analyte particles which then pass through a beam of light. The analyte particles disperse the light eliciting a signal.

A wide range of preparative HPLC columns are available. The surface modification of the column packing material determines the kind of interactions between the sample analytes and the stationary phase. For the isolation of secondary metabolites, RP columns are most frequently utilized, because most drug-like compounds can be purified using RP-HPLC [36]. Among the available RP column packing material surface modifications, octadecyl (C18), bonded silica is most widely used. In addition, a wide range of other RP column packing material surface modifications exists like phenyl-hexyl, fluorophenyl, and dihydroxypropane [35]. The isolation process is often initiated by trial and error, where various HPLC columns and elution gradients are tested for their ability to separate the desired compound from the rest of the sample matrix.

If two compounds show different retention time values (t_R (HPLC)) on similar chromatographic systems, they are never the same and can be regarded as two different compounds. If more than one peak is detected, then the compounds are entirely different. However, in rare cases different compounds show the same retention time; in such cases, additional identification methods should be adopted to verify the result.

7.5.7   STRUCTURAL ELUCIDATION

Elucidating the structure of secondary metabolites often involves the accumulation of data from numerous sources. A wide range of spectroscopic instrumentation (such as FT-IR spectroscopy, NMR, and Mass Spectrophotometry) currently form the backbone of modern structural analysis. Usually, more than one analysis is necessary for accurate determination of the compound structure.

7.5.7.1   MASS SPECTROMETRY (MS)

Mass spectrophotometry (MS) is a sensitive and extremely accurate method for analyzing molecules. The technique is highly sensitive and needs only a microgram amount of the compound for detection. The main disadvantage regarding MS is that for ionization of any unknown compound a universal ionization type is lacking which restricts generalization of results. Fortunately, for many of the marine natural products, various ionization techniques have been introduced, such as electrospray ionization (ESI) for polar extracts [333].

MS is only useful for characterizing molecules that can be ionized to a positive or negative charged state, and most secondary metabolites can be ionized by means of variety of these techniques. For replication and structural elucidation purpose, liquid chromatography coupled with MS (LC-MS) is the most convenient technique for identification of molecular ions of compounds. Further, it is possible to find out the molecular weight of compounds to the closest atomic mass unit (amu) [255]. Modern MS comes with database that has molecular structural information of previously detected compounds. Nowadays more researchers rely on the LC-MS-MS, which has the advantage of a second round of fragmentation of certain molecular ions, which were already separated [242].

7.5.7.2   NMR SPECTROSCOPY

Nuclear NMR is the study of the interaction between electromagnetic radiation and matter. NMR spectroscopy utilizes the physical phenomenon, where a magnetic nucleus in a fixed external magnetic field absorbs and re-emits measurable electromagnetic radiation. The two most commonly examined nuclei are 1H and 13C. A parallel alignment of the previously randomly-oriented nuclei will occur when they are subjected to an external magnetic field. The nuclei will align either with it or against the magnetic field, with the latter being the alignment requiring least energy. The difference in energy between the two spin states will increase with increased strength of the external magnetic field.

NMR spectroscopic analysis is a useful technique for the structural elucidation of pure compounds. In case of unknown compounds, NMR can either be used to match against spectral libraries or to infer the basic structure directly. When the nuclei are exposed to electromagnetic radiation with a frequency matching its Larmor frequency, nucleus transitions occur from lower-energy spin state to higher energy spin state. When the radio frequency is switched off, the nuclei relax back to the lower energy state by re-emitting the absorbed energy. This emitted energy is of a particular resonance frequency, which depends upon the magnetic field and the magnetic properties of the isotope and produces a measurable radio frequency signal accordingly. This signal, called the resonance frequency, aids to recover structural details of the analyzed molecule. It depends on its atomic properties, such as nature of nucleus, its hybridization state, and the electronic atmosphere surrounding the nucleus (bonds, conjugation network, etc.) [219]. The transmitted "fro" frequencies are processed to NMR spectrum. In NMR spectrum, each atom is represented with a peak noted by a chemical shift.

More than 1000 different NMR experiments have been developed to provide spectra to deliver information about the examined nuclei and can either be one-dimensional (1D) or two-dimensional (2D) [179]. The 1D experiment is spectroscopic analysis of a single nucleus. The most commonly used NMR experiment is 1H-NMR, providing information about chemical shifts, multiple structures, homonuclear coupling constants and integrations of all protons present in the sample.

When the 13C nucleus is examined using a 1D-NMR experiment, each peak in the resulting NMR spectrum identifies a carbon atom in a different environment within the molecule. The phenomenon of spin-spin coupling explains that the nuclei as they resonate behave like small magnets, thereby influencing each other and alter the energy of nearby nuclei to form the basis for 2D-NMR experiments [219].

The most common 2D-NMR experiment for structure elucidation includes nuclear over Hauser enhancement spectroscopy (NOESY), correlated spectroscopy (COSY), hetero-nuclear multiple quantum correlation (HMQC) and heteronuclear multiple bond correlation (HMBC) [155]. All the different 2D-NMR experiments are designed to ascertain a different type of physical information about the molecule being studied.

NMR spectroscopy coupled with HPLC systems has radically increased the detection of trace amounts of analytes. LC-MS data is insufficient for the identification of a compound like isomers having same molecular weights. This technique is very successful in identifying the alkaloid aaptamine [31].

7.5.8   NEW TECHNOLOGIES IN SEPARATION SCIENCE

Recently, many methods have been put forward creating vast and diversified natural products libraries for HTS [2] to avoid the time-consuming procedures for isolation. One method is the creating a broad spectrum of semi-purified fractions having the advantage of increased reliability of results from biological testing and subsequent reduction in workload for detection and dereplication. Another approach is the construction of a library of pure natural molecules. Two main technologies include 8X parallel HPLC and SEPBOX: HPLC-SPE-HPLC-SPE arrangements.

Using these eight complex, extract mixtures can be simultaneously fractionated. It is capable of separating hundreds of natural product extracts a day. A series of HPLC columns are used for sample separation with one column for each sample. Most samples can be completely dissolved in dimethyl sulfoxide (DMSO) [296]. Samples dissolved in DMSO are inoculated in the module with the help of an autosampler [100]. As the gradient polarity is increased, the extract components will get adsorbed onto the SPE column.

7.6   MARINE NATURAL PRODUCTS AND THEIR POTENTIAL APPLICATIONS IN MEDICINE

During 2009 to 2013, there has been 20% increase in the number of identified marine natural biocompounds. Since 2009, more than 4400 new marine natural products have been identified [26 to 30]. Out of these, 4% (188) of them are derived from deep-sea (50 to 5000 m) fauna, including Bryozoans, Coelenterates, Echinodermata, Molluscans, sponges, and microorganisms.

Research in the field of marine natural products has resulted in the discovery of 20,000 new substances, including compounds with high bioactivity. The success rate of discovery of natural products from the marine ecosystem (record 7 clinically approved drugs out of 28,175 molecular entities discovered) is approximately 1.2 to 2.5 folds greater than the average (1 in 5000–10,000) compounds tested [19, 94, 189, 194, 224, 261, 284] marine natural products (MNP). Examples for some of the marine natural products currently in market are shown in Figure 7.1.

During the early 1950s, the discovery and identification of the two nucleosides (spongothymidine and spongouridine) from the Caribbean marine sponge (*Cryptotethia crypta*) made the marine natural products as promising candidates with a high therapeutic value [184]. A few noteworthy findings of compounds and products pertaining to marine natural products towards the development of possible drug candidates for the future are listed in Table 7.4.

Yondelis® (Trabectedin)

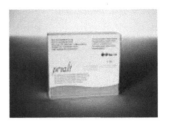

Prialt® (Zinconotide)

Halaven® (Eribulin mesylate)

Adcetris ® (Brentuximab Vedotin)

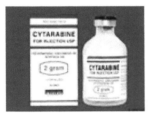

Cytarabine ® (Ara- C)

Viroptic ® (Vidarabine)

FIGURE 7.1   Pharmaceutical drugs in commercial use from marine natural products. Source compounds are given in brackets.

7.6.1   MARINE NATURAL PRODUCTS AND CANCER

Cancer is one of the major public health issues [313]. Hence, it is the need of the hour to identify new compounds with anticancer activity from plant and marine sources. A global initiative has led to the discovery of new and effective anti-cancer agents derived from marine organisms, due to extensive research findings. A significant rise has been seen in the number of preclinical marine anticancer compounds that have entered human clinical trials since early 1990s [185, 204].

In recent years, more, and more researchers are involved in the research on marine organisms with potential as a source of novel molecules and new anticancer agents [306]. According to recent reports, more than 10 novel marine anti-tumor agents have entered clinical trials [330], such as

TABLE 7.4 List of Some Marine Natural Products for Treatment of Chronic Diseases

Therapeutic Area	Compound	Biological Source	Chemical Nature	Mechanism of Action
Cancer	Dolastatin-10	Cyanobacteria; *Symploca hydnoides*	Peptide	Inhibits microtubule formation
	Didemnin B	Tunicate; *Trididemnum solidum*	Depsipeptide	Immune modulatory
	Bryostatin-1	Bryozoan; *Bugula neritina*	Polyketide	Inhibits a protein kinase
	HTI-286	Sponge; *Cymbastella* sp.	Tripeptide	Inhibits microtubule formation.
	Discodermolid	Sponge; *Discodermia dissoluta*	Tripeptide	
	Cryptophycin	Cyanobacterium; *Nostoc* sp.	Cyclic depsipeptide	
	Aplidine	Ascidium; *Aplidium albicans*	Cyclic cepsipeptide	Causes oxidative stress in cells
	Eribulin mesylate	Sponge; *Halichondria okadai*	Polyester derivative	Inhibits microtubule formation
	Squalamine	Shark; *Squalus acanthias*	Steroid	Inhibits angiogenesis
	Kahalalide F	Mollusk; *Elysia rufescans*	Cyclic cepsipeptide	Lysosome-tropic effect
	Ecteinascidin 743	Tunicate; *Ecteinacidia turbinate*	Alkaloid	Non p-53 mediated apoptosis
	Halichondrin	Sponge; *Halichondria ckadai*	Polyether macrolide	Tubulin depolymerization
HIV	Avarol	Sponge; *Dysidea avara*	Sesquiterpene hydroquinone	UAG suppressor glutamine transfer tRNA
	Clathsterol	Sponge; *Clathria* sp	Depsipeptide	Inhibit HIV-1 RTase
	Neamphamide	Sponge; *Neamphius huxleyi*	Depsipeptide	Inhibit cytopathic effect of HIV-1
	Lamellarins	Ascidin; *Didemnid ascidians*	alkaloid	Inhibits leukemia cancer cells
	Tachyplesin I & II	Horse shoe crab; *Tachypleus tridentatus*	Peptide	HIV cell fusion inhibitors

TABLE 7.4 *(Continued)*

Therapeutic Area	Compound	Biological Source	Chemical Nature	Mechanism of Action
Inflammation	Ascidiathiazone	Ascidian; *Aplidium* sp.	Alkaloids	Anti-inflammatory action in human neutrophils
	Manzamine	*Sponge* sp; *Acanthostrongylophora* sp.	Alkaloids	Inhibitors of thromboxane B2
	Cembranolides	Soft coral (*Lobophytum crassum*)	Cembranoids	Inhibitors of COX-2
	Plakortide P	Sponge; *Plakortis angulospiculatus*)	Polyketide	Anti-neuroinflammatory
	Manoalide	Sponge; *Luffariella variabilis*	Sesterterpene	Inhibition of Phospholipase A_2
	Pseudopterosin	*Pseudopterogonia elisabathe*	Diterpene glycoside	Inhibit PMN-PLA_2
	Fucoside	Gorgonian; *Eunicea fusca*	Glycoside of fucose	Inhibits PMA
Atherosclerosis	Eryloside F	Sponge; *Erylus formosus*	Penasterol	Potent thrombin receptor antagonist
	Halichlorine	Sponge; *Halichondria okadai*	Alkaloid	Reduces monocyte adhesion
Pain	Zinconotide (Prialt)	Cone snail; *Conus magus*	Peptide	Block Ca^{2+} channels
	ω conotoxin	Cone snail; *Conus magus*	Peptide	Block Ca^{2+} channels
	CGX-1007	Cone snail; *Conus Geographus*	Peptide	NMDA receptor antagonist

bryostatin-1, aplidine, ecteinascidin-743 (ET-743), Kahalalide F as well as derivatives of dolastatin (such as TZT-1027 and LU 103793). A synthetic analog of C-nucleoside (cytarabine or Ara-C) from the Caribbean sponge (*Cryptothethya crypta*) was the first marine derived anticancer agent developed for clinical use. It was approved in 1969 and currently in use for the treatment of acute myelocytic leukemia and non-Hodgkin's lymphoma [250, 258]. A third discovery from marine source was halichondrin A (a polyether metabolite from the sponge, *Halichondria okadai*) [119].

Very recently, a marine anticancer compound, brentuximab vedotin (Adcetris), was approved for human use. It is a chimeric antibody, attached to a derivative of the potent anti-tubulin agent dolastatin 10, through a protease-cleavable linker [145]. In addition, there are many biocompounds, such as, discodermolide, eleutherobin, and sarcodictyin A, didemnin B, dehydrodidemnin B, bryostatin 1, dolastatin 10, ectcinascidin 743, halichondrin B, isohomohalichondin B, curacin A, girolline, jaspamide, and thiocoraline, which were discovered using traditional screening methods.

Dolastatin 10is one of the early discovered antitumor compound isolated (very low yield, 10.6 to 10.7%) during an expedition to the Mauritius Island in 1972 by Pettit. It was isolated from *Dolabella auricularia* (marine nudibranch) [223]. In the case of the dolastatins, studies have proven that the peptides were produced by the cyanobacterium *Lyngbya majuscula* or its epiphytes [112] and have higher yields than were found in the sea. In 1989, researchers made a synthetic Dolastatin 10, which was found to be exceptionally toxic to tumor cells, having an IC_{50} of 4.5 × 10^{-5} µg/ml against lymphocytic leukemia P388. TZT-1027 (soblidotine), asynthetic derivative of dolastatin (dolaphenine amino acid was replaced with the phenylalanine group), inhibited tubulin polymerization and angiogenesis similar to dolastatin [259].

Didemnin B is perhaps the most studied of marine cyclic peptides isolated from the Caribbean tunicate (*Trididemnum solidum*) [243] and has shown anti-viral and immunosuppressive activities and is an effective agent for the treatment of leukemia and melanoma. It was developed by NCI (US National Cancer Institute) and went through phase II clinical trials but was withdrawn. Later, an analog of didemnin, named aplidin (Dehydrodidemnin B), was isolated from the Mediterranean tunicate (*Aplidium albicans*), with superior anticancer properties than its predecessor, and was found to initiate oxidative stress leading to apoptosis in tumor cells [251]. Dehydrodidemnin B, which can also be prepared by oxidation of didemnin B or by total synthesis, is being developed by Pharma Mar S. A. in Europe and has been scheduled for phase I clinical trials [194].

Another anti-cancer metabolite, Bryostatin 1, a 26-numbered macrocyclic lactone was isolated [225]. Bryostatin 1 is a partial agonist of protein kinase C and showed potent activity on human tumor xenografts *in vivo*. It has been developed predominantly by the NCI for the treatment of melanoma, non-Hodgkins lymphoma, and renal cancer and is currently in Phase II clinical trials in the United States [117].

Wender et al. [310] prepared a simplified analog of bryostatin. Phase II clinical trial on patients with recurrent ovarian cancer treated with a combination of bryostatin-1 and cisplatin (cDDP) showed moderate results; however, this combination was excluded from a further investigation [197] due to severe myalgia. It is the most potent of the halichondrins against P-388 leukemia, B-16 melanoma and L-1210 leukemia *in vivo*, and is in preclinical trials at the NCI. A new source of halichondrins from the deep-water sponge (*Lissodendoryx* sp.) was discovered by some New Zealand scientists. Later, an effort was made to cultivate *Lissodendoryx* in shallow waters of New Zealand; however, the amount of active component was much lower than in the wild species [201] to attain 310 mg of halichondrin [56, 194, 241, 278].

Squalamine was also found useful in treating diseases of aged-people and also impairments related to vision (macular degeneration) [53, 101]. Kahalalide F was discovered along with several new highly active depsipeptides from the mollusk (*Elysia rufescens*) by Scheuer of Hawai University, USA [120]. The actual sources of kahalalide were the algae *Bryopsis* sp, which this mollusk feeds on. The structure and relative stereochemistry of kahalalide-F was corrected after the solid-phase synthesis [176] followed by preclinical and clinical trials by the PharmaMar company.

Kahalalide seems to induce vacuole formation in tumor cells and stimulates lysosomes and was found to be several times toxic to tumor cells than against healthy cells [230, 233, 252, 264, 283]. Curacin A, which was isolated from the cyanobacterium (*Lyngbya majuscula*) from Curacao and can inhibit microtubule assembly by binding at the colchicine site [95]; and it is a good example with anti-tumor activity *in vitro*, but has yet to be successfully formulated for use *in vivo*.

Marine corals with Eleutherobin from a small Australian soft coral of the genus *Eleutherobia* sp. [173] and sarcodictyin A from the Mediterranean stoloniferan coral (*Sarcodictyon roseum*) [61] are good sources of anticancer drugs. They are closely related diterpenoids that mimic Taxol by stabilizing microtubules. Eleutherobin showed selective cytotoxicity toward breast, renal, ovarian, and lung cancer cell lines and is generally more inhibitory than sarcodictyin A. Despite some difficulties in the beginning, soon after the reporting of its molecular structure and issuing of patent, these two biocompounds were successfully synthesized [48, 206, 207]. Eleutherobin underwent some preclinical trials at Bristol-Myers Squibb but is no longer being pursued, due to the difficulty in obtaining sufficient material.

Two alkaloids (which were isolated from ascidians), polycarpin [236], and varacin C [180], are characterized by high toxicity against tumor cells. The Varacin C is selectively cytotoxic to tumor cells than normal cells, because of its activity in the acidic environment, and the activity was higher than that of well-known doxorubicin [165]. Another alkaloid from a prosobranch mollusk, *Lamellaria* named as Lamellarin D (LAM-D), showed cytotoxicity against various tumors [16, 263, 292]. Chitosan is produced commercially by deacetylation of chitin, which is the major component of exoskeleton of crustaceans [168].

Agelasines was first reported by Nakamura et al. and it was isolated from marine sponges [203]. In 2011, two analogs of Agelasine (2F and 2G) were reported to have high cytotoxicity [247, 257]. In 2012, Pimentel et al. purified agelasine B from the marine sponge *Agelas clathrodes* and reported its cytotoxicity and probable mechanism of action [226]. This compound was comparatively more toxic to cancer cells (IC_{50} value for MCF-7 = 3.22 μM, SKBr3 = 2.99 μM and PC-3 = 6.86 μM) than the normal cells (IC_{50} value for fibroblasts = 32.91 μM).

According to the recent data, around 87 molecules (62%) were reported in 2012, along with 53 known biocompounds with promising results as anti-cancer drugs. Neamphamides and the derivatives of kulokekahilide-2 (5 and 5a) are other promising cytotoxic compounds against wide range of cancer cells with significant anti-cancer activity at nanomolar concentrations [306].

7.6.2 MARINE NATURAL PRODUCTS AS ANTI-INFLAMMATORY AGENTS

Inflammation is a pathophysiological response of mammalian tissues to a variety of hostile agents (including infectious organisms, toxic chemical substances, physical injury, or tumor growth leading to local accumulation of plasma and blood cells) [276]. Inflammation is characterized by redness, and sometimes loss of function. When persistent inflammation on immune system gets activated, it can lead to chronic diseases and to the condition known as chronic inflammation [319]. However, if the reaction gets elaborated and misdirected, sustains for a longer period, then it can adversely affect the health leading to many other disorders [177, 192].

Phytochemicals derived from marine organisms have provided several lead biocompounds [60, 334], which are anti-inflammatory. The marine metabolites (such as pacifenol, stypotriol triacetate, and epitaondiol) have been tested for their effects on number of inflammatory responses [97]. Biocompounds, such as, manoalide, pseudopterosins, topsentins (e.g., debromohymenialdisine (DBH)), have been investigated by both SmithKline Beecham and Osteoarthritis Sciences Inc. for the treatment of rheumatoid arthritis and osteoarthritis, respectively. Manoalide, which was isolated from the Palauan sponge *Luffariella variabilis* [63], irreversibly inhibits the release of arachidonic acid (AA) from membrane phospholipids by the enzyme phospholipase A2 (PLA2), thus inhibiting inflammation [98, 99]. The University of California patented Manoalide and licensed to Allergan Pharmaceuticals, for Phase I clinical trials for the treatment of psoriasis. But to date, no drug that is apparently based on manoalide has entered the market. However, manoalide is used as a standard drug for PLA2 inhibition and is commercially available [84].

Mayer et al. [183] isolated group of diterpene glycosides (pseudo-pterosins from the gorgonian (sea whip)) and pseudo-pterogorgia (*elisabethae*). These compounds were effective in inhibiting PMA induced mouse ear edema. Pseudopterosin A, the most active of the first group of pseudopterosins so far studied [175], possessed potent anti-inflammatory and analgesic activities. Further research [248] revealed that pseudopterosin E and a related compound were superior anti-inflammatory agents than pseudopterosin A and was nontoxic even at 300 mg/kg in mice. Ata et al. [12] isolated and identified hydroxyquinone, elisabethadione, and pseudopterosins and seco-pseudopterosins from marine gorgonian (*Pseudopterogorgia elisabethae*). Anti-inflammatory assays indicated that elisabethadione was more potent than the pseudopterosin A and E.

DBH was first described as a metabolite from *Phakellia flabellata* [265] in the Great Barrier Reef and was later rediscovered as a metabolite of Okinawan marine sponge (*Hymeniacidon aldis*) [152]. It is also one of several constituents of common Palauan shallow-water sponge (*Stylotella aurantia*) [216, 315]. DBH has been patented as a Protein Kinase C inhibitor by SmithKline Beecham for the treatment of osteoarthritis by University of California and Osteoarthritis Sciences Inc. [83].

Fuscoside, an anti-inflammatory marine natural product isolated from the Caribbean gorgonian (*Eunicea fusca*) [132], significantly inhibited PMA (phorbol myristate acetate) induced edema inmouse ears at levels comparable to indomethacin. Scytonemin is a sheath pigment from cyanobacteria [232] that has been patented as an anti-inflammatory agent. In some systemic assays, it was effective than topsentin although its mechanism of action is not still known. It is one of the most potent inhibitors of neurogenic inflammation and can be used for treating burns.

Several Indole alkaloids from marine invertebrates have anti-inflammatory potential. Cembranoids and crassumolides A and C isolated from soft coral (*Lobophytum crissum*) repressed the expression of iNOS and COX-2 (IC_{50}> 10 μM) [47, 49]. El Sayed et al. in 2008 reported the inhibition of thromboxane (TXB2) generation in brain microglia [154].

Several authors have reported the anti-inflammatory activity of omega-3 polyunsaturated fatty acids (ω-3 PUFA) from New Zealand green-lipped mussel. Halpern [108] evaluated anti-inflammatory activity of a lipid-rich extract of freeze-dried mussel powder (Lyprinol) isolated from a supercritical-CO_2 lipid extract of the tartaric acid-stabilized freeze-dried mussel powder. The study showed significant anti-inflammatory activity in animals and humans. In contrast to NSAIDs, Lyprinol was non-gastro toxic and does not seem to affect platelet aggregation (human and rat). Treschow et al. [298] reported omega-3 polyunsaturated fatty acids (omega-3 PUFA) purified in the same way as above with significant *in vitro* anti-inflammatory (AI) activity. Related studies were also carried out by Singh et al. [271] in rat adjuvant arthritis model and proved significant anti-inflammatory activity of CO_2-SFE crude lipid extract and its FFA (free fatty acid) components with no adverse side effects.

During 2009–2011, the preclinical pharmacological studies were reported on several marine natural products involving their molecular mechanism of action. Steroid callysterol acting on pro-inflammatory mediators produced by activated brain microglia, from the Red Sea sponge (*Callyspongia siphonella*), was reported by Youssef and his coworkers [326]. Callysterol effectively suppressed rat hind paw edema and decreased release of TXB2 [326]. Villa et al. reported the inhibition of nitric oxide production in LPS-primed RAW 264.7 macrophage cells by malyngamide Facetate isolated from marine cyanobacterium (*Lyngbya majuscula*) with IC_{50} = 7.1 μM [304].

Kim and his colleagues [151] showed inhibition of nitric oxide and PGE2 generation by downregulation [151]. Anti-inflammatory properties of dermatan sulfate, isolated from the Brazilian ascidian (*Styela plicata*) were investigated on rat colitis model. The compound effectively reduced the infiltration of lymphocytes and macrophages [18].

7.6.3   MARINE NATURAL PRODUCTS AS ANALGESIC AGENTS

Ziconotide was the first analgesic drug of marine origin to obtain approval from the U.S Food and Drug Administration (USFDA) to treat the pain. The analgesic property of ziconotide is found to have role in blocking of N-type calcium channels on the primary nociceptive nerves of the spinal cord in animal studies [274]. At the end of 2004 after twenty years of research on the toxins from predatory molluscan gastropods of genus *Conus,* the toxin got approval for making and clinical use in the USA. After few months, its production was started in Europe under the commercial name "prialt." Elan Pharmaceuticals launched Prialt® (Ziconotide) as a remedy for chronic pain due to its noteworthy anti-nociceptive action even in patients not responsive to morphine.

Ziconotide is a peptide consisting of 25-amino acid derived from the ω-conotoxin, a toxin from cone snail (*Conus magus*) found in tropical sea, [187]. The snails produce conotoxin for fish hunting to immobilize the prey by acting on the neuromuscular system [213]. In patients, ω-conotoxin acts by blocking the N-type voltage-sensitive calcium channels and inhibiting the neurotransmitters release thereby interrupting the nerve signal conduction leading to pain relief [212]. Since the large-scale production is limited due to the lack of enough biological specimens; and ziconotide was produced by peptide synthesis [5, 148].

Conotoxin MVIIA, being a natural toxin, was of top priority as a prospective drug compared to their synthetic derivatives. The ω-conotoxin is a linear peptide consisting of 25 amino acid residues with six-cysteine residues forming three disulfide bridges [52, 214]. Neurex (branch Elan Pharmaceuticals) company carried out clinical investigations of synthetic ω-conotoxin [307]. These investigations showed that ziconotide does not have hallucinogenic effects and does not cause addiction, as in morphine [188]. Many conotoxins are still in different phases of investigation as analgesic drugs [114]. Tetrodotoxin (TTX) 24 (a guanidine alkaloid isolated from fish, algae, and bacteria) was found to block the voltage-dependent sodium channels [325] and has shown analgesic property in cancer patients [123, 125].

7.6.4   MARINE NATURAL PRODUCTS AGAINST CARDIO VASCULAR DISEASES

The risk factors related to CVD and atherosclerosis are the quantity and type of fatty acids (FAs) in our daily diet [177]. According to the World Health Organization (WHO), CVD, and stroke will remain the leading causes of death and about 23.6 million people will die due to CVD in 2030 [316].

Significant discoveries have resulted in drugs from marine organisms against CVD [279]. Activation of thrombin receptor plays a key role in arterial thrombosis and atherosclerosis [45]. Atherosclerosis begins with damage of endothelium and ensuing deposition of fats, cholesterol, platelets, cellular waste products, calcium, and other substances on the arterial wall [272]. Studies show that halichlorine from sponge (*Halichondria okadai*) inhibits the vascular cell adhesion molecule 1 expression [162]. Two polysaccharides isolated from sea cucumber (a fucosylated chondroitin sulfate and a sulfated fucan) showed anti-thrombic activity [143, 199].

7.6.5   MARINE NATURAL PRODUCTS AGAINST HYPERTENSION

Hypertension or high blood pressure is a chronic medical condition resulting in elevated blood pressure in the arteries [72]. Thus, ACE inhibitors can be used for the treatment of hypertension; and there is a need for increased demand for natural or food-derived inhibitors [303] due to the adverse side effects [13]. The peptides showed IC_{50} values 3.09 and 4.22 μM, respectively for the anti-hypertensive activities [205]. An anti-angiotensin I converting enzyme (ACE) peptide (Ala-His-Ile-Ile-Ile, with MW: 565.3Da) was isolated from *Styela clava*. The induction of vasorelaxation in the rat aortas was observed with the isolated fractions and the peptide [153].

However, due to complex nature of constituents and the amino acid in seaweed hydrolysates, sequencing of bioactive peptides in ACE-active hydrolysates is rarely determined. Antihypertensive properties of peptides from macroalgae have also been shown in humans. Diet with a tridecapeptide obtained from *Palmaria palmata* after papain digestion has shown to be active against renin [92] and could potentially reduce the systolic blood pressure of about 33 mmHg in spontaneously-hypertensive rats [91]. In addition, many di- and tetrapeptides from *Undaria pinnafita* have been characterized by Suetsuna and coworkers [285, 286]. Among them, diet supplementation with dipetides and the tetrapeptides significantly lowered the blood pressure in spontaneously hypertensive rats [285].

7.6.6   MARINE NATURAL PRODUCTS AGAINST HIV

By early 2003, screening of drugs against HIV from marine sources established more than 150 highly active marine metabolites [105, 299]. Peptides from some marine invertebrates are especially good sources of antiviral substances. Peptides are composed of 17–18 amino-acid residues [193, 198]. Four new cyclic depsipeptides (termed mirabamides A, B, C, D) have been isolated from the sponge *Siliquariaspongia mirabilis* with antiviral properties [227]. Anti-HIV activity of Mirabamide A, C, and D was performed using neutralization and fusion assays. The IC_{50} values of Mirabamide A was between 40–140 nM, and for Mirabamide C, IC_{50} value ranged between 140 nM-1.3 µM and 190 nM and 3.9 µM for Mirabamide D, showing that these peptides can prevent HIV-entry in the early stages [238]. Another cyclodepsipeptide homophymine A, isolated from the marine sponge *Homophymia* sp., exhibited cyto-protective activity against HIV infestation (IC_{50} of 75 nM) with MTT assay [327].

It was found to inhibit protein tyrosine phosphatase 1B. Another new C22 furano-terpene (dehydrofurodendin) isolated from the sponge (*Madagascan Lendenfeldia*) was found to be active against HIV-1 RT with an IC_{50} value of 3.2–5.6 μM [50]. Bioassay-guided fractionation of extracts of the Palauan ascidian (*Didemnum guttatum*) led to the isolation of a sulfated serinolipid, cyclodidemniserinol trisulfate, as an inhibitor of HIV-1 integrase [191].

Metabolites from sponges (such as avarol, avarone, ilimaquinone, and several phloroglucinols) exhibited anti-HIV activity [299, 324]. Avarol inhibited HIV almost completely by blocking the synthesis of the natural UAG suppressor glutamine transfer tRNA, synthesis of which is upregulated after viral infection and is important for the synthesis of a viral protease, which is necessary for viral proliferation [200]. A novel and active sulfated sterol, Clathsterol from the sponge *Clathria* sp., inhibited HIV-1 RT at a concentration of 10 μM [249].

Lamellarin, an alkaloid, showed anti-HIV activity by inhibiting the activity of strands transfer and integrase terminal cleavage [239]. Lamellarin was earlier purified from the mollusk Lamellaria and later was also found in *Didemnid ascidians* [6]. Neamphamide A, a new HIV-inhibitory depsipeptide from the Papua New Guinea marine sponge *Neamphius huxleyi,* repressed cytopathic effect (CPE) of HIV-1 infection [210]. Fan and his coworkers in 2010 reported some sulphated alkaloids as anti-HIV agents, named as baculiferins A-O from the Chinese marine sponge *Iotrochota baculifera* [82].

7.6.7   BIOCOMPOUNDS AGAINST OTHER INFECTIOUS AGENTS

During 2009–2011, antibacterial studies on marine natural products reported many new compounds isolated from a varied group of marine bacteria, ascidians, soft corals, algae, bryozoans, and sponges [185], and these efforts continue to contribute for more compounds in future in this field.

Many bromophenol compounds have frequently been found in algae and bacteria. The antimicrobial activity of crude extracts of *Odonthalia corymbifera* against various microorganisms was reported by Oh et al. [209]. Bioassay-guided isolation of its crude extract resulted in many bromophenol compounds; and among these, the 2,2,'3,3'-tetrabromo-4,4,'5,5'-tetrahydroxy diphenylmethane was most effective against *Aspergillus fumigatus, Candida albicans, Trichophyton rubrum* and *Trichophyton mentagrophytes*. Antimicrobial activity against Gram +ve, Gram –ve bacteria and fungi was reported for ALAA 2000 [77].

Desbois and his colleagues isolated some antimicrobial FAs from marine algae; and among them (9Z)-hexadecenoic acid and (6Z, 9Z, 12Z)-hexadecatrienoic acid from the marine diatom *Phaeodactylum tricornutum* [64], displayed anti-bacterial property against Gram (+) bacteria and also suppressed the progression of Gram (-) marine pathogen *Listonella anguillarum*.

Another potent inhibitor of gram-positive methicillin-resistant *Staphylococcus aureus* was isolated from alga *Chrysophaeum taylori*, bisdiaryl butene macrocycle chrysophaentin A, which also repressed *Enterococcus faecium* (vancomycin-resistant) [228, 282]. Marine sponges are exceptional sources of many interesting antimicrobial compounds isolated from marine environment. A review on presently available chemical data suggests that marine sesterterpenoids have prominent bioactivities, which occur frequently in marine sponges with antimicrobial and antiviral properties [75]. *Candida an albicans* inhibitory property was shown by seven sesterterpene sulphates extracted from the tropical sponge *Dysidea* sp. [166].

Puupehanol, a sesquiterpene-dihydroquinone derivative isolated from the marine sponge *Hyrtios* sp., was responsible for the antifungal activity of the sponge extract [320]. Another compound, puupehenone exhibited more potent inhibitory activity than the two compounds tested, against *Cryptococcus neoformans* and *Candida krusei* with MIC of 1.25 and 2.5 µg/ml, respectively. Zhang and his coworkers isolated bioactive meroterpene (Fascioquinols A) from *Fasciospongia* sp, a southern Australian sponge [328]. Fascioquinol A on an acid-mediated hydrolysis/cyclization process resulted in new products, such as Fascioquinols B, C, and D [67], which inhibited malaria on two enzymatic targets (plasmodial kinase Pfnek-1 and a protein farnesyltransferase) in one µM range activity [317].

Two novel sulfated sesterterpene alkaloids were isolated by Yao and Chang. Fasciospongins A and B exhibited strong inhibitory activity against *Streptomyces* 85E. Many antimicrobial phenolic compounds extracted from marine sponges have been reported recently [269]. Bioactive polybrominated diphenyl ethers were isolated from *Dysidea* species [329]. The same compounds were also reported from *Lamellodysidea herbacea,* an Indonesian sponge [111], which displayed antimicrobial activity against *Bacillus subtilis.*

Many antimicrobial nitrogen-containing heterocyclic compounds have been isolated from marine sponges, such as alkyl-piperidine, bromopyrrole, and pyrrolo iminoquinone alkaloids. An anti-tuberculosis agent, Halicyclamine A, was re-discovered from the marine sponge *Haliclona* sp. against dormant *Mycobacterium tuberculosis* [10]. The compound showed growth inhibition against many species of *Mycobacterium,* such as, *M. smegmatis*; and alkaloids (e.g., 22-hydroxyhaliclonacyclamine B and haliclonacyclamine A and B from *Haliclona* sp.) inhibited Mycobacterium *tuberculosis* [9].

From the Turkish sponge *Agelas oroides,* oroidin (a bromopyrrole alkaloid) was isolated [293]; and it inhibited enoyl reductases in *Plasmodium falciparum* and *Mycobacterium.* Ceratinadins A-, B-, C-, Bromotyrosine alkaloids possessing N-imidazolyl-quinolinone moiety having antifungal property were purified from an Okinawan marine sponge *Pseudoceratina* sp. [156].

Jang et al. isolated [133] two bicyclic bromotyrosine-derived metabolites (pseudoceratins A and B) from *Pseudoceratina* sponge species (*Pseudoceratina purpurea*); and both were active against *Candida albicans* [137]. Herpes simplex virus type 1 penetration was inhibited by 4-methylaaptamine, an alkaloid from the marine sponge *Aaptos aaptos* (EC_{50} = 2.4 µM) [277]. This sponge also yielded four aaptamines, which inhibited sortase A (an enzyme responsible for virulence and anchoring of cell wall proteins) in *Staphylococcus aureus* [134]. Bioactivity-guided fractionation of a marine sponge *Spongosorites* sp. yielded antimicrobial alkaloids in the class topsentin and hamacanthin [17]. Other anti-microbial alkaloids of sponge origin were found to have inhibitory activity against isocitrate lyase in *Candida albicans* [167].

Some peptides of sponge origin also showed antimicrobial and antiviral activity such as callyaerins -A to -F, and H from the *Callyspongia aerizusa* off Indonesian coast [130], and the antifungal bicyclic peptide-theonellamides [208]. Other sponge derived antimicrobial peptides are amino-lipopeptides [234]. Furthermore, two novel cyclic hexapeptides had antifungal and antibacterial activity [332]. The fungal species also produced some novel aspochracin-type cyclic tripeptides, named as sclerotiotides -A to -K [331].

Studies have reported many sponge derived polyketides showing potent antimicrobial and antiviral activities [174]. Orholquinone 8 significantly inhibited farnesyl transferase enzymes in both human and yeast [318]. The compound showed antimicrobial and antiviral properties and inhibited *Candida albicans* (MIC of 0.62 µg/ml). A novel acetylenic fatty acid purified from the calcareous sponge *Paragrantiawaguensis* [297] inhibited both *Staphylococcus aureus* and *Siliquaria spongia* [146].

7.7 CHALLENGES FACED BY INDUSTRIES IN DRUGS FROM MARINE NATURAL PRODUCTS

The marine natural product drug development process faces various obstacles in each step starting from the selection of organism up to the marketing of novel compounds as drugs. The serious obstacle to the ultimate development of most marine natural products is the supply problem. Development of a new drug must always include a plan to supply enough biocompounds (raw material) for the preclinical and clinical phases. It is particularly important to address supply issues especially when developing drugs from marine invertebrates or algae that are found below the intertidal zone.

Collection of marine invertebrates for screening and chemical studies is a good practice not only with respect to sustaining biological diversity but also adopting an optimal strategy for maximizing the diversity. If collectors cannot find more than 100 g of an organism, this usually indicates that the specimen is too rare and will be very difficult to re-collect in sufficient quantities for subsequent development. Moreover, the concentrations of many highly active compounds in marine invertebrates are often minute, sometimes accounting for less that 10^{-6}% of the wet weight, necessitating the collection of more amounts of raw materials. For example, nearly one metric ton (wet wt.) of the tunicate *Ecteinascidia turbinate* is needed to be harvested and extracted [94].

Adequate supply of the marine organism always poses great challenge. Often gram amounts of a compound are usually needed. Moreover, for large collections involving more than 5 kg of rare organism or 100 kg of a very common species require the cooperation and consent of the country in which the collection is made. Such collections should involve the preparation of an environmental impact report and careful monitoring of the impact of collecting; hence, it is the need of the hour to find alternative approaches for an ecologically and economically feasible source of marine natural products [84].

Mariculture of invertebrates-like sponges, tunicates, and bryozoans for getting a steady supply of compounds-has not been into consideration. However, in the case of bryozoan *Bugula neritina* (the source of the bryostatins) and Tunicate *E. turbinate,* it was made possible through mariculture, though they obtained yields of biomass was less than the required. Mariculture of invertebrates in tank or in the sea is rather unpredictive due to the loss by storms or diseases. It is hoped that future developments in the field of mariculture will make it possible for marine invertebrates to be cultured as a part of community-based conservation project in the developing nations, thereby providing an economic incentive for the restoration of coral reef environments. An attractive alternative to mariculture of sponges would be to grow sponge cell tissue culture but research in progress suggests that this approach is not easy [84].

The commercial source of choice for the pharmaceutical industry is synthesis, which allows the company to control all aspects of production. This is the best solution for relatively simple compounds, but many bioactive marine natural products are extremely complex and require multi-step synthesis for heroic proportions. If synthesis is not economically viable, then mariculture should be considered as an alternative to harvest wild specimens [140].

7.8 SUMMARY

During the past 20 years, the chemical industry for marine-based natural products has discovered more than their fair share of promising pharmaceuticals. There are many serious obstacles faced for the development of marine natural products, both in academic and industrial pipelines. There have been difficulties in supplying sufficient raw materials for the early stages of development and clinical trials. It must be acknowledged that many of the problems experienced in the past seem close to being solved. Synthetic methods are constantly improving so that even complex molecules or preferably (simpler analogs based on marine metabolites) can be synthesized on an industrial useful scale. Aquaculture of marine invertebrates is a reality and an attractive alternative to mariculture would be to grow cells in tissue culture but research in progress suggests that this approach will be very difficult to achieve. Gene transfer technology is another novel technology to solve the supply problem in the future. Further comprehensive research in this field is needed to develop alternative technologies, which are more cost-effective and produce less pollution.

KEYWORDS

- bioassay-guided fractionation
- marine bioprospecting
- marine natural products
- marine secondary metabolites
- structural elucidation
- supercritical fluid extraction

REFERENCES

1. Abrantes, J. L., Barbosa, J., Cavalcanti, D., & Pereira, R. C., (2010). The effects of the diterpenes isolated from the Brazilian brown algae *Dictyota pfaffii* and *Dictyota menstrualis* against the herpes simplex type-1 cycle. *Planta Medicinal*, pp. 339–344.

2. Adel, U., Kock, C., Speitling, M., & Hansske, F. G., (2002). Modern methods to produce natural-product libraries. *Current Opinion in Chemical Biology*, pp. 453–458.

3. Aicher, T. D., (1992). Total synthesis of halichondrin B and nor-halochondrin B. *Journal of the American Chemical Society, 114*, 3162–3164.

4. Aiello, A., Borrelli, F., & Capasso, R., (2003). Conicamin: Novel histamine antagonist from the Mediterranean tunicate *Aplidium conicum*. *Bioorganic and Medicinal Chemistry Letters, 13*(24), 4481–4483.

5. Allen, J. W., Hofer, K., McCumber, D., & Wagstaff, J. D., (2007). An assessment of the antinociceptive efficacy of intrathecal and epidural contulakin-G in rats and dogs. *Anesthesia and Analgesia, 104*, 1505–1513.

6. Andersen, R. J., Faulkner, D. J., & He, C. H., (1985). Metabolites of the marine prosobranch mollusk *Lamellaria sp. Journal of the American Chemical Society, 107*(19), 5492–5495.

7. Anderson, H. J., Coleman, J. E., Andersen, R. J., & Roberge, M., (1997). Cytotoxic peptides hemiasterlin, hemiasterlin-A and hemiasterlin-B induce mitotic arrest and abnormal spindle formation. *Cancer Chemotherapy and Pharmacology, 39*(3), 223–226.

8. Aoki, S., Ye, Y., & Higuchi, K., (2001). Novel neuronal nitric-oxidesynthase (nNOS) selective inhibitor, aplysinopsin-type indole alkaloid, from marine sponge *Hyrtios erecta*. *Chemical and Pharmaceutical Bulletin, 49*(10), 1372–1374.

9. Arai, M., Ishida, S., Setiawan, A., & Kobayashi, M., (2009). Haliclonacyclamines, tetracyclic alkyl-piperidine alkaloids, as anti-dormant mycobacterial substances from a marine sponge of *Haliclona* sp. *Chemical and Pharmaceutical Bulletin, 57*, 1136–1138.

10. Arai, M., Sobou, M., Vilcheze, C., & Baughn, A., (2008). Marine spongean alkaloid as a lead for anti-tuberculosis agent. *Bioorganic and Medicinal Chemistry, 16*, 6732–6736.

11. Asolkar, R. N., Freel, K. C., & Jensen, P. R., (2009). Cytotoxic NFκB inhibitors from the marine actinomycete *Salinispora arenicola. Journal of Natural Products, 72*, 396–402.

12. Ata, A., Kerr, R. G., Moya, C. E., & Jacobs, R. S., (2003). Identification of anti-inflammatory diterpenes from the marine gorgonian *Pseudopterogorgia elisabethae*. *Tetrahedron, 59*, 4215–4222.

13. Atkinson, A. B., & Robertson, J. I., (1979). Captopril in the treatment of clinical hypertension and cardiac failure. *Lancet, 2*, 836–839.

14. Atta-Ur, R., Choudhary, M. I., & Ata, A., (1994). Microbial transformations of 7-a-hydroxyfrullanolide. *Journal of Natural Products, 57*, 1251–1255.

15. Ausubel, J., Crist, D. T., & Waggoner, P. E., (2010). *First Census of Marine Life 2010: Highlights of a Decade of Discovery* (p. 68). Washington, DC: Census of Marine Life (CoML)-International Secretariat Consortium for Ocean Leadership.

16. Ballot, C., Kluza, J., & Martoriati, M., (2009). Essential role of mitochondria in apoptosis of cancer cells induced by the marine alkaloid Lamellarin D. *Molecular Cancer Therapeutics, 8*(12), 3307–3317.

17. Bao, B., Sun, Q., Yao, X., & Hong, J., (2007). Bisindole alkaloids of the topsentin and hamacanthin classes from a marine sponge *Spongosorites* sp. *Journal of Natural Products, 70*, 2–8.

18. Belmiro, C. L., & Castelo-Branco, M. T., (2009). Unfractionated heparin and new heparin analogs from ascidians (chordate-tunicate) ameliorate colitis in rats. *Journal of Biological Chemistry, 284*, 11267–11278.

19. Berdy, J., (2005). Bioactive microbial metabolites. *Journal of Antibiotics, 58*, 1–26.

20. Bergmann, W., & Burke, D. C., (1956). Contributions to the study of marine products: Nucleosides of Sponges-1 and Spongosine-2. *Journal of Organic Chemistry, 21*(2), 226–228.

21. Bergmann, W., & Feeney, R. J., (1951). Contributions to the study of marine products: Nucleosides of sponges. *Journal of Organic Chemistry, 16*, 981–987.

22. Bergmann, W., & Feeney, R. J., (1950). The isolation of a new thymine pentoside from sponges. *Journal of the American Chemical Society, 72*, 2809–2810.

23. Bergmann, W., & Stempien, M. F., (1957). Contributions to the study of marine products: Nucleosides of sponges-V, synthesis of spongosine-1. *Journal of Organic Chemistry, 22*(12), 1575–1577.

24. Bhakuni, D. S., & Rawat, D. S., (2005). *Bioactive Marine Natural Products* (pp. 278–328). Springer: New York.

25. Bitencourt, F. S., Figueiredo, J. G., & Mota, M. R. L., (2008). Antinociceptive and anti-inflammatory effects of a mucin-binding agglutinin isolated from the red marine alga *Hypnea cervicornis*. *Naunyn-Schmiedeberg's Archives of Pharmacology, 377*, 139–148.

26. Blunt, J. W., Copp, B. R., & Munro, M. H. G., (2011). Marine natural products. *Natural Product Reports, 28*, 196–268.

27. Blunt, J. W., Copp, B. R., Keyzers, R. A., & Munro, M. H. G., (2014). Marine natural products-II. *Natural Product Reports, 31*, 160–258.

28. Blunt, J. W., Copp, B. R., Keyzers, R. A., & Munro, M. H. G., (2013). Marine natural products-III. *Natural Product Reports, 30*, 237–323.

29. Blunt, J. W., Copp, B. R., Keyzers, R. A., & Munro, M. H. G., (2012). Marine natural products-IV. *Natural Product Reports, 29*, 144–222.

30. Blunt, J. W., Copp, B. R., & Munro, M. H. G., (2010). Marine natural products-V. *Natural Product Reports, 27*, 165–237.

31. Bobzin, S. C., Yang, S., & Kasten, T. P., (2000). LC-NMR: New tool to expedite the dereplication and identification of natural products. *Journal of Industrial Microbiology and Biotechnology*, *25*, 342–345.

32. Boonlarppradab, C., & Faulkner, D. J., (2007). Eurysterols -A and -B, cytotoxic and antifungal steroidal sulfates from a marine sponge of the genus *Euryspongia*. *Journal of Natural Products*, *70*, 846–848.

33. Bramley, A. M., Langlands, J. M., & Jones, A. K., (1995). Effects of IZP-94005 (contignasterol) on antigen induced bronchial responsiveness in ovalbumin-sensitized guinea-pigs. *British Journal of Pharmacology*, *115*, 1433–1438.

34. Bruneton, J., (1999). *Pharmacognosy, Phytochemistry, Medicinal Plants* (2nd edn., p. 23). Intercept. Ltd., Paris and New York.

35. Bucar, F., Wube, A., & Schmid, M., (2013). Natural product isolation: How to get from biological material to pure compounds. *Natural Product Reports*, *30*, 525–545.

36. Bugni, T. S., & Harper, M. K., (2009). Advances in instrumentation, automation, dereplication and prefractionation. In: *Natural Product Chemistry for Drug Discovery* (pp. 272–298). London: The Royal Society of Chemistry.

37. Bugni, T. S., & Ireland, C. M., (2004). Marine-derived fungi: A chemically and biologically diverse group of microorganisms. *Natural Product Reports*, *21*, 143–163.

38. Burres, N. S., Hunter, J. E., & Wright, A. E., (1989). Mammalian cell agar-diffusion assay for the detection of toxic compounds. *Journal of Natural Products*, *52*, 522–527.

39. Camp, D., Davis, R. A., Evans-Illidge, E. A., & Quinn, R. J., (2012). Guiding principles for natural product drug discovery. *Future Medicinal Chemistry*, *4*, 1067–1084.

40. Cannell, R. J. P., (1998). How to approach the isolation of a natural product? In: Cannell, R. J. P., (eds.), *Methods in Biotechnology: Natural Products Isolation* (Vol. 4, pp. 1–51). Totowa, NJ: Humana.

41. Carbone, A., Parrino, B., & Barraja, P., (2013). Synthesis and anti-proliferative activity of 2,5-bis(3'-Indolyl)pyrroles, analogs of the marine alkaloid nortopsentin. *Marine Drugs*, *11*, 643–654.

42. Carter, N. J., & Keam, S. J., (2010). Trabectedin: A review of its use in soft tissue sarcoma and ovarian cancer. *Drugs*, *70*(3), 355–376.

43. Cavalier-Smith, T., (1992). Origins of secondary metabolism. *Ciba Foundation Symposium*, *171*, 64–87.

44. Cha, S. H., Lee, K. W., & Jeon, Y. J., (2006). Screening of extracts from red algae in juju for potentials marine angiotensin-I converting enzyme (ACE) inhibitory activity. *Algae*, *21*, 343–348.

45. Chackalamannil, S., (2001). Thrombin receptor antagonists as novel therapeutic targets. *Current Opinion in Drug Discovery and Development*, *4*(4), 417–427.

46. Chang, L., Whittaker, N. F., & Bewley, C. A., (2003). Crambescidin 826 and dehydrocrambine A: New polycyclic guanidine alkaloids from the marine sponge *Monanchora sp.* that inhibit HIV-1 fusion. *Journal of Natural Products*, *66*(11), 1490–1494.

47. Chao, C. H., Wen, Z. H., Wu, Y. C., Yeh, H. C., & Sheu, J. H., (2008). Cytotoxic and anti-inflammatory cembranoids from the soft coral *Lobophytum crissum*. *Journal of Natural Products*, *71*(11), 1819–1824.

48. Chen, X. T., & Gutteridge, C. E., (1998). A convergent route for the total synthesis of the eleuthesides. *Angewandte Chemie International Edition (German Chemical Society)*, *37*, 185–187.

49. Cheng, S. Y., Wen, Z. H., & Chiou, S. F., (2008). Durumolides AE: Anti-inflammatory and antibacterial cembranolides from the soft coral *Lobophytum durum*. *Tetrahedron*, *64*(41), 9698–9704.

50. Chill, L., Rudi, A., Aknin, M., Loya, S., Hizi, A., & Kashman, Y., (2004). New sesterterpenes from Madagascan *Lendenfeldia* sponges. *Tetrahedron*, *60*(47), 10619–10626.

51. Chin, Y., Balunas, M., & Chai, H., (2006). Drug discovery from natural sources. *The AAPS (Am. Assoc. of Pharma. Scientists)*, *8*, E239–E253.

52. Chung, D., Guar, S., Bell, J. R., Ramachandran, J., & Nadasci, L., (1995). Determination of disulfide bridge pattern in omega-conopeptides. *International Journal of Peptide and Protein Research*, *46*, 320–325.

53. Ciulla, T. A., Criswell, M. H., Danis, R. P., & Williams, J. I., (2003). Squalamine lactate reduces choroidal neovascularization in laser-injury model in the rat. *Retina*, *23*, 808–814.

54. Claeson, P., & Bohlin, L., (1997). Some aspects of bioassay methods in natural-product research aimed at drug lead discovery. *Trends in Biotechnology*, *15*, 245–248.

55. Coleman, J. E., Dilip, E., & De-Silva, E. D., (1995). Cytotoxic peptides from the marine sponge *Cymbastela sp. Tetrahedron*, *51*, 10653–10662.

56. Cortes, J., Vahdat, L., Blum, J. L., & Twelves, C., (2010). Phase-II study of the halichondrin B analog eribulin mesylate in patients with locally advanced or metastatic breast cancer previously treated with an anthracycline, a taxane, and capecitabine. *Journal of Clinical Oncology*, *28*, 3922–3928.

57. Costantino, V., Fattorusso, E., & Mangoni, A., (2009). Tedanol: Potent anti-inflammatory ent-pimarane diterpene from the Caribbean sponge *Tedania ignis*. *Bioorganic and Medicinal Chemistry*, *17*, 7542–7547.

58. Costello, M. J., Bouchet, P., Emblow, C. S., & Legakis, A., (2006). European marine biodiversity inventory and taxonomic resources: State of the art and gaps in knowledge. *Marine Ecology Progress Series*, *316*, 257–268.

59. Cragg, G. M., Katz, F., Newman, D. J., & Rosenthal, J., (2012). Legal and ethical issues involving marine biodiscovery and development. In: *Handbook of Marine Natural Products* (pp. 1314–1342) New York: Springer.

60. Croft, J. E., (1979). *Relief from Arthritis: A Safe and Effective Treatment from the Ocean* (p. 128). Welling borough, UK: Thorsons Publishers.

61. D'Ambrosio, M., Guerriero, A., & Pietra, F., (1987). *Sarcodictyin* A and *sarcodictyin* B, novel diterpenoidic alcohols esterified by (*E*)-*N* (1)-methylurocanic acid: Isolation from the Mediterranean stolonifer *Sarcodictyon roseum*. *Helvetica Chimica Acta*, *70*, 2019–2027.

62. Dassonneville, L., Wattez, N., & Baldeyrou, B., (2000). Inhibition of topoisomerase II by the marine alkaloid ascididemin and induction of apoptosis in leukemia cells. *Biochemical Pharmacology*, *60*(4), 527–537.

63. De-Silva, E. D., & Scheuer, P. J., (1980). Manoalide: An antibiotic sesterterpenoid from the marine sponge *Luffariella variabilis* (Polejaeff). *Tetrahedron Letters*, *21*, 1611–1614.

64. Desbois, A. P., Lebl, T., Yan, L., & Smith, V. J., (2008). Isolation and structural characterization of two antibacterial free fatty acids fromthe marine diatom *Phaeodactylum tricornutun*. *Applied Microbiology and Biotechnology, 81*, 755–764.

65. Desbois, A. P., Meams-Spragg, A., & Smith, V. J., (2009). Fatty acid from the diatom *Phaeodactylum tricornutum* is antibacterial against diverse bacteria including multi-resistant *Staphylococcus aureus* (MRSA). *Marine Biotechnology, 11*, 45–52.

66. Deslandes, S., Lamoral-Theys, D., & Frongia, C., (2012). Synthesis and biological evaluation of analogs of the marine alkaloids granulatimide and isogranulatimide. *European Journal of Medicinal Chemistry, 54*, 626–636.

67. Desoubzdanne, D., Marcourt, L., & Raux, R., (2008). Alisiaquinones and alisiaquinol: Dual inhibitors of *Plasmodium falciparum* enzyme targets from a New Caledonian deep water sponge. *Journal of Natural Products, 71*, 1189–1192.

68. Diana, P., Carbone, A., & Barraja, P., (2010). Synthesis and antitumor activity of 2,5-bis(3'-indolyl)-furans and 3,5-bis(3'-indolyl)-isoxazoles, nortopsentin analogs. *Bioorganic and Medicinal Chemistry, 18*, 4524–4529.

69. Diaz, N., Vezin, H., Lansiaux, A., & Bailly, C., (2005). Topoisomerase inhibitors of marine origin and their potential use as anticancer agents. In: *DNA Binders and Related Subjects, Topics in Current Chemistry* (Vol 253, pp 89–108). Berlin, Germany: Springer.

70. Digirolamo, J. A., Li, X. C., Jacob, M. R., Clark, A. M., & Ferreira, D., (2009). Reversal of fluconazole resistance by sulphated sterols from the marine sponge *Topsentia* sp. *Journal of Natural Products, 72*, 1524–1528.

71. Donia, M., & Hamann, M., (2003). Marine natural products and their potential applications as anti-infective agents. *The Lancet Infectious Diseases, 3*, 338–348.

72. Dostal, D. E., & Baker, K. M., (1999). The cardiac renin-angiotensin system: Conceptual or a regulator of cardiac function. *Circulation Research, 85*, 643–650.

73. Du, L., Shen, L., Yu, Z., Chen, J., & Guo, Y., (2008). Hyrtiosal from the marine sponge *Hyrtios erectus* inhibits HIV-1 integrase binding to viral DNA by a new inhibitor binding site. *Chem. Med. Chem., 3*, 173–180.

74. Duarte, K., Rocha-Santos, T. A. P., Freitas, A. C., & Duarte, A. C., (2012). Analytical techniques for discovery of bioactive compounds from marine fungi. *TrAC Trends in Analytical Chemistry, 34*, 97–110.

75. Ebada, S. S., Lin, W., & Proksch, P., (2010). Bioactive sester-terpenes and triterpenes from marine sponges: Occurrence and pharmacological significance. *Marine Drugs, 8*, 313–346.

76. Edwards, V., Benkendorff, K., & Young, F., (2012). Marine compounds selectively induce apoptosis in female reproductive cancer cells but not in primary-derived human reproductive granulosa cells, *Marine Drugs, 10*(1), 64–83.

77. El Gendy, M. M., Hawas, U. W., & Jaspars, M., (2008). Novel bioactive metabolites from a marine derived bacterium *Nocardia* sp. ALAA2000. *Journal of Antibiotics, 61*, 379–386.

78. El Gendy, M. M., Shaaban, M., El Bonkly, A. M., & Shaaban, K. A., (2008). Bioactive benzopyrone derivatives from new recombinant fusant of marine *Streptomyces*. *Applied Biochemistry and Biotechnology, 150*, 85–96.

79. El Sayed, K. A., Khalil, A. A., & Yousaf, M., (2008). Semisynthetic studies on the manzamine alkaloids. *Journal of Natural Products, 71*(3), 300–308.

80. Epifanio, R. D. A., Gabriel, R., Martins, D. L., & Muricy, G., (1999). The sester-terpene-variabilin as a fish-predation deterrent in the western Atlantic sponge *Irciniastrobilina*. *Journal of Chemical Ecology, 25*, 2247–2254.

81. Epifanio, R. D. A., Martins, D. L., Villac, R., & Gabriel, R., (1999). Chemical defenses against fish predation in three Brazilian octocorals: 11b, 12b-epoxypukalide as a feeding deterrent in *Phyllogorgiadilatata. Journal of Chemical Ecology, 25*, 2255–2265.

82. Fan, G., Li, Z., Shen, S., Zeng, Y., & Yang, Y., (2010). Baculiferins A-O,O-sulfated pyrrole alkaloids with anti-HIV-1 activity from the Chinese marine sponge *Iotrochota baculifera. Bioorganic and Medicinal Chemistry, 18*, 5466–5474.

83. Faulkner, D. J., (1993). Academic chemistry and the discovery of bioactive marine natural products. In: Attaway, D. H., & Zaborsky, O. R., (eds.), *Marine Biotechnology, Pharmaceutical and Bioactive Natural Products* (Vol. 1, pp. 459–474). Plenum Press, New York.

84. Faulkner, D. J., (2000). Marine natural products. *Natural Product Reports, 17*, 7–55.

85. Faulkner, D. J., (2002). Marine natural products. *Natural Product Reports, 19*, 1–48.

86. Faulkner, D. J., (1977). Interesting aspects of marine natural products chemistry. *Tetrahedron, 33*, 1421–1443.

87. Faulkner, D. J., (2000). Marine pharmacology. *Antonie Van Leeuwenhoek, 77*, 135–145.

88. Fedorov, S. N., Bode, A. M., & Stonik, V. A., (2004). Marine alkaloid polycarpine and its synthetic derivative dimethylpolyearpine induce apoptosis in JB6 cells through p53-and caspase 3-dependent pathways. *Pharmaceutical Research, 21*, 2307–2319.

89. Fenical, W., (1993). Chemical studies of marine bacteria: Developing a new resource. *Chemical Reviews, 93*, 1673–1683.

90. Firn, R. D., & Jones, C. G., (2000). The evolution of secondary metabolism-a unifying model. *Molecular Microbiology, 37*, 989–994.

91. Fitzgerald, C., Aluko, R. E., Hossain, M., Rai, D. K., & Hayes, M., (2014). Potential of a renin inhibitory peptide from the red seaweed *Palmaria palmata* as a functional food ingredient following confirmation and characterization of a hypotensive effect in spontaneously hypertensive rats. *Journal of Agricultural and Food Chemistry, 62*, 8352–8356.

92. Fitzgerald, C., Mora-Soler, L., & Gallagher, E., (2012). Isolation and characterization of bioactive pro-peptides with *in vitro* renin inhibitory activities from the macroalga *Palmaria palmata. Journal of Agricultural and Food Chemistry, 60*, 7421–7427.

93. Gamble, W. R., Durso, N. A., & Fuller, R. W., (1999). Cytotoxic and tubulin-interactive hemiasterlins from *Auletta sp.* and *Siphonochalina spp.* sponges. ⌈ *Bioorganic and Medicinal Chemistry, 7*(8), 1611–1615.

94. Gerwick, W. H., & Moore, B. S., (2012). Lessons from the past and charting the future of marine natural products drug discovery and chemical biology. *Chemical Biology, 19*, 85–98.

95. Gerwick, W. H., Proteau, P. J., Nagle, D. G., & Hamel, E., (1994). Structure of curacin A, a novel antimitotic, antiproliferative, and brine shrimp toxic natural product from the marine cyanobacterium *Lyngbya majuscule. Journal of Organic Chemistry, 59*, 1243–1245.

96. Gibbons, S., (2006). Introduction to planar chromatography methods. In: Satyajit, D., (ed.), *Biotechnology Natural Products Isolation* (2nd edn., pp. 77–116). New York: Humana Press Inc.

97. Gil, B., Ferrándiz, M. L., & Sanz, M. J., (1995). Inhibition of inflammatory responses by epitaondiol and other marine natural products. *Life Sciences, 57*(2), 25–30.

98. Glaser, K. B., & Jacobs, R. S., (1987). Inactivation of bee venom phospholipase A2 by manoalide: A model based on the reactivity of manoalide with amino acids and peptide sequences. *Biochemical Pharmacology, 36*, 2079–2086.

99. Glaser, K. B., & Jacobs, R. S., (1986). Molecular pharmacology of manoalide: Inactivation of bee venom phospholipase A2. *Biochemical Pharmacology, 35*, 449–453.

100. God, R., Gumm, H., Heuer, C., & Juschka, M., (1999). Online coupling of HPLC and solid-phase extraction for preparative chromatography. *GitLab Journal, 3*, 188–191.

101. Gross, H., & Konig, G. H., (2006). Terpenoids from marine organisms: Unique structures and their pharmacological potential. *Phytochemistry Reviews, 5*, 115–141.

102. Gulavita, N. K., Gunasekera, S. P., & Pomponi, S. A., (1992). Bioactive depsipeptide from the marine sponge *Discodermia* sp. *Journal of Organic Chemistry, 57*, 1767–1772.

103. Gunasekera, S. P., Gunasekera, M., Longley, R. E., & Schulte, G. K., (1990). Discodermolide: New bioactive polyhydroxylated lactone from the marine sponge *Discodermia dissoluta. Journal of Organic Chemistry, 55*, 4912–4915.

104. Gu-Ping, H., Jie, Y., Li, S., & Zhi-Gang, S., (2011). Statistical research on marine natural products based on data obtained between 1985 and 2008. *Marine Drugs, 9*, 514–525.

105. Gustafson, K. R., Oku, N., & Milanowski, D. J., (2004). Antiviral marine natural products. *Current Medicinal Chemistry, 3*, 233–249.

106. Haefner, B., (2003). Drugs from the deep: Marine natural products as drug candidates. *Drug Discover Today (DDT), 8*(12), 536–544.

107. Hale, K. J., & Manaviazar, S., (2010). New approaches to the total synthesis of the bryostatin antitumor macrolides. *Chemistry: An Asian Journal, 5*(4), 704–754.

108. Halpern, G. M., (2000). Anti-inflammatory effects of a stabilized lipid extract of *Perna canaliculus* (Lyprinol). *Clinical Reviews in Allergy and Immunology, 32*(7), 272–278.

109. Hamburger, M. O., & Cordell, G. A., (1987). Direct bioautographic TLC assay for compounds possessing antibacterial activity. *Journal of Natural Products, 50*, 19–22.

110. Hanif, N., Ohno, O., Kitamura, M., Yamada, K., & Uemura, D., (2010). Symbiopolyol, VCAM-1 inhibitor from a symbiotic dinoflagellate of the jellyfish *Mastigias papua. Journal of Natural Products, 73*, 1318–1322.

111. Hanif, N., Tanaka, J., Setiawan, A., & Trianto, A., (2007). Polybrominated diphenyl ethers from the Indonesian sponge *Lamellodysidea herbacea. Journal of Natural Products, 70*, 432–435.

112. Harrigan, G. G., Yoshida, W. Y., & Moore, R. E., (1998). Isolation, structure determination, and biological activity of dolastatin and lyngbyastatin-1 from *Lyngbya majuscula/Schizothrix calcicola* cyanobacterial assemblages. *Journal of Natural Products, 61*, 1221–1225.

113. Harvell, C. D., Fenical, W., & Roussis, V., (1993). Local and geographic variation in the defensive chemistry of a West Indian gorgonian coral (*Briareum asbestinum*). *Marine Ecology Progress Series, 93*, 165–173.

114. Harvey, A. L., (2002). Toxins 'R': More pharmacological tools from nature's superstore. *Trends in Pharmacological Sciences, 5*, 201–203.

115. Hay, M. E., (1996). Marine chemical ecology: What's known and what's next? *Journal of Experimental Marine Biology and Ecology, 200*, 103–134.

116. Hay, M. E., (1992). The role of seaweed chemical defenses in the evolution of feeding specialization and in the mediation of complex interactions. In: Paul, V. J., (ed.), *Ecological Roles for Marine Natural Products* (pp. 93–118) Comstock Press, Ithaca, NY.

117. Haygood, M. G., & Davidson, S. K., (1997). Small-subunit rRNA genes and *in situ* hybridization with oligonucleotides specific for the bacterial symbionts in the larvae of the bryozoan *Bugula neritina* and proposal of *Candidatus Endobugula sertula*. *Applied and Environmental Microbiology, 63*, 4612–4616.

118. Hilton, M. D., (2003). Natural products: Discovery and screening. In: Vinci, V. A., & Parekh, S. R., (eds.), *Handbook of Industrial Cell Cultures* (pp. 107–136.). Humana, Totowa, NY.

119. Hirata, Y., & Uemura, D., (1986). Halichondrins: Antitumor polyether macrolides from a marine sponge. *Pure Applied Chemistry, 58*, 701–710.

120. Homann, M. T., & Scheuer, P. J., (1993). Kahalalide: Bioactive depsipeptide from the sacoglossan mollusk *Elysia rufescens* and the green alga *Bryopsis* sp. *Journal of the American Chemical Society, 115*, 5825–5826.

121. Houghton, P. J., Raman, A., & Chapman, H., (1998). *Laboratory Handbook for the Fractionation of Natural Extracts* (p. 199). Springer, London, UK.

122. Houssen, W. E., & Jaspars, M., (2006). Isolation of marine natural products. In: Sarker, S. D., (ed.), *Methods in Biotechnology Natural Products Isolation* (2nd edn., pp 353–390). Totowa, NY: Humana Press.

123. https://wexpharma.com/clinical-trials/clinical-experience/ (accessed on 18 May 2020).

124. https://www.midwestern.edu/departments/marinepharmacology/clinical-pipeline.xml (accessed on 18 May 2020).

125. http://www.Clinicaltrials.gov (accessed on 18 May 2020).

126. Hua, H. M., Peng, J., Dunbar, D. C., & Schinazi, R. F., (2007). Batzelladine alkaloids from the Caribbean sponge *Monanchora unguifera* and the significant activities against HIV-1 and AIDS opportunistics infections pathogens. *Tetrahedron, 63*, 11179–11188.

127. Hughes, C. C., & Fenical, W., (2010). Antibacterial from the sea. *Chemistry, 16*, 12512–12525.

128. Hugon, B., Anizon, F., Bailly, C., & Golsteyn, R. M., (2007). Synthesis and biological activities of isogranulatimide analogs. *Bioorganic and Medicinal Chemistry, 15*, 5965–5980.

129. Huntley, M., Sykes, P., Rohan, S., & Martin, V., (1986). Chemically- mediated rejection of dinoflagellate prey by the copepods *Calanus pacificus* and *Paracalanus purvus*: Mechanism, occurrence and significance. *Marine Ecology Progress Series, 28*, 105–120.

130. Ibrahim, S. R., Min, C. C., & Teuscher, F., (2010). Callyaerins: New cytotoxic cyclic peptides from the Indonesian marine sponge *Callyspongia aerizusa*. *Bioorganic and Medicinal Chemistry, 18*, 4947–4956.

131. Izzo, I., Pironti, V., Della, M. C., Sodano, G., & De Riccardis, F., (2001). Stereo-controlled synthesis of contignasterol's side chain. *Tetrahedron Letters, 42*, 8977–8980.

132. Jacobson, P. B., & Jacobs, R. S., (1992). Fuscoside: An anti-inflammatory marine natural product which selectively inhibits 5-lipoxygenase, Part I: Physiological and biochemical studies in murine inflammatory models. *Journal of Pharmacology and Experimental Therapeutics, 262*(2), 866–873.

133. Jang, J. H., Van, S. R. W., Fusetani, N., & Matsunaga, S., (2007). Pseudoceratins A and B, antifungal bicyclic bromotyrosine-derived metabolites from the marine sponge *Pseudoceratina purpurea*. *Journal of Organic Chemistry, 72*, 1211–1217.

134. Jang, K. H., Chung, S. C., Shin, J., & Lee, S. H., (2007). Aaptamines as sortase A inhibitors from the tropical sponge *Aaptos aaptos*. *Bioorganic and Medicinal Chemistry Letters, 17*, 5366–5369.

135. Jang, K. H., Kang, G. W., & Jeon, J. E., (2009). Haliclonin A: New macrocyclic diamide from the sponge *Haliclona* sp. *Organic Letters, 11*, 1713–1716.

136. Jean, Y. H., Chen, W. F., Sung, C. S., & Duh, C. Y., (2009). Capnellene: Natural marine compound derived from soft coral, attenuates chronic constriction injury-induced neuropathic pain in rats. *British Journal of Pharmacology, 158*, 713–725.

137. Jeon, J. E., Na, Z., Jung, M., & Lee, H. S., (2010). Discorhabdins from the Korean marine sponge *Sceptrella* sp. *Journal of Natural Products, 73*, 258–262.

138. Ji, N. Y., Li, X. M., Li, K., Ding, L. P., Gloer, J. B., & Wang, B. G., (2007). Diterpenes, sesquiterpenes and a C15-acetogenin from the marine redalga *Laurencia mariannensis*. *Journal of Natural Products, 70*, 1901–1905.

139. Jiang, X., Zhao, B., Britton, R., & Lim, L. Y., (2004). Inhibition of Chk1 by the G2 DNA damage checkpoint inhibitor isogranulatimide. *Molecular Cancer Therapeutics, 3*, 1221–1227.

140. Jianga, X., Xionga, J., & Songa, Z., (2013). Is coarse taxonomy sufficient for detecting macroinvertebrate patterns in floodplain lakes? *Ecological Indicators, 27*, 48–55.

141. Jimeno, A., (2009). Eribulin: Rediscovering tubulin as an anticancer target. *Clinical Cancer Research, 15*(12), 3903–3905.

142. Jones, D. O., Wigham, B. D., Hudson, I. R., & Bett, B. J., (2007). Anthropogenic disturbance of deep-sea megabenthic assemblages. A study with remotely operated vehicles in the Faroe-Shetland Channel, NE Atlantic. *Marine Biology, 151*, 1731–1741.

143. Jung, W. K., & Kim, S. K., (2009). Isolation and characterization of an anticoagulant oligopeptide from blue mussel (*Mytilus edulis*). *Food Chemistry, 117*, 687–692.

144. Justino, K., & Duarte, A. C., (2014). Classical methodologies for preparation of extracts and fractions. In: Rocha, S., & Duarte, A. C., (eds.), *Analysis of Marine Samples in Search of Bioactive Compounds* (Vol. 65, pp. 35–57). New York: Elsevier.

145. Katz, J., Janik, J. E., & Younes, A., (2011). Brentuximab Vedotin (SGN-35). *Clinical Cancer Research, 17*, 6428–6436.

146. Keffer, J. L., Plaza, A., & Bewley, C. A., (2009). Motualevic acids A-F, antimicrobial acids from the sponge *Siliquaria spongia*. *Organic Letters, 11*, 1087–1090.

147. Kelecom, A., (2002). Secondary metabolites from marine microorganisms. *Anais da Academia Brasileria de Ciencias (Annals of the Brazilian Academy of Sciences), 74*, 151–170.

148. Kern, S. E., Allen, J., Wagstaff, J., Shafer, S. L., & Yaksh, T., (2007). The pharmaco kinetics of the conopeptide contulakin-G (CGX-1160) after intrathecal administration: An analysis of data from studies in beagles. *Anesthesia and Analgesia, 104*, 1514–1520.

149. Kerr, R. G., & Kerr, S. S., (1999). Marine natural products as therapeutic agents. *Expert Opinion on Therapeutic Patents, 9*, 1207–1222.

150. Kicklighter, C. E., Kubanek, J., & Hay, M. E., (2004). Do brominated natural products defend marine worms from consumers? Some do, most don't. *Limnology and Oceanography, 49*, 430–444.

151. Kim, A. R., Shin, T. S., & Lee, M. S., (2009). Isolation and identification of phlorotannins from *Ecklonia stolonifera* with antioxidant and anti-inflammatory properties. *Journal of Agricultural and Food Chemistry, 57*, 3483–3489.

152. Kitagawa, I., Kobayashi, M., Kitanaka, K., & Kido, M., (1983). Marine natural products, XII: On the chemical constituents of the Okinawan marine sponge *Hymeniacidon aldis*. *Chemical and Pharmaceutical Bulletin, 31*, 2321–2328.

153. Ko, S. C., Kim, D. G., & Han, C. H., (2012). Nitric oxide-mediated vasorelaxation effects of anti-angiotensin I-converting enzyme (ACE) peptide from *Styela clava* flesh tissue and its anti-hypertensive effect in spontaneously hypertensive rats. *Food Chemistry, 134*, 1141–1145.

154. Kobayashi, H., Kitamura, K., & Nagai, K., (2007). Carteramine A: An inhibitor of neutrophil chemotaxis, from the marine sponge *Stylissa carteri*. *Tetrahedron Letters, 48*(12), 2127–2129.

155. Koehn, F. E., & Carter, G. T., (2005). The evolving role of natural products in drug discovery. *Nature Reviews Drug Discovery, 4*, 206–220.

156. Kon, Y., Kuboto, T., Shibazaki, A., Gonoi, T., & Kobayashi, J., (2010). Ceratinadins-A-C: New bromotyrosine alkaloids from Okinawan marine sponge *Pseudoceratina* sp. *Bioorganic and Medicinal Chemistry Letters, 20*, 4569–4572.

157. Korhonen, H., (2009). Milk-derived bioactive peptides: From science to applications. *Journal of Functional Foods, 1*, 177–187.

158. Kossuga, M. H., Nascimento, A. M., & Reimao, J. Q., (2008). Antiparasitic, antineuroinflammatory, and cytotoxic polyketides from the marine sponge *Plakortis angulospiculatus* collected in Brazil. *Journal of Natural Products, 71*(3), 334–339.

159. Kuboto, T., Araki, A., & Yasuda, T., (2009). Benzosceptrin C: New dimeric bromopyrrole alkaloid from sponge *Agelas* sp. *Tetrahedron Letters, 50*, 7268–7270.

160. Kumar, D., Maruthi, N., Ghosh, S., & Shah, K., (2012). Novel bis(indolyl)hydrazide-hydrazones as potent cytotoxic agents. *Bioorganic and Medicinal Chemistry Letters, 22*, 212–215.

161. Kupchan, S. M., Britton, R. W., Lacadie, J. A., Ziegler, M. F., & Sigel, C. W., (1975). Isolation and structural elucidation of bruceantin and bruceantinol: New potent antileukemic quassinoids from *Brucea antidysenterica*. *Journal of Organic Chemistry, 40*, 648–654.

162. Kuramoto, M., Tong, C., Yamada, K., & Chiba, T., (1996). Halichlorine, an inhibitor of VCAM-1 induction from the marine sponge *Halichondria okadai* Kadota. *Tetrahedron Letters, 37*(22), 3867–3870.

163. Lane, A. L., Mulac, L., & Drenkarrd, E. J., (2010). Ecological leads for natural product discovery: Novel sesquiterpene hydroquinones from the red macroalga *Peysonnelia* sp. *Tetrahedron Letters, 66*, 455–461.

164. Leal, M. C., Madeira, C., Brandão, C. A., Puga, J., & Calado, R., (2012). Bioprospecting of marine invertebrates for new natural products: Chemical and zoogeographical perspective. *Molecules, 17*, 9842–9854.

165. Lee, A. H., Chen, J., Liu, D., Leung, T. Y., Chan, A. S., & Li, T., (2002). Acid-promoted DNA-cleaving activities and total synthesis of varacin C. *Journal of the American Chemical Society, 124*, 13972–13973.

166. Lee, D., Shin, J., Yoon, K. M., Kim, T. I., Lee, S. H., Lee, H. S., & Oh, K. B., (2008). Inhibition of *Candida albicans* isocitrate lyase activity by sesterterpene sulfates from the tropical sponge *Dysidea*. *Bioorganic and Medicinal Chemistry Letters, 18*, 5377–5380.

167. Lee, H. S., Yoon, K. M., Han, Y. R., & Lee, R. J., (2009). The 5-hydroxyindole-type alkaloids as *Candida albicans* isocitrate lyase inhibitors, from the tropical sponge *Hyrtios*. *Bioorganic and Medicinal Chemistry Letters, 19*, 1051–1053.

168. Lee, S. H., Ryu, B., Je, J. Y., & Kim, S. K., (2011). Diethylaminoethyl chitosan induces apoptosis in HeLa cells via activation of caspase-3 and p53 expression. *Carbohydrate Polymers, 84*(1), 571–578.

169. Li, G. H., Le, G. W., Shi, Y. H., & Shrestha, S., (2004). Angiotensin I-converting enzyme inhibitory peptides derived from food proteins and their physiological and pharmacological effects. *Nutrition Research, 24*, 469–486.

170. Li, G. H., Wan, J. Z., Le, G. W., & Shi, Y. H., (2006). Novel angiotensin I-converting enzyme inhibitory peptides isolated from alcalase hydrolysate of mung bean protein. *Journal of Peptide Science, 12*, 509–514.

171. Li, Y. X., Li, Y., Lee, S. H., Qian, Z. J., & Kim, S. K., (2010). Inhibitors of oxidation and matrix metalloproteinases, floridoside, and D-isofloridoside from marine red alga *Laurencia undulata. Journal of Agricultural and Food Chemistry, 58*, 578–586.

172. Lin, A. S., Stout, E. P., & Prudhomme, J., (2010). Bioactive bromophycolides R-U from the Fijian red alga *Callophycus serratus. Journal of Natural Products, 73*, 275–278.

173. Lindel, T., Jensen, P. R., & Fenical, W., (1997). Eleutherobin, a new cytotoxin that mimics Paclitaxel (Taxol) by stabilizing microtubules. *Journal of the American Chemical Society, 119*, 8744–8745.

174. Longeon, A., Copp, B. R., Rove, M., & Dubois, J., (2010). New bioactive halenaquinone derivatives from South Pacific marine sponges of the genus *Xestospongia. Bioorganic and Medicinal Chemistry, 18*, 6006–6011.

175. Look, S. A., Fenical, W., Jacobs, R. S., & Clardy, J., (1986). The pseudopterosins: Anti-inflammatory and analgesic natural products from the sea whip, *Pseudopterogorgia elisabethae*. In: *Proc. Natl. Acad. Sci. USA*, p. 8.

176. Lopez-Macia, A., Jimenez, J. C., Royo, M., Giraet, E., & Albericio, F., (2001). Synthesis and structure determination of kahalalide F (1,2). *Journal of the American Chemical Society, 123*, 11398–11401.

177. Lunn, J., & Theobald, H. E., (2006). The health effects of dietary unsaturated fatty acids. *Nutrition Bulletin, 31*(3), 178–224.

178. Mackay, H. J., & Twelves, C. J., (2007). Targeting the protein kinase C family: Are we there yet? *Nature Reviews Cancer, 7*(7), 554–562.

179. Macomber, R. S., (1997). *A Complete Introduction to Modern NMR Spectroscopy* (p. 400). New York: Wiley.

180. Makarieva, T. N., Stonik, V. A., Dmitrenok, A. S., & Grebnev, B. B., (1995). Anthocyanins and other flavonoids. *Journal of Natural Products, 58*, 254–258.

181. Malloy, K. L., Villa, F. A., Engene, N., & Matainaho, T., (2011). Malyngamide 2, an oxidized lipopeptide with nitric oxide inhibiting activity from a Papua New Guinea marine cyanobacterium. *Journal of Natural Products, 74*, 95–98.

182. Marin, L., (2003). *A Marine Literature Database*. Department of Chemistry, University of Canterbury, New Zealand.

183. Mayer, A. M. S., Jacobson, P. B., Fenical, W., Robert, R. S., & Glaser, K. B., (1998). Pharmacological characterization of the pseudopterosins: Novel anti-inflammatory natural products isolated from the Caribbean soft coral, *Pseudopterogorgia elisabethae. Life Sciences, 624*, 401–407.

184. Mayer, A. M. S., Rodriguez, A. D., Berlinck, R. G. S., & Hamann, M. T., (2009). Marine pharmacology in 2005–2006: Marine compounds with anthelmintic, antibacterial, anticoagulant, antifungal, anti-inflammatory, antimalarial, antiprotozoal, antituberculosis, and antiviral activities, affecting the cardiovascular, immune and nervous systems, and other miscellaneous mechanisms of action. *Biochimica et Biophysica Acta, 1790*, 283–308.

185. Mayer, A. M. S., Rodríguez, A. D., Berlinck, R. G., & Fusetani, N., (2011). Marine pharmacology in 2007–2008: Marine compounds with antibacterial, anticoagulant, antifungal, anti-inflammatory, antimalarial, antiprotozoal, antituberculosis, and antiviral activities, affecting the immune and nervous system, and other miscellaneous mechanisms of action. *Comparative Biochemistry and Physiology-Part C, 153*, 191–222.

186. McClintock, J. B., (1994). *An Overview of the Chemical Ecology of Antarctic Marine Invertebrates, the Ireland Lecture 1993* (p. 24). University of Alabama, Birmingham, USA.

187. McIntosh, M., Cruz, L. J., Hunkapiller, M. W., Gray, W. R., & Olivera, B. M., (1982). Isolation and structure of a peptide toxin from the marine snail *Conus magus*. *Archives of Biochemistry and Biophysics, 218*, 329–334.

188. Miljanich, G. P., (2004). Ziconotide: Neuronal calcium channel blocker for treating severe chronic pain. *Current Medicinal Chemistry, 11*(23), 3029–3040.

189. Mishra, B. B., & Tiwari, V. K., (2011). Natural products: An evolving role in future drug discovery. *European Journal of Medicinal Chemistry, 46*, 4769–4807.

190. Mishra, K. P., Ganju, L., Sairam, M., & Banerjee, P. K., (2008). Review of high throughput technology for the screening of natural products. *Biomedicine and Pharmacotherapy, 62*, 94–98.

191. Mitchell, S. S., Rhodes, D., Bushman, F. D., & Faulkner, D. J., (2000). Cyclodidemniserinol trisulfate: A sulfated serinolipid from the Palauan ascidian *Didemnum guttatum* that inhibits HIV-1 integrase. *Organic Letters, 2*(11), 1605–1607.

192. Miyashita, K., (2009). Function of marine carotenoids. *Forum of Nutrition, 61*, 136–146.

193. Miyata, T., Tokunaga, F., & Yoneya, T., (1989). Antimicrobial peptides, isolated from *Horseshoe crab hemocytes*, tachyplesin II, and polyphemusins I and II: Chemical structures and biological activity. *The Journal of Biochemistry, 106*(4), 663–668.

194. Molinski, T. F., Dalisay, D. S., Lievens, S. L., & Saludes, J. P., (2009). Drug development from marine natural products. *Nature Reviews Drug Discovery, 8*, 69–85.

195. Montalvão, S. I. G. H. M., Singh, V., & Haque, S., (2014). Bioassays for bioactivity screening. In: Rocha-Santos, T., & Duarte, A. C., (eds.), *Analysis of Marine Samples in Search of Bioactive Compounds* (Vol. 65, pp. 79–114). New York: Elsevier.

196. Moore, K. S., Wehrli, S., & Roder, H., (1993). Squalamine: An aminosterol antibiotic from the shark. *Proceedings of the National Academy of Sciences of the United States of America, 90*, 1354–1358.

197. Morgan, R. J. Jr., Leong, L., & Chow, W., (2012). Phase II trial of bryostatin-1 in combination with cisplatin in patients with recurrent or persistent epithelial ovarian cancer: A California cancer consortium study. *Investig. New Drugs, 30*, 723–728.

198. Morimoto, M., Mori, H., & Otake, T., (1991). Inhibitory effect of tachyplesin I on the proliferation of human immunodeficiency virus *in vitro*. *Chemotherapy, 37*(3), 206–211.

199. Mourão, P. A. S., Pereira, M. S., & Pavão, M. S. G., (1996). Structure and anticoagulant activity of a fucosylated chondroitin sulfate from echinoderm: Sulfated fucose branches on the polysaccharide account for its high anticoagulant activity. *The Journal of Biological Chemistry, 271*, 23973–23984.

200. Muller, W. E. G., & Schroder, H. C., (1991). Cell biological aspects of HIV-1 infection: Effect of the anti-HIV-1 agent avarol. *Journal of Sports Medicine, 12*(1), S43–S49.

201. Munro, M. H., Blunt, J. W., & Dumdei, E. J., (1999). The discovery and development of marine compounds with pharmaceutical potential. *Journal of Biotechnology, 70*, 15–25.

202. Najafian, L., & Babji, A. S., (2012). Review of fish-derived antioxidant and antimicrobial peptides: Their production, assessment, and applications. *Peptide, 33*, 178–185.

203. Nakamura, H., Wu, H., Ohizumi, Y., & Hirata, Y., (1984). Agelasine-A, -B, -C and -D, novel bicyclic diterpenoids with a 9-methyladeninium unit possessing inhibitory effects on Na, K-ATPase from the Okinawa sea sponge Agelas sp.1. *Tetrahedron Letter, 25*, 2989–2992.

204. Newman, D. J., & Cragg, G. M., (2012). Natural products as sources of new drugs over the 30 years from 1981 to 2010. *Journal of Natural Products, 75*(3), 311–335.

205. Ngo, D. H., Ryu, B., & Kim, S. K., (2014). Active peptides from skate (*Okamejei kenojei*) skin gelatin diminish angiotensin-I converting enzyme activity and intracellular free radical-mediated oxidation. *Food Chemistry, 143*, 246–255.

206. Nicolau, K. C., Xu, J. Y., & Kim, S., (1997). Synthesis of the tricyclic core of eleutherobin and sarcodictyins and total synthesis of sarcodictyin *A. Journal of the American Chemical Society, 19*, 11353–11354.

207. Nicolau, K. C., Van, D. F., & Ohshima, T., (1997). Total synthesis of eleutherobin. *Angewandte Chemie International Edition in English, 36*, 2520–2524.

208. Nishimura, S., Arita, Y., Honda, M., & Iwamoto, K., (2010). Marine antifungal theonellamides target 3β-hydroxysterol to activate Rho1 signaling. *Nature Chemical Biology, 6*, 519–526.

209. Oh, K. B., Lee, J. H., & Chung, S. C., (2008). Antimicrobial activities of the bromophenols from the redalga *Odonthalia corymbifera* and some synthetic derivatives. *Bioorganic and Medicinal Chemistry Letters, 18*, 104–108.

210. Oku, N., Gustafson, K. R., & Cartner, L. K., (2004). Neamphamide A: New HIV-inhibitory depsipeptide from the Papua New Guinea marine sponge *Neamphius huxleyi. Journal of Natural Products, 67*(8), 1407–1411.

211. Olivera, B. M., & Cruz, L. J., (2001). Conotoxins in retrospect. *Toxicon, 39*, 7–14.

212. Olivera, B. M., & Cruz, L. J., (1987). Neuronal calcium channel antagonists. Discrimination between calcium channel subtypes using omega-conotoxin from *Conus magus* venom. *Biochemistry, 26*(8), 2086–2090.

213. Olivera, B. M., Gray, W. R., & Zeikus, R., (1985). Peptide neurotoxins from fish-hunting cone snails. *Science, 230*, 1338–1343.

214. Olivera, B. M., Miljanich, G. P., Ramachandran, J., & Adams, M. E., (1994). Calcium channel diversity and neurotransmitter release: The omega-conotoxins and omega-agatoxins. *Annual Review of Biochemistry, 63*, 823–867.

215. Paerl, H. W., (1988). Nuisance phytoplankton blooms in coastal, estuarine, and inland waters. *Limnology and Oceanography, 33*, 824–846.

216. Patil, A. D., Freyer, A. J., & Killmer, L., (1997). Z-Axinohydantoin and debromo-Z-axinohydantoin from the sponge *Stylotella aurantium*: Inhibitors of protein kinase C. *Natural Products Letters, 9*, 201–207.

217. Paul, V. J., (1992). *Ecological Roles J or Marine Natural Products* (p. 245). Comstock Publishing Associates, Ithaca, NY.

218. Paul, V. J., Ritson-Williams, R., & Sharp, K., (2011). Marine chemical ecology in benthic environments. *Natural Products Letters, 28*, 345–388.

219. Pauli, G. F., Jaki, B. U., & Lankin, D. C., (2004). Quantitative 1H NMR: Development and potential of a method for natural products analysis. *Journal of Natural Products, 68*, 133–149.

220. Pawlik, J. R., (1993). Marine invertebrate chemical defenses. *Chemical Reviews, 93,* 1911–1922.
221. Pearce, A. N., Chia, E. W., & Berridge, M. V., (2007). Anti-inflammatory thiazine alkaloids isolated from the New Zealand ascidian *Aplidium sp.*: Inhibitors of the neutrophil respiratory burst in a model of gouty arthritis. *Journal of Natural Products, 70*(6), 936–940.
222. Pearce, A. N., Chia, E. W., & Berridge, M. V., (2007). E/Z-rubrolide O, an anti-inflammatory halogenated furanone from the New Zealand ascidian *Synoicumn.* Sp. *Journal of Natural Products, 70*(1), 111–113.
223. Pettit, G. R., (1987). Isolation and structure of a remarkable marine animal atineoplastic constituent: Dolastatin 10. *Journal of the American Chemical Society, 109,* 6883–6885.
224. Pettit, G. R., Fujii, Y., Hasler, J. A., & Schmidt, J. M., (1982). Isolation and characterization of palystatins A-D. *Journal of Natural Products, 45,* 272–276.
225. Pettit, G. R., Herald, C. L., Doubek, D. L., Herald, D. L., Arnold, E., & Clardy, J., (1982). Isolation and structure of bryostatin 1. *Journal of the American Chemical Society, 104,* 6846–6848.
226. Pimentel, A. A., Felibertt, P., & Sojo, F., (2012). The marine sponge toxin agelasine B increases the intracellular Ca^{2+} concentration and induces apoptosis in human breast cancer cells (MCF-7). *Cancer Chemotherapy and Pharmacology, 69,* 71–83.
227. Plaza, A., Gustchina, E., Baker, H. L., Kelly, M., & Bewley, C. A., (2007). Mirabamides A-D, depsipeptides from the sponge *Siliquaria spongia mirabilis* that inhibit HIV-1 fusion. *Journal of Natural Products, 70,* 1753–1760.
228. Plaza, A., Keffer, J. L., Bifulco, G., Lloyd, J. R., & Bewley, C. A., (2010). Chrysophaentins A-H, antibacterial bisdiarylbutene macrocycles that inhibit the bacterial cell division protein FtsZ. *Journal of the American Chemical Society, 132,* 9069–9077.
229. Potts, B. C. M., Faulkner, D. J., & Chan, J. A., (1991). Didemnaketals A and B, HIV-1 protease inhibitors from the ascidian *Didemnum sp. Journal of the American Chemical Society, 113*(16), 6321–6322.
230. Prado, M. P., Torres, Y. R., & Berlinck, R. G. S., (2004). Effects of marine organisms extracts on microtubule integrity and cell cycle progression in cultured cells. *Journal of Experimental Marine Biology and Ecology, 313*(1), 125–137.
231. Proksch, P., (1994). Defensive roles for secondary metabolites from marine sponges and sponge-feeding nudibranchs. *Toxicon, 32,* 639–655.
232. Proteau, P. J., Gerwick, W. H., Garcia-Pichel, F., & Castenholz, R. W., (1993). The structure of scytonemin, an ultraviolet sunscreen pigment from the sheaths of cyanobacteria. *Experientia, 49,* 825–829.
233. Provencio, M., Sanchez, A., Gasent, J., Gomez, P., & Rosell, R., (2009). Cancer treatments: Can we find treasures at the bottom of the sea? *Clinical Lung Cancer, 10*(4), 295–300.
234. Pruksakorn, P., Arai, M., & Kotoku, N., (2010). Trichoderins, novel aminolipopeptides from a marine sponge-derived *Trichoderma* sp. are active against dormant mycobacteria. *Bioorganic and Medicinal Chemistry Letters, 20,* 3658–3663.
235. Pullen, F. S., Swanson, A. G., Newman, M. J., & Richards, D. S., (1995). Online liquid chromatography/nuclear magnetic resonance/mass spectrometry: A powerful spectrometric tool for the analysis of mixtures of pharmaceutical interest. *Rapid Communivation in Mass Spectrometry, 9,* 1003–1006.

236. Radchenko, O. S., (1997). Marine sulfur-containing natural products. *Tetrahedron Letters*, *38*, 3581–3584.

237. Radjasa, O. K., Vaske, Y. M., & Navarro, G., (2011). Highlights of marine invertebrate-derived biosynthetic products: Their biomedical potential and possible production by microbial associants. *Bioorganic and Medicinal Chemistry Letters, 19*, 6658–6674.

238. Rashid, M. A., Gustafson, K. R., & Cartner, L. K., (2001). Microspinosamide: New HIV inhibitory cyclic depsipeptide from the marine sponge *Sidonops microspinosa*. *Journal of Natural Products*, *64*(1), 117–121.

239. Reddy, M. V. R., Rao, M. R., & Rhodes, D., (1991). Lamellarin 20-sulfate, an inhibitor of HIV-1 integrase active against HIV-1 virus in cell culture. *Journal of Medicinal Chemistry*, *42*(11), 1901–1907.

240. Ren, S., Ma, W., Xu, T., & Lin, X., (2010). Two novel alkaloidsfrom the South China sea marine sponge *Dysidea* sp. *The Journal of Antibiotics, 63*, 699–701.

241. Renouf, D. J., Tang, P. A., & Major, P., (2012). Phase II study of the halichondrin B analog eribulin mesylate in gemcitabine refractory advanced pancreatic cancer. *Investigational New Drugs, 30*, 1203–1207.

242. Riguera, R., (1997). Isolating bioactive compounds from marine organisms *Journal of Marine Biotechnology, 5*, 187–193.

243. Rinehart, K. L. Jr., & Gloer, J. B., (1981). Didemnins: Antiviral and antitumor depsipeptides froma Caribbean tunicate. *Science*, *212*, 933–935.

244. Rinehart, K. L., (1992). Secondary metabolites from marine organisms. *Ciba Foundation Symposium*, *171*, 236–254.

245. Roberge, M., Berlinck, R. G., & Xu, L., (1998). High-throughput assay for G2 checkpoint inhibitors and identification of the structurally novel compound isogranulatimide. *Cancer Research, 58*, 5701–5706.

246. Rodriguez, J., Nieto, R. M., & Crews, P., (1993). New structures and bioactivity patterns of bengazole alkaloids from a Choristid marine sponge. *Journal of Natural Products*, *56*, 2034–2040.

247. Roggen, H., Charnock, C., & Burman, R., (2011). Antimicrobial and antineoplastic activities of agelasine analogs modified in the purine 2-position. *Archiv Der Pharmazie (Wiley), 344*, 50–55.

248. Roussis, R., Zhongde, W., & Fenical, W., (1990). New anti-inflammatory pseudopterosins from the marine octocoral *Pseudopterogorgia elisabethae*. *The Journal of Organic Chemistry*, *55*, 4916–4922.

249. Rudi, A., Yosief, T., Loya, S., Hizi, A., Schleyer, M., & Kashman, Y., (2001). Clathsterol, a novel anti-HIV-1 RT sulfated sterol from the sponge *Clathria* species. *Journal of Natural Products, 64*(11), 1451–1453.

250. Sagar, S., Kaur, M., & Minneman, K. P., (2010). Antiviral lead compounds from marine sponges. *Marine Drugs, 8*, 2619–2638.

251. Sakai, R., Rinehart, K. L., & Kishore, V., (1996). Structure–activity relationships of the didemnins. *Journal of Medicinal Chemistry*, *39*, 2819–2934.

252. Salazar, A., & Cortes-Funes, A., (2013). Phase I study of weekly kahalalide -F as prolonged infusion in patients with advanced solid tumors. *Cancer Chemotherapy and Pharmacology*, *72*(1), 1–9.

253. Sams-Dodd, F., (2005). Target-based drug discovery: Is something wrong? *Drug Discovery Today*, 139–147.

254. Samuelsson, G., (2004). *Drugs of Natural Origin: A Textbook of Pharmacognosy* (p. 244). Stockholm, Swedish Academy of Pharmaceutical Sciences.
255. Sarker, S. D., & Nahar, L., (2006). Hyphenated techniques. In: Sarker, S. D., Nahar, L., & Gray, A. I., (eds.), *Methods in Biotechnology Natural Products Isolation* (pp. 233–267). Totowa, NJ: Humana Press.
256. Sauviat, M. P., Vercauteren, J., & Grimaud, N., (2006). Sensitivity of cardiac background inward rectifying K^+ outward current (IK1) to the alkaloids lepadiformines A, B, and C. *Journal of Natural Products, 69*(4), 558–562.
257. Sawadogo, W. R., Schumacher, M., & Teiten, M. H., (2013). Survey of marine natural compounds and their derivatives with anti-cancer activity reported in 2011. *Molecules, 18*, 3641–3673.
258. Schoffski, P., Dumez, H., Wolter, P., & Stefan, C., (2008). Clinical impact of trabectedin (ecteinascidin-743) in advanced/metastatic soft tissue sarcoma. *Expert Opinion on Pharmacotherapy, 9*, 1609–1618.
259. Schoffski, P., Thate, B., & Beutel, G., (2004). Phase I and pharmacokinetic study of TZT-1027, a novel synthetic dolastatin 10 derivatives, administered as a 1-hour intravenous infusion every 3 weeks in patients with advanced refractory cancer. *Annals of Oncology, 15*, 671–679.
260. Schumacher, M., Kelkel, M., Dicato, M., & Diederich, M., (2011). Survey of marine natural compounds and their derivatives with anti-cancer activity reported in 2010. *Molecules, 16*(7), 5629–5646.
261. Schumacher, M., Kelkel, M., Dicato, M., & Diederich, M., (2011). Gold from the sea: Marine compounds as inhibitors of the hallmarks of cancer. *Biotechnology Advances, 29*, 531–547.
262. Schupp, P., Eder, C., Paul, V., & Proksch, P., (1999). Distribution of secondary metabolites in the sponge *Oceanapia* sp. and its ecological implications. *Marine Biology, 135*, 573–580.
263. Schyschka, L., Rudy, A., & Jeremias, I., (2008). Spongistatin 1: A new chemo sensitizing marine compound that degrades XIAP. *Leukemia, 22*(9), 1737–1745.
264. Serova, M., De Gramont, M., & Bieche, I., (2013). Predictive factors of sensitivity to elisidepsin, a novel Kahalalide F-derived marine compound. *Marine Drugs, 11*(3), 944–959.
265. Sharma, G. W., & Buyer, J. S., (1980). Characterization of a yellow compound isolated from the marine sponge *Phakellia flabellata*. *Journal of the Chemical Society, Chemical Communication*, 435–456.
266. Shen, Y. C., & Chen, Y. H., (2007). Four new briarane diterpenoids from the gorgonian coral *Junceella fragilis*. *Helvetica Chimica Acta, 90*(7), 139.
267. Sherif, S., (2008). Methods for isolation, purification, and structural elucidation of bioactive secondary metabolites from marine invertebrates. *Nature Protocols, 3*, 12–15.
268. Shi, Y. P., Wei, X., & Rodríguez, I. I., (2009). New terpenoid constituents of the Southwestern Caribbean-sea whip *Pseudopterogorgia elisabethae* (Bayer) including a unique pentanorditerpene. *European Journal of Organic Chemistry, 4*, 493–502.
269. Shridhar, D. M., & Mahajan, G. B., (2009). Antibacterial activity of 2-(2',4'-dibromophenoxy)-4,6-dibromophenol from *Dysidea granulosa*. *Marine Drugs, 7*, 464–471.

270. Sills, A. K. Jr., Williams, J. I., & Tyler, B. M., (1998). Squalamine inhibits angiogenesis and solid tumor growth *in vivo* and perturbs embryonic vasculature. *Cancer Research, 58*, 2784–2792.

271. Singh, M., & Hodges, L. D., (2008). The CO_2-SFE crude lipid extract and the free fatty acid extract from *Perna canaliculus* have anti-inflammatory effects on adjuvant-induced arthritis in rats. *Comparative Biochemistry and Physiology-Part B: Biochemistry and Molecular Biology, 149*(2), 251–258.

272. Sipkema, D., Franssen, M. C. R., Osinga, R., Tramper, J., & Wijffels, R. H., (2005). Marine sponges as pharmacy. *Marine Biotechnology, 7*(3), 142–162.

273. Sittampalam, G. S., Coussens, N. P., & Nelson, H., (2004). *Assay Guidance Manual.* Eli Lilly & Company and the National Center for Advancing Translational Sciences, (ebook), https://www.ncbi.nlm.nih.gov/pubmed/22553861 (accessed on 18 May 2020).

274. Skov, M. J., Beck, J. C., Kater, A. W., & Shopp, G. M., (2007). Nonclinical safety of Ziconotide: An intrathecal analgesic of a new pharmaceutical class. *International Journal of Toxicology, 26*, 411–421.

275. Smith, V. J., Desbois, A. P., & Dyrynda, E. A., (2010). Conventional and nonconventional antimicrobials from fish, marine invertebrates, and micro-algae. *Marine Drugs, 8*, 1213–1262.

276. Sobota, R., Szwed, M., Kasza, A., Bugno, M., & Kordula, T., (2000). Arthenolide inhibits activation of signal transducers and activators of transcription (STATs) induced by cytokines of the IL-6 family. *Biochemical and Biophysical Research Communications, 267*, 329–333.

277. Souza, T. M., Abrantes, J. L., Epifanio, R., Leite, F. C. F., & Frugulhetti, I. C., (2007). The alkaloid 4-methylaaptamine isolated from thesponge *Aaptos aaptos* impairs herpes simplex virus type 1 penetration and immediate-early protein synthesis. *Planta Medica, 73*, 200–205.

278. Spira, A. I., Iannotti, N. O., & Savin, M. A., (2012). Phase II study of eribulin mesylate (E7389) in patients with advanced, previously treated non-small-cell lung cancer. *Clinical Lung Cancer, 13*, 31–38.

279. Stead, P., Hiscox, S., Robinson, P. S., & Eryloside, F., (2000). Novel penasterol disaccharide possessing potent thrombin receptor antagonist activity. *Bioorganic and Medicinal Chemistry Letters, 10*(7), 661–664.

280. Stead, P., (1998). Isolation by preparative HPLC. In: Cannell, R. J. P., (ed.), *Methods in Biotechnology, Natural Products Isolation* (Vol. 4, pp. 165–208). Humana, Totowa, NJ.

281. Steinberg, P. D., (1992). Geographical variation in the interaction between marine herbivores and brown algal secondary metabolites. In: Paul, V. J., (ed.), *Ecological Roles or Marine Natural Products* (pp. 51–92). Comstock Press, Ithaca, NY.

282. Stout, E. P., Hasemeyer, A. P., & Lane, A. L., (2009). Antibacterial neuromenolides from the Fijian red alga *Neurymenia fraxinifolia*. *Organic Letters, 11*, 225–228.

283. Suárez, Y., González, L., & Cuadrado, A., (2003). New marine-derived compound induces oncosis in human prostate and breast cancer cells. *Molecular Cancer Therapeutics, 2*, 863–872.

284. Sudek, S., & Lopanik, N. B., (2007). Identification of the putative bryostatin polyketide synthase gene cluster from *Candidatus Endobugula sertula*, the uncultivated microbial symbiont of the marine bryozoan *Bugula neritina*. *Journal of Natural Products, 70*, 67–74.

285. Suetsuna, K., Maekawa, K., & Chen, J. R., (2004). Antihypertensive effects of *Undaria pinnatifida* (wakame) peptide on blood pressure in spontaneously hypertensive rats. *The Journal of Nutritional Biochemistry, 15*, 267–272.

286. Suetsuna, K., & Nakano, T., (2000). Identification of an antihypertensive peptide from peptic digest of wakame (*Undaria pinnatifida*). *The Journal of Nutritional Biochemistry, 11*, 450–454.

287. Suffness, M., & Pezzuto, J. M., (1991). Assays related to cancer drug discovery. In: Hostettmann, K., (ed.), *Methods in Plant Biochemistry, Assays for Bioactivity* (Vol. 6, pp. 71–153). London: Academic Press.

288. Swinney, D. C., & Anthony, J., (2011). How were new medicines discovered? *Nature Reviews Drug Discovery*, 507–519.

289. Takei, M., Burgoyne, D. L., & Andersen, R. J., (1994). Effect of contignasterol on histamine release induced by anti-immunoglobulin E from rat peritoneal mast cells. *Journal of Pharmaceutical Sciences, 83*, 1234–1235.

290. Takishima, S., Ishiyama, A., & Iwatsuki, M., (2009). Merobatzelladines A and B, anti-infective tricyclic guanidines from a marine sponge *Monanchora* sp. *Organic Letters, 11*, 2655–2658.

291. Taniguchi, M., Uchio, Y., Yasumoto, K., Kusumi, T., & Ooi, T., (2008). Brominated unsaturated fatty acids from marine sponge collected in Papua New Guinea. *Chemical and Pharmaceutical Bulletin, 56*, 378–382.

292. Tardy, C., Facompre, M., & Laine, W., (2004). Topoisomerase I mediated DNA cleavage as a guide to the development of antitumor agents derived from the marine alkaloid lamellarin D: Triester derivatives incorporating amino acid residues. *Bioorganic and Medicinal Chemistry, 12*(7), 1697–1712.

293. Tasdemir, D., Topaloglu, B., & Perozzo, R., (2007). Marinenatural products from the Turkish sponge *Agelas oroides* that inhibit the enoyl reductases from *Plasmodium falciparum, Mycobacterium tuberculosis,* and *Escherichia coli. Bioorganic and Medicinal Chemistry, 15*, 5834–5845.

294. Ter-Haar, E., & Kowalski, R. J., (1996). Discodermolide: A cytotoxic marine agent that stabilizes microtubules more potently than Taxol. *Biochemistry, 35*, 243–250.

295. Terracciano, S., Aquino, M., & Rodriguez, M., (2006). Chemistry and biology of anti-inflammatory marine natural products: Molecule interfering with cyclooxygenase, NF-κB and other unidentified targets. *Current Medicinal Chemistry, 13*, 1947–1969.

296. Thurman, E. M., & Mills, M. S., (1998). *Solid Phase Extraction: Principles and Practice* (1st edn., p. 384). John Wiley & Sons, New York.

297. Tianero, M. D., Hanif, N., DeWoogd, N. J., Van, S. R. W., & Tanaka, J., (2009). New antimicrobial fatty acids from the calcareous sponge *Paragrantia* cf. *waguensis. Chemistry and Biodiversity, 6*, 1374–1377.

298. Treschow, A. P., & Hodges, L. D., (2007). Novel anti-inflammatory ω-3 PUFAs from the New Zealand green-lipped mussel, *Perna canaliculus. Comparative Biochemistry and Physiology-Part B: Biochemistry and Molecular Biology, 147*(2), 645–656.

299. Tziveleka, L. A., Vagias, C., & Roussis, V., (2003). Natural products with anti-HIV activity from marine organisms. *Current Topics in Medicinal Chemistry, 3*(13), 1512–1535.

300. Vairappan, C. S., Suzuki, M., Ishii, T., Okino, T., Abe, T., & Masuda, M., (2008). Antibacterial activity of halogenated sesquiterpenes from Malaysian *Laurencia* sp. *Phytochemistry, 69*, 2490–2494.

301. Van-Middlesworth, F., & Cannell, R. J. P., (1998). Dereplication and partial identification of natural products. *Methods in Biotechnology, Isolation of Marine Natural Products, 4*, 387–390.

302. Venkateshwar-Goud, T., Srinivasa-Reddy, N., & Raghavendra-Swamy, N., (2003). Anti-HIV activepetrosins from the marine sponge *Petrosia similis*. *Biological and Pharmaceutical. Bulletin, 26*(10), 1498–1501.

303. Vermeirssen, V., Van, C. J., & Verstraete, W., (2004). Bioavailability of angiotensin I converting enzyme inhibitory peptides. *British Journal of Nutrition, 92*, 357–366.

304. Villa, F. A., Lieske, K., & Gerwick, L., (2010). Selective MyD88-dependent pathway inhibition by the cyanobacterial natural product malyngamide F acetate. *European Journal of Pharmacology, 629*, 140–146.

305. Volcho, K. P., (2009). Use of natural compounds in catalytic synthesis of chiral biologically active substances. *Thesis for Degree of Doctor of Chemical Sciences* (p. 223). Gainseville, FL: University of Florida.

306. Wamtinga, K., & Sawadogo, R., (2015). Survey of marine natural compounds and their derivatives with anti-cancer activity reported in 2012. *Molecules, 20*, 7097–7142.

307. Wang, Y. X., Pettus, M., Gao, D., & Phillips, C., (2000). Effects of intrathecal administration of ziconotide, a selective neuronal N-type calcium channel blocker, on mechanical allodynia and heat hyperalgesia in a rat model of postoperative pain. *Pain, 84*, 151–158.

308. Wehrli, S. L., Moore, K. C., Roder, H., Durell, S., & Zasloff, M., (1993). Structure of the novel steroidal antibiotic squalamine determined by two-dimensional NMR spectroscopy. *Steroids, 58*, 370–378.

309. Weinheimer, A. J., & Spraggins, R. L., (1969). The occurrence of two new prostaglandin derivatives (15-epi-PGA2 and its acetate, methyl ester) in the gorgonian chemistry of coelenterates. XV. *Tetrahedron Letters, 10*(59), 5185–5188.

310. Wender, P. A., & De Brabander, J., (1998). Synthesis of the first members of a new class of biologically active bryostatin analogs. *Journal of the American Chemical Society, 120*, 4534–4535.

311. West, L. M., & Northcote, P. T., (2000). Peloruside A: Potent cytotoxic macrolide isolated from the New Zealand marine sponge Mycale sp. *The Journal of Organic Chemistry, 65*, 445–449.

312. West, L. M., Northcote, P. T., Hood, K. A., Miller, J. H., & Page, M. J., (2000). Mycalamide D: New cytotoxic amide from the New Zealand marine sponge Mycale species. *Journal of Natural Products, 63*, 707–709.

313. WHO., (2018). *Cancer: Fact Sheet 297*. http://www.who.int/mediacentre/factsheets/fs297/en/ (accessed on 18 May 2020).

314. Wijffels, R. H., (2008). Potential of sponges and microalgae for marine biotechnology. *Trends in Biotechnology, 26*, 26–31.

315. Williams, D. H., & Faulkner, D. J., (1996). Isomers and tautomers of hymenialdisine and debromohymenialdisine. *Natural Product Letters, 9*, 57–64.

316. World Health Organization., (2017). *Cardiovascular Diseases*. Online: http://www.who.int/cardiovascular_diseases/about_cvd/en/ (accessed on 18 May 2020).

317. Wright, A. D., & McCluskey, A., (2011). Anti-malarial, anti-algal and antitubercular, anti-bacterial, anti-photosynthetic and anti-fouling activity of diterpene and diterpene isonitriles from the tropical marine sponge *Cymbastela hooperi*. *Organic and Biomolecular Chemistry, 9*, 400–407.

318. Wright, A. E., & Boterho, J. C., (2007). Neopeltolide, amacrolide from a lithistid sponge of the family *Neopeltidae*. *Journal of Natural Products, 70*, 412–416.
319. Wu, M. L., Ho, Y., Lin, C. Y., & Yet, S. F., (2011). Heme oxygenase-1 in inflammation and CV disease. *American Journal of Cardiovascular Disease, 1*(2), 150–158.
320. Xu, W. H., & Ding, Y., (2009). Puupehanol, a sesquiterpene hydroquinone derivative from the marine sponge *Hyrtios* sp. *Bioorganic and Medicinal Chemistry Letters, 19*, 6140–6143.
321. Yamanokuchi, R., Imada, K., & Miyazaki, M., (2012). Hyrtioreticulins A-E, indole alkaloids inhibiting the ubiquitin-activating enzyme, from the marine sponge *Hyrtios reticulatus*. *Bioorganic and Medicinal Chemistry Letters, 20*, 4437–4442.
322. Yao, G., & Chang, L. C., (2007). Novel sulfated sesterterpene alkaloids from the marine sponge *Fasciospongia* sp. *Organic Letters, 9*, 3037–3040.
323. Yao, G., Kondratyuk, T. P., Tan, G. T., Pezzuto, J. M., & Chang, L. C., (2009). Bioactive sulfated sesterterpene-alkaloids and sesterterpene sulfates from the marine sponge *Fasciospongia* sp. *Journal of Natural Products, 72*, 319–323.
324. Yasuhara-Bell, J., & Lu, Y., (2010). Marine compounds and their antiviral activities. *Antiviral Research, 86*(3), 231–240.
325. Yasumoto, T., & Yotsu, M., (1989). Inter-species distribution and biogenetic origin of tetrodotoxin and its derivatives. *Pure and Applied Chemistry, 61*, 505–508.
326. Youssef, D. T., Ibrahim, A. K., & Khalifa, S. I., (2010). New anti-inflammatory sterols from the red sea sponges, *Scalarispongia aqabaensis* and *Callyspongia siphonella*. *Natural Product Communications, 5*, 27–31.
327. Zampella, A., Sepe, V., & Luciano, P., (2008). Homophymine A, an anti-HIV cyclodepsipeptide from the sponge *Homophymia* sp. *The Journal of Organic Chemistry, 73*, 5319–5327.
328. Zhang, H., Khalil, Z. G., & Capon, R. J., (2011). Fascioquinols A-F: Bioactive meroterpenes from a deep-water Southern Australian marine sponge, *Fasciospongia* sp. *Tetrahedron, 67*, 2591–2595.
329. Zhang, H., Skildum, A., Stromquist, E., Rose-Hellekant, T., & Chang, L. C., (2008). Bioactive polybrominated diphenyl ethers from the marine sponge *Dysidea* sp. *Journal of Natural Products, 71*, 262–264.
330. Zhang, Y. P., Xu, F., & Wang, M., (2006). Progress on marine anticancer drugs in R&D. *Program Details-Pharmaceutical Science, 30*, 433–442.
331. Zheng, J., Xu, Z., Wang, Y., Hong, K., Liu, P., & Zhu, W., (2010). Cyclic tripeptides from the halotolerant fungus *Aspergillus sclerotiorum* PT06-1. *Journal of Natural Products, 73*, 1133–1137.
332. Zheng, J., Zhu, H., Hong, K., Wang, Y., Liu, P., Wang, X., et al., (2009). Novel cyclic hexapeptides from marine-derived fungus *Aspergillus sclerotiorum* PT06-1. *Organic Letters, 11*, 5262–5265.
333. Zhou, S., & Hamburger, M., (1996). Application of liquid chromatography atmospheric pressure ionization mass spectrometry in natural products analysis: Evaluation and optimization of electrospray and heated nebulizer interfaces. *Journal of Chromatography A, 755*, 189–204.
334. Zwar, D., (1994). *The Magic Mussel-Arthritis Another Way?* (2nd edn., p. 108). Ideas Unlimited, Cairns, Australia.

CHAPTER 8

MARINE FUNGI-DERIVED SECONDARY METABOLITES: POTENTIAL AS FUTURE DRUGS FOR HEALTH CARE

SYED SHAMS UL HASSAN, HUI-ZI JIN, ABDUR RAUF, SAUD BAWAZEER, and HAFIZ ANSAR RASUL SULERIA

ABSTRACT

About half of all drugs are dependent on the support of natural products. In current years, a substantial number of structurally unique compounds with interesting therapeutic activities have been isolated from the marine-derived fungus (alkaloids, polyketides, peptides, terpenoids, lactones, and steroids) with pharmacological activities, such as antibacterial, antiviral, anticancer, anti-inflammatory, Immunosuppressive, cardiovascular, antioxidant, antidiabetic, and enzymatic. This review chapter focuses on therapeutically active biocompounds from marine-derived fungus with classification according to the sources of isolation of fungus and their pharmacological activities. Novel biocompounds without pharmacological activity have not been reported. The chapter includes 111 bioactive compounds isolated from marine fungi during 2016 to 2017.

8.1 INTRODUCTION

Plant and marine-based natural products have complex chemical structures and these cooperate effectively with their biological targets [1 to 4.6, 7, 17]. Despite the facts that mostly all existing natural product-derived therapeutics has global origins, yet untapping of marine environment will make the road for chemical and biological innovations [22]. Today, thousands of biocompounds with their unique structures with their bioactivities have been reported from marine fungi [9, 10, 12, 15].

The research review in this chapter was aimed to overview sources, chemistry, biological activities, and mechanisms of action of bioactive secondary metabolites derived from marine-derived fungi. This chapter has focused on the period from 2016 to 2017.

In this review chapter, the marine-derived biomolecules (secondary metabolites) were classified according to the sources of marine fungi, such as mangroves, alga, sediment, sponges, corals, sea cucumber, sea worms, and seawater (Table 8.1).

8.2   MARINE FUNGI AS SOURCE OF FUTURE DRUGS

8.2.1   MANGROVES

Mangroves are small trees or shrubs that grow out in coastal saline or brackish water. According to literature, there are around 80 pronounced species of mangroves, 60 of which grow absolutely on coasts between the high-tide and low-tide lines. Mangroves are now a major concern for bioactive compounds studies. Almost all body parts of it have been studied. Some biocompounds are isolated from leaves [5], and some belong to its fresh tissue [60], etc. Mangroves derived fungus is one of the largest sources of the bioactive secondary metabolites with promising pharmacological properties, such as antibacterial, anticancer, anti-inflammatory, antiviral, etc. [8].

A novel marine chromone derivate (2'S*)-2-(2'-hydroxypropyl)-5-methyl-7,8-dihydroxy (1) (Figure 8.1) chromone was isolated from fungus (*Penicillium aculeatum*) that was obtained from mangrove leaves (*Kandeliacandel*) collected in Yangjiang, Guangdong province of South China Sea. The compound exhibited potent antibacterial activity against *Salmonella* with an MIC value of 2.00 ± 0.02 μM [20]. Six novel alkaloidal compounds (penicisulfuranols A to F: Figure 8.1 (2 to 7)) were isolated from the mangrove endophytic fungi (*Penicillium janthinellum*) that were obtained from the roots of mangrove plant *Sonneratia caseolaris,* in the South China Sea (Note: In this chapter, numbers 2 to 7, etc. refer to the images in this figure, etc.).

The compounds 2 to 4 exhibited remarkable cytotoxic activity against both cell lines with its IC_{50} values ranging from 0.1 to 3.9 μM, while the compounds 5 to 7 did not exhibit activity ($IC_{50} > 30$ μM) as indicated in Table 8.1 [64].

TABLE 8.1 Bioactive Compounds Derived from Marine Fungi

Sampling Region	Biocompound	Species	Activity	Cell Line	Isolation Area	IC$_{50}$/MIC*/ EC$_{50}$**	References
Mangroves (Figure 8.1)							
South China Sea	(2'S*)-2-(2'-hydroxypropyl)-5-methyl-7, 8-dihydroxy (1***)	Penicillium aculeatum	Anti-bacterial	Salmo-nella	Leaves/ Kandelia candel sp.	2.00 ± 0.02 µM*	[20]
South China Sea	Penicisulfuranols A to F (2–7)	Penicillium janthinellum	Cytotoxic	HL-60	Roots/ Sonneratia caseolaris	0.1 to 3.9 µM	[64]
South China Sea	Asperiso-coumarins A, B, E & F (8–11)	Aspergillus sp.	Anti-diabetic	—	Roots/ Acanthus ilicifolius	87.8; 52.3; and 95.6 µM	[60]
South China Sea	Nectriacids B, C (12, 13)	Nectria sp.	Anti-diabetic	—	Branches/ Sonneratia ovata	23.5; and 42.3 µM	[8]
South China Sea	Peniphenone (14); methyl peniphenone (15)	Penicillium sp	Immuno-suppressive	LPS-induced (B cell)	Leaves/ Sonneratia apetala	9.3; and 23.7 µM	[34]
South China Sea	Penicopeptide A (16)	Penicillium commune	anti-adipogenic	11β-HSD1) enzyme	Leaves/Vitis vinifera	9.07 ± 0.61 µM	[54]

TABLE 8.1 *(Continued)*

Sampling Region	Biocompound	Species	Activity	Cell Line	Isolation Area	IC$_{50}$/MIC*/ EC$_{50}$**	References
Alga (Figure 8.2)							
South China Sea	8-hydroxy-coniothyrinone B (17), 8,11-dihydroxy-coniothyrinone B (18), 4S,8-dihydroxy-10-Omethyldendryol E (19)	*Talaromyces islandicus*	Anti-bacterial	*Staphylococcus aureus; E. coli*	Inner tissue/ *Laurencia okamurai*	2–8 μg/ml*	[30]
East coast of China	Isovariecolorin I (20), 30-Hydroxy-echinulin (21)	*Eurotium cristatum*	Anti-helminths	—	*Sargassum thunbergii*	19.4 μg/ml	[10]
Northern Taiwan	Phomaketides A, C (22, 23)	*Phoma sp.*	Anti-inflammatory	RAW264.7	*Pterocladiella capillacea*	8.8 μM	[28]
South sea of China	Varioloid A (24)	*Paecilomyces variotii*	Cytotoxic	HCT116; A549	*Grateloupia turuturu*	2.5–6.4 μg/ mL	[62]
Egyptian Red Sea	Griseoxanthone C (25)	*Fusarium equiseti,*	Anti-Viral	HCV protease	*Padina pavonica*	19.8 μM	[16]
Kongdong Island, China	Sesteralterin (26)	*Alternaria alternate*	Anti-microbial	*Heterosigma akashiwo*	*Lomentaria hakodatensis*	41–69%	[49]

Sampling Region	Biocompound	Species	Activity	Cell Line	Isolation Area	IC₅₀/MIC*/ EC₅₀**	References
Shandong, China	Chermesins A, B (27, 28) Chermesins C, D (no activity)	*Penicillium chermesinum*	Anti-microbial	*C. albicans; E. coli*	*Pterocladiella tenuis*	32 µg/mL*	[35]
Qingdao, China	2,2′,3,5′-tetrahydroxy-3′-methyl-benzophenone (29);2,2′,5′-trihydroxy-3-methoxy-3′-methyl-benzophenone (30); 1,4,7-trihydroxy-6-methyl-xanthone (31)	*Talaromyces islandicus*	Anti-oxidant	ABTS	*Laurencia okamurai*	0.69 µg/mL; 0.58 µg/mL; 2.35 µg/ mL.	[31]
South sea of China	Aspernigrin C (32)	*Aspergillus Niger*	Anti-Viral	HIV-1 SF162	Un-identified	4.7 ± 4 µM	[63]
Sediment (Figure 8.3)							
South sea of China	Aspergilols H, I (33, 34)	*Aspergillus versicolor*	Anti-Viral	HSV-1	Depth of 2326 m	4.68; 6.25 µM**	[19]
South sea of China	4-carbglyceryl-3,30-dihydroxy-5,50 dimethyl diphenyl ether (35)	*Aspergillus versicolor*	Anti-fouling	*Bugula neriina*	Depth of 2326 m	1.28 µg/ ml**	[18]
North east coast of Brazil	Ester furan derivate (36)	*Aspergillus Niger*	Cytotoxic	HCT-116 cell	Depth of 1820 m	2.9 µg/ml	[57]

TABLE 8.1 *(Continued)*

Sampling Region	Biocompound	Species	Activity	Cell Line	Isolation Area	IC$_{50}$/MIC*/ EC$_{50}$**	References
Indian ocean	Gliotoxin (37)	*Aspergillus sp.*	Anti-T.B	*M. Tuberculosis*	Depth of 4530 m	0.030 μM*	[37]
Indian ocean	Penicitrinone F (38),	*Penicillium chrysogenum*	Anti-Viral	EV71	—	14.50 μM	[6]
Indian ocean	Bipenicilisorin (39)	*Penicillium chrysogenum*	Cytotoxic	A549	—	2.59 μM	[6]
Indian ocean	Butyrolactone derivatives (40, 41)	*Penicillium chrysogenum*	Anti-inflammatory	—	—	1.09 μM; 1.97 μM	[6]
Korea	Citrinin H1 (42)	*Penicillium sp*	Anti-inflammatory	—	—	8.1 μM	[42]
South China Sea	Aspewentins D, F, G, H (43, 44, 45, 46)	*Aspergillus wentii*	Anti-microbial	*M. luteus; F. graminearum; P. aeruginosa*	depth of 2038 m	2–32 μg/mL	[32]
Sponges (Figure 8.4)							
South sea of China	Asteltoxin E (47) and F (48)	*Aspergillus sp*	Anti-Viral	H3N2	*Cally-spongia sp.*	6.2 ± 0.08 μM and 8.9 ± 0.3 μM	[56]
East sea Korea	Tanzawaic acid Q (49)	*Penicillium steckii*	Anti-inflammatory	—	—	—	[50]

Sampling Region	Biocompound	Species	Activity	Cell Line	Isolation Area	IC$_{50}$/MIC*/ EC$_{50}$**	References
South sea of China	Chartarenes A to D (50, 51, 52, 53)	Stachybotrys chartarum	Anti-tumor	HepG2, HCT-116; BGC-823	Niphates recondite	3.39 μM; 5.58 μM; 0.74 μM; 1.48 μM.	[30]
South sea of China	Pestaloisocoumarins A & B (54, 55)	Pestalotiopsis heterocornis	Anti-bacterial	B. Subtilis; S. aureus	Phakellia fusca	50 μM*; 25 μM*	[29]
South sea of China	Austalide S & U (56, 57)	Aspergillus aureolatus	Anti-Viral	H1N1	—	90 μM	[46]
—	Diasteltoxins A, B (58, 59)	Emericella variecolor	Anti-cancer	MCF	—	73 μM; 127 μM	[36]
—	Trichodermanin C (60)	Trichoderma harzianum	Cytotoxic	HL-60; P388; L1210	Halichondria okadai	6.8 μM; 7.9 μM; 7.6 μM	[61]
Corals (Figure 8.5)							
South sea of China	Aspergivones A & B (61, 62)	Aspergillus candidus	Anti-diabetic	—	Anthogorgia ochracea	244 μg/ml	[38]
South sea of China	Versiquin-azoline (A, B, G, K) (63, 64, 65, 66)	Aspergillus versicolor	Anti-cancer	A2780; A549;	Pseudo pterogorgiasp	20 ± 1 μM; 2 ± 2 μM; 13 μM; and 13 μM	[7]
—	Chondrosterins K, L, M (67, 68, 69)	Chondro-stereum sp.	Cytotoxic	CNE1, CNE2, A549	Sarcophyton tortuosum	12.0 μM; 22.5 μM; 44.1 μM	[18]

TABLE 8.1 *(Continued)*

Sampling Region	Biocompound	Species	Activity	Cell Line	Isolation Area	IC$_{50}$/MIC*/EC$_{50}$**	References
South sea of China	Aspergillipeptides D, E (70, 71)	*Aspergillus sp.*	Anti-Viral	HSV-1	*Melitodes squamata*	9.5 μM; 19.8 μM	[39]
South sea of China	Cytosporin L (72)	*Eutypella sp.*	Anti-bacterial	*Micrococcus lysodeikticus*	*Dichotella gemmacea*	3.12 μM*	[33]
—	Pseuboydones A, C (73, 74)	*Pseudallescheria boydii*	Cytotoxic	Sf9	*Lobophytum crissum*	2.2 ± 0.2 μM*; 0.7 ± 0.1 μM*.	[27]
South sea of China	Ochramide B (75)	*Aspergillus ochraceus*	Anti-microbial	*Enterobacter aerogenes*	*Dichotella gemmacea*	40 μM*	[47]
South sea of China	Ochralate (76)	*Aspergillus ochraceus*	Anti-microbial	*Enterobacter aerogenes*	*Dichotella gemmacea*	18.9 μM*	[47]
South sea of China	Phomaethers A, C (77, 78)	*Phoma sp.*	Anti-bacterial	*S. aureus; E. coli; V. parahaemolyticus.*	*Dichotella gemmacea*	0.63 μM*, 0.32 μM*	[49]
Tunicates (Figure 8.6)							
Red Sea, Egypt	Penicillosides A, B (79, 80)	*Penicillium sp*	Anti-Bacterial; Antifungal	*Candida albicans*	*Didemnum sp*	≥ 50 μg/ml	[40]
Suez Canal, Egypt	Terretrion D (81)	*Penicillium sp*	Anti-migratory	MDA-MB-231	*Didemnum sp.*	16.5 μM	[48]

Sampling Region	Biocompound	Species	Activity	Cell Line	Isolation Area	IC$_{50}$/MIC*/ EC$_{50}$**	References
Manado, Indonesia	2-hydroxy-6-(2'-hydroxy-3'-hydroxy-methyl-5-methyl-phenoxy)-benzoic acid (82)	*Penicillium aibobi-verticillium*	Anti-enzymatic	CD45	Un-identified	43 μM	[53]
Crabs (Figure 8.7)							
Kueishantao Hydro thermal vent, Taiwan	Methyl isoverrucosidinol (83)	*Penicillium sp*	Anti-bacterial	*Bacillus subtilis*	*Xenograpsus testudinatus*	32 μg/ml*	[43]
Kueishantao hydrothermal vent, Taiwan	7-methoxycyclopeptin (84); 7-methoxycyclopenin (85); 9-hydroxy-3-methoxyviridicatin (86)	*Aspergillus versicolor*	Anti-bacterial	*E. coli*	*Xenograpsus testudinatus*	32 μg/ml*	[44]
Kueishantao hydrothermal vent, Taiwan	Penicilliso-coumarin A, B, D (87, 88, 89)	*Penicillium sp*	Anti-bacterial	*E. coli*	*Xenograpsus testudinatus*	32 μg/ml*	[45]

TABLE 8.1 (Continued)

Sampling Region	Biocompound	Species	Activity	Cell Line	Isolation Area	IC$_{50}$/MIC*/EC$_{50}$**	References
Sea Worms (Figure 8.8)							
Haikou Bay, China	Chrodrimanins K, N (90, 91)	Penicillium sp	Anti-viral	H1N1	Sipun-culus nudus	74 µM; 58 µM	[24]
Haikou Bay, China	Verruculide B2 (92)	Penicillium sp	Anti-bacterial	Staphylococcus aureus	Sipun-culus nudus	32 µg/mL*	[24]
Haikou Bay, China	Chrodrimanins O,R,S (93, 94, 95)	Penicillium sp	Anti-diabetic	PTP1B	Sipun-culus nudus	71.6 ±12.3 µM; 62.5 ±10.32 µM; 63.1 ±15.6 µM.	[23]
Haikou Bay, China	Penicillars B, C (96, 97)	Penicillium sp	Anti-enzymetic	Acetylcholi-nesterase	Sipun-culus nudus	19.5 % and 21.3%.	[26]
Haikou Bay, China	Aculene E (98); Penicitor B (99)	Penicillium sp	Anti-QS	C. violaceum	Sipun-culus nudus	300 µM	[25]
Sea Cucumber (Figure 8.9)							
—	Isopimarane diterpene (100)	Epicoccum sp	Anti-diabetic	—	Apostichopus japonicas	4.6 ± 0.1 µM	[59]

Sampling Region	Biocompound	Activity	Species	Cell Line	Isolation Area	IC$_{50}$/MIC*/ EC$_{50}$**	References
–	Virescenosides R1, R2, R3, R4 (101 to 104)	Anti-enzymatic	*Acremonium striatisporum*	Esterase enzyme	*Holothurian Eupentacta Fraudatrix*	56, 58, 36 and 40%	[1]
Sao Pao Plateau, Brazil	Cladomarine (105)	Anti-parasitic	*Penicillium coralligerum*	*Saprolegnia parasitica*	Un-identified	64 µg/ml*	[55]
Seawater (Figure 8.9)							
Depth of 30 m in the Indian Ocean.	Communesin I; Fumiquinazoline Q; Protuboxepin E (106, 107, 108)	Cardio-vascular	*Penicillium expansum*	Zebrafish model	–	Potent activity at 20 µg/mL; 50 µg/mL; 100 µg/mL	[11]
East China Sea.	Flavichalasines A, E, F (109–111),	Cytotoxic	*Aspergillus flavipes*	–	–	10.5–26.0 µM	[58]

Legend:

* MIC value;

** EC$_{50}$ Va;

*** Each number refers to the corresponding image in Figures 8.1 to 8.9.

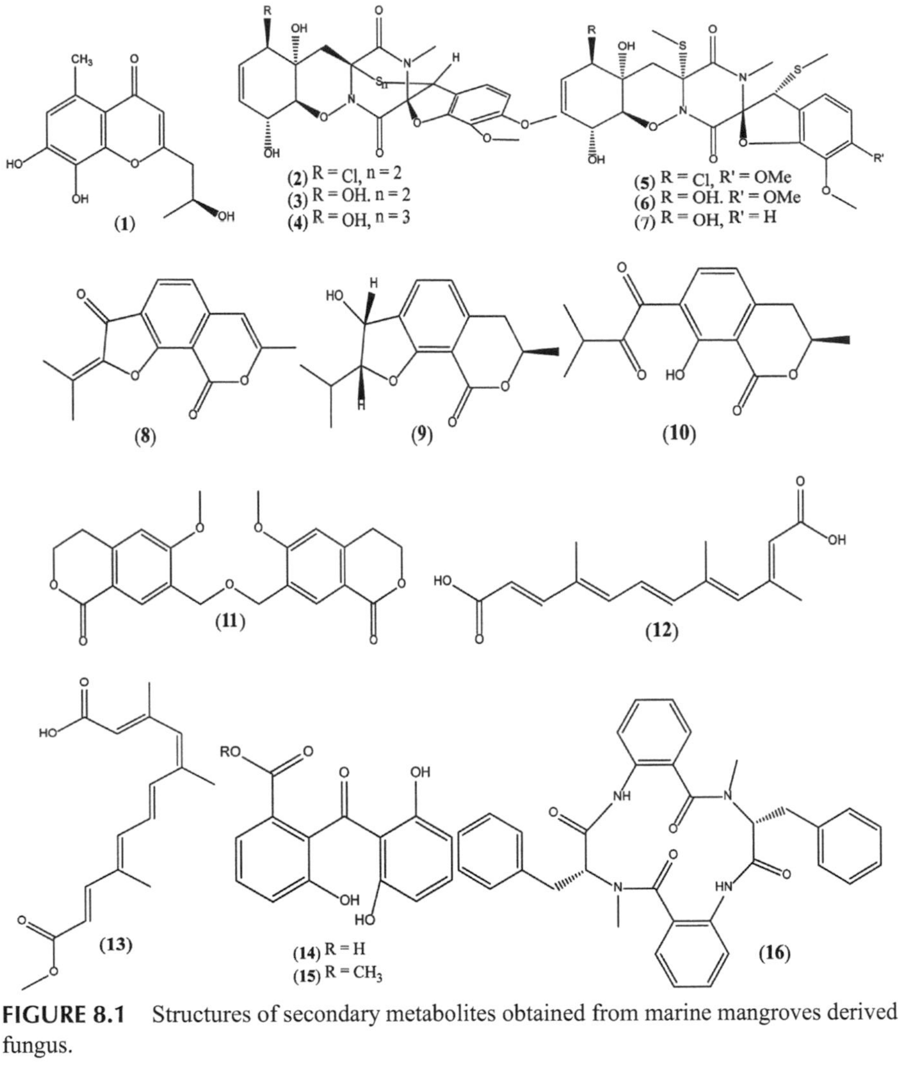

FIGURE 8.1 Structures of secondary metabolites obtained from marine mangroves derived fungus.

During the search for novel α-glucosidase inhibitors from marine environment, four novel isocoumarin derivatives (asperisocoumarins A, B, E, and F: 8 to 11) (Figure 8.1: images 8 to 11) were isolated from mangrove endophytic fungus (*Aspergillus* sp.) that was obtained from the roots of mangrove plants (*Acanthus ilicifolius*) in the South China Sea. The compounds 9 to 11 displayed a potent activity as α-glucosidase inhibitors with IC_{50} values of 87.8, 52.3, and 95.6 μM, respectively. Furthermore, the compound in image

8 showed a weak DPPH radical scavenging activity with EC_{50} values of 125 µM [60].

Two polyketides (nectriacids B, C): images 12, 13 in Figure 8.1) were isolated from the endophytic fungus (*Nectria* sp.) derived from the branches of the mangrove plant (*Sonneratia ovata*) obtained in the South China Sea. The compounds were evaluated for their α-glucosidase inhibitory activity, and the compounds displayed potent activity with IC_{50} values of 23.5 and 42.3 µM, demonstrating stronger efficacy compared to positive control (acarbose: IC_{50} of 815.3 µM) [8].

Two polyketides benzophenone derivatives (peniphenonein Figure 8.1: 14; and methyl peniphenonein Figure 8.1: 15) were derived from the endophytic fungi (*Penicillium* sp.). That was isolated from the leaves of mangrove plant (*Sonneratia apetala*) in the South China Sea. The compounds displayed a potent immunosuppressive activity against LPS-induced (B-cells) and Con A-induced (T-cells) proliferations of mouse splenic lymphocytes with IC_{50} values of 9.3 and 8.1 µM and 23.7 and 17.5 µM, respectively. The carboxylic acid group at C-1 (image 14) enhanced the immunosuppressive activity compared with the compound in image 15 containing a methyl ester group [34].

The cyclic tetrapeptide penicopeptide A (image 16 in Figure 8.1) was isolated from the culture broth of the endophytic fungus (*Penicillium commune*) that was derived from the leaves of mangrove plant (*Vitis vinifera*) in Gansu province of China. The compound penicopeptide A was examined against 11β-hydroxysteroid dehydrogenase type 1 (11β-HSD1) enzyme; and it exhibited potent inhibitory activity against human 11β-HSD1, with an IC_{50} value of 9.07 ± 0.61µM. From these studies, it has been revealed that the novel structure of the compound penicopeptide A with its potent inhibitory activities against 11β-HSD1 opens the gates for more 11β-HSD1 inhibitors from marine natural products [54].

8.2.2 ALGA

Three hydroanthraquinone derivatives (8-hydroxyconiothyrinone B (image 17), 8,11-dihydroxyconiothyrinone B (image 18), and 4S,8-dihydroxy-10-Omethyldendryol E (image 19) in Figure 8.2)) were examined for its pharmacological potential as cytotoxic, antimicrobial, and antioxidant agents. During the assay for its antimicrobial potency against the gram-positive and gram-negative microorganisms *E. coli* and *Staphylococcus aureus*, the compounds in images 17 to 19 exhibited their antimicrobial efficacy with the MIC value ranging from 2 to 8 μg/ml [30].

Two indole diketopiperazine derivatives (isovariecolorin I (image 20) and 30-Hydroxyechinulin (image 21 in Figure 8.2) were isolated from the culture extract of endophytic fungus (*Eurotium cristatum*) that were derived from brown alga (*Sargassum thunbergii*) in Qingdao, East coast of China.

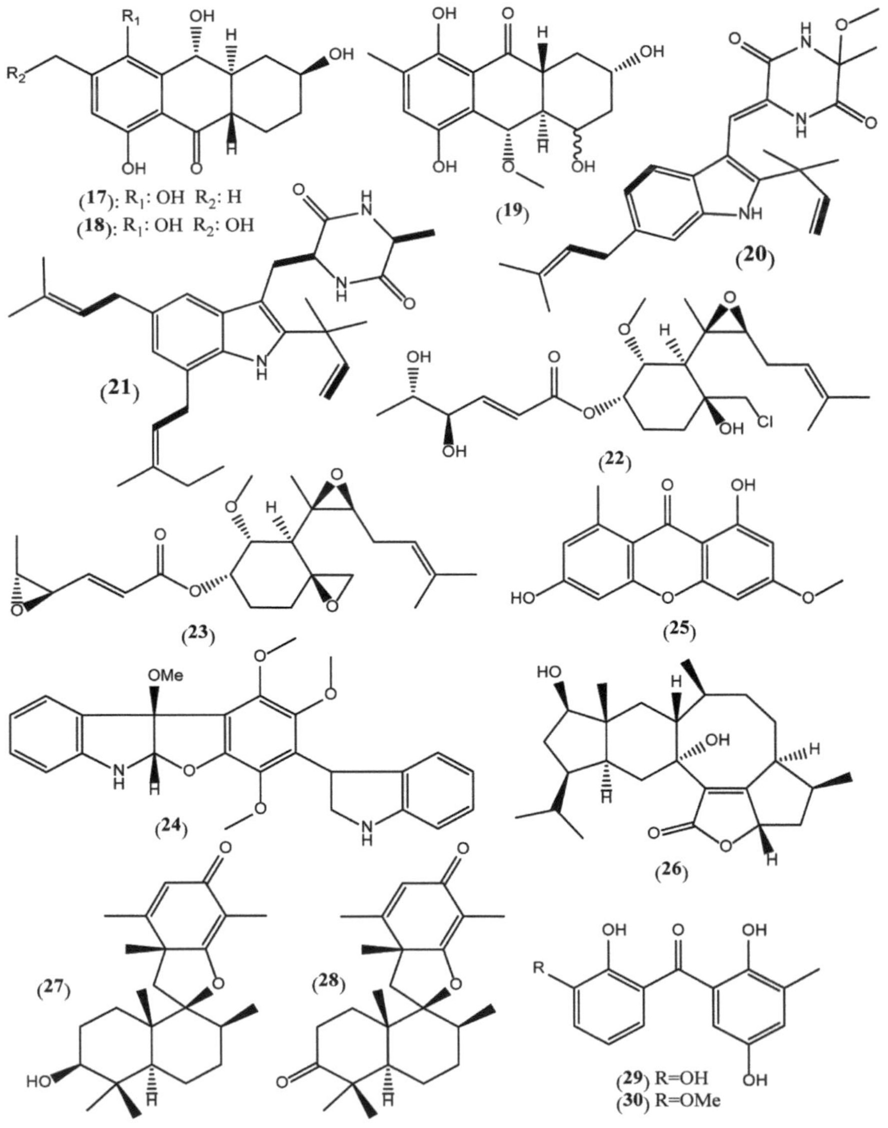

(17): R₁: OH R₂: H
(18): R₁: OH R₂: OH
(19)
(20)
(21)
(22)
(23)
(24)
(25)
(26)
(27)
(28)
(29) R=OH
(30) R=OMe

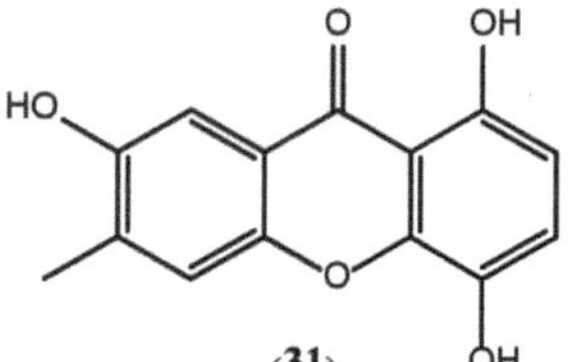

FIGURE 8.2 Structures of secondary metabolites obtained from marine alga derived fungus.

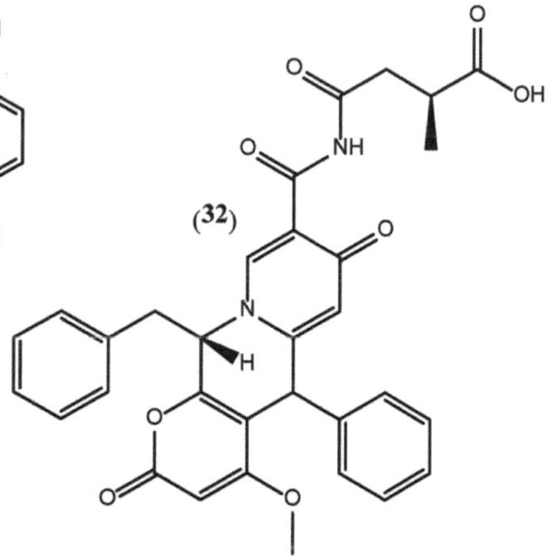

For enriching the bioactivity of these compounds, these were evaluated for their multiple pharmacological activities, such as nematicidal activity, brine shrimp lethality, antioxidative potency. For the evaluation of nematicidal efficacy, only the compound in image 20 exhibited weak activities with LD_{50} value of 110.3 µg/ml, while the compound in image 21 did not displayed any of these activities. For the brine shrimp lethality, the compound in image 20 exhibited potent lethal activity with LD_{50} value of 19.4 µg/ml compared to weak activity of compound in image 21. For the antioxidant potential, only the compound in image 20 exhibited radical scavenging activity against DPPH with IC_{50} value of 20.6 µg/ml [10].

Two polyketides (phomaketides A and phomaketides C (images 22, 23 in Figure 8.2) were isolated and purified from the fermented mycelium and broth of the endophytic fungus (*Phoma* sp.) that were derived from the marine red alga (*Pterocladiella capillacea*) in Northern Taiwan. Both compounds (images 22 and 23) exhibited selective inhibitory efficacy of LPS-induced NO production in RAW264.7 macrophages without any cytotoxicity with an IC_{50} value of 8.8 µM. The compound in image 23 displayed the potent antiangiogenic efficacy by inducing inhibitory effects on the tube formation of EPCs with an IC_{50} value of 8.1 µM. Therefore, it is suggested to evaluate the extracts of *Phoma* sp. anti-inflammatory and antiangiogenic agents [28].

The varioloid A (image 24 in Figure 8.2) was assayed for its cytotoxic activities against a panel of cell lines including HCT116 and A549 (human lungadenocarcinoma cells) and HepG2 (human hepatoma cells). This compound exhibited a remarkable activity against all three cell lines with IC_{50} values of 6.4, 3.5, and 2.5 µg/mL [62], respectively. The secondary metabolites were evaluated for their anti-viral efficacy against hepatitis C virus (HCV). Among all compounds, the compound griseoxanthone C (image 25 in Figure 8.2) exhibited a remarkable activity against HCV protease inhibition with IC_{50} value of 19.8 µM; and potent Trypsin inhibitory activity with the IC_{50} value of 48.5 µM [16].

Sesteralterin (image 26 in Figure 8.2) was isolated from the culture extract of marine fungus (*Alternaria alternate*) that were derived from marine red alga (*Lomentaria hakodatensis*) in Kongdong Island of China. The compound sesteralterin represents the first "nitidasanesesterterpene" naturally produced by fungus. This compound was examined for arresting the growth of three marine plankton (*Heterosigma akashiwo*, *Chattonella marina and Prorocentrum donghaiense*); and it indicated 41–69% of inhibition against these three pathogens [49].

Two spiromeroterpenoids (Chermesins A in image 27 and Chermesins B in image 28 (Figure 8.2)) were evaluated for their antimicrobial activities against human pathogens and aquatic bacteria. Both compounds displayed antimicrobial activity against *C. albicans*, *E. coli* and *V. alginolyticus*. While compounds Chermesins C and Chermesins D did not show any antimicrobial activity because of the absence of double bond at C-5′ or the cyclohexa-2,5 dienone unit, which is essential for antimicrobial efficacy [35].

Two diphenylketones (2,2′,3,5′-tetrahydroxy-3′-methylbenzophenone (image 29 in Figure 8.2); and 2,2′,5′-trihydroxy-3-methoxy-3′-methylbenzophenone (image 30 in Figure 8.2) and third biocompound (xanthone 1,4,7-trihydroxy-6-methylxanthone (image 31 in Figure 8.2) were isolated from marine fungus (*Talaromyces islandicus*) that were derived as an endophytic fungi from marine red alga (*Laurencia okamurai*) in coastal region of Qingdao, China. These three compounds were evaluated for their antimicrobial and antioxidant efficacies; and these displayed DPPH (1,1-diphenyl-2-picrylhydrazyl) activity with an IC_{50} value of 1.26, 1.33 and 6.92 µg/mL and showed more potent ABTS (2,2′-azino-bis(3-ethylbenzothiazoline-6-sulfonate) radical scavenging activity with an IC_{50} value of 0.69, 0.58 and 2.35 µg/mL, respectively. During the antimicrobial assay, the compound in image 30 exhibited low antimicrobial efficacy compared to the compounds in images 29 and 31 that exhibited antibacterial activity against *E. coli*, *P. aeruginosa* and the aquatic bacteria *V. alginolyticus* [31].

The 2-benzylpyridin-4-one containing metabolite (aspernigrin C (image 32 in Figure 8.2)) was isolated from the marine-derived fungus (*Aspergillus Niger*) that was derived from the unidentified marine alga. This compound aspernigrin C was evaluated its potent anti-viral efficacy against HIV-1 SF162 with an IC_{50} value of $4.7\pm.4$ µM. The other new compounds from aspernigrin class did not exhibit any anti-viral activity because 2-methylsuccinic acid moiety may remarkably improve the anti-HIV-1 activity of aspernigrins [63].

8.2.3 SEDIMENT

Two anthraquinones (aspergilol H (image 33 in Figure 8.3), anthraquinones aspergilol I (image 34 in Figure 8.3)), and one diphenyl ether (4-carbglyccryl-3,3'-dihydroxy-5,5'dimethyldiphenyl ether (image 35 in Figure 8.3)) were isolated from deep marine-derived fungi (*Aspergillus versicolor*).

These three compounds were evaluated for multiple pharmacologic activities. The compounds were first tested towards HSV-1 using a plaque reduction assay. The results exhibited that under their non-cytotoxic concentrations (CC_0) against a Vero cell-line, compounds (33) and (34) had evident antiviral activity towards HSV-1 with EC_{50} values of 4.68 and 6.25 µM and CC_{50} values of 108.6 and 50.7 µM, respectively [19].

The ester furan derivate (image 36 in Figure 8.3) was isolated from a fungus (*Aspergillus Niger*) that was derived from marine sediment collected from North-east coast of Brazil at a depth of 1820 m. This compound exhibited its cytotoxic activities against HCT-116 cell line with IC_{50} value of 2.9 µg/ml [57].

An alkaloid gliotoxin (image 37 in Figure 8.3) was isolated from the marine fungus (*Aspergillus* sp.) that was obtained from the marine sediment collected from an Indian ocean from the depth of 4530 m. The compound (37) was isolated by liquid rotary fermentation based on OSMAC technique to get different compounds; and it was evaluated for pharmacological activities, such as anti-tubercular, antibacterial, and cytotoxic. The compound (37) displayed potent anti-tubercular activity against *Mycobacterium Tuberculosis* with MIC_{50} value of 0.030 µM. Moreover, the anti-tubercular efficacy in terms of inhibition was superior to positive control INH suggesting that Sulphur-bridge may improve the efficacy. The compound (37) also exhibited significant selective *in vitro* cytotoxicity against A549, K562, and Huh-7 cell lines with IC_{50} values of 0.015, 0.191, and 95.4 µM, respectively. Additionally, deep-sea-derived fungus *Aspergillus* sp. SCSIO Ind09F01 exhibited moderate antibacterial activity against Gram-positive and Gram-negative microorganisms, such as *S. aureus*, *E. coli* and *Salmonella* [37].

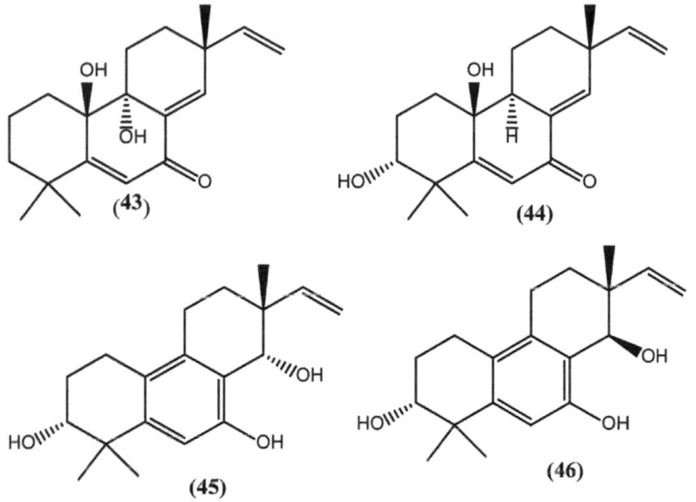

FIGURF 8.3 Structures of secondary metabolites obtained from marine sediment derived fungus.

The citrinin dimer (penicitrinone F (image 38 in Figure 8.3)) and dimeric isocoumarin (bipenicilisorin (image 39 in Figure 8.3), and two butyrolactone derivatives (images 40, 41 in Figure 8.3) were isolated from marine fungus (*Penicillium chrysogenum*) that were obtained from an Indian Ocean. All four compounds were evaluated for pharmacological activities, such as Antiviral, cytotoxic, and anti-inflammatory. The penicitrinone F (image 38) exhibited moderate antiviral activity against EV71 with IC_{50} value of 14.50 µM. The bipenicilisorin (image 39) demonstrated potent cytotoxic activity against A549, K562 and Huh-7 cell lines with IC_{50} value of 6.94, 6.78 and 2.59 µM, respectively. Furthermore, the compounds butyrolactone derivatives (images 40, 41) exhibited specific COX-2 inhibitory activity with IC_{50} values of 1.09 µM and 1.97 µM [6]. The compound citrinin H1 (image 42 in Figure 8.3) along with seven other metabolites were tested for their anti-inflammation activities. Among these 8 compounds, only citrinin H displayed potent anti-inflammatory activity [42].

Four 20-nor-isopimarane diterpenoids (Aspewentins D (Image 43 in Figure 8.3), Aspewentins F (Image 44 in Figure 8.3), Aspewentins G (Image 45 in Figure 8.3), and Aspewentins H (Image 46 in Figure 8.3)) were isolated from cultural extract of marine fungus (*Aspergillus wentii*). The aspewentin D (43) displayed moderate activity against *E. coli*, *P. aeruginosa* and *G. graminis* with an MIC value of 32 µg/mL and displayed strong activity against *M. luteus* and *F. graminearum* with an MIC value of 4 µg/mL and 2 µg/mL. The aspewentin F (44) displayed moderate activity against *S. aureus, V. anguillarum* and *V. parahemolyticus* with an MIC value of 32 µg/mL and displayed strong activity against *E. tarda* with an MIC value of 4 µg/mL. The aspewentin G (45) displayed moderate activity against *S. aureus, C. gloeosprioides*, and *G. graminis* and strong activity against *V. harveyi* with an MIC value of 4 µg/mL. The aspewentin H (46) showed moderate activity against *E. coli, V. anguillarum, V. parahemolyticus, C. gloeosprioides*, and *G. graminis* with an MIC value of 32 µg/mL and displayed strong activity against *P. aeruginosa* and *F. graminearum* with an MIC value of 4 µg/mL [32].

8.2.4 SPONGES

Sponge-derived microorganisms associated pharmacologically active compounds have gained significance as a source of antimicrobial, cytotoxic, antioxidant, anti-inflammatory, and immunosuppressive agents. Marine sponges-oriented fungus has proved to produce unique chemical diversity and potent biologically activities [4].

The asteltoxin E (image 47 in Figure 8.4) and asteltoxin F (image 48 in Figure 8.4) were evaluated for antiviral efficacy against (H1N1 and H3N2) through cytopathic effect (CPE) inhibition assay [56]. Tanzawaic acid Q (image 49 in Figure 8.4) was also isolated from marine-derived fungus (*Penicillium steckii*) that was obtained from a marine sponge in East Sea Korea, for evaluation as anti-inflammatory agent. The Tanzawaic acid Q (49) reduced the mRNA levels of inflammatory cytokines (such as IL-1β and TNF-α); and inhibited LPS-induced inflammation. This is the first report on the anti-inflammatory activity of Tanzawaic acid Q [50].

Four sesquiterpenes (Chartarenes A (image 50 in Figure 8.4), Chartarenes B (image 51 in Figure 8.4), Chartarenes C (image 52 in Figure 8.4), Chartarenes D (image 53 in Figure 8.4)) were isolated from marine sponge (*Niphates recondite*) that was obtained from marine fungus (*Stachybotrys chartarum*) collected at a depth of 10 m from Guangxi province in China. All four compounds were evaluated for anti-tumor efficacy. The compounds chartarenes A to D displayed potent activity against HepG2, HCT-116, BGC-823, A2780, and NCI-H1650 cancer cell lines (Table 8.2).

Also, all four compounds revealed inhibitory activity against tumor related-kinases, such as FGFR3, PDGFRb, IGF1R, WT, and TRKB (Table 8.3). FGF signaling plays crucial role in several aspects of cancer biology, e.g., anti-apoptosis, involvement in proliferation, angiogenesis, drug resistance, and EMT; thus, FGFR-targeted therapeutics is an attractive target in clinical oncology. These four compounds reveal that these compounds can be considered for further investigations as future drugs in clinical oncology [30].

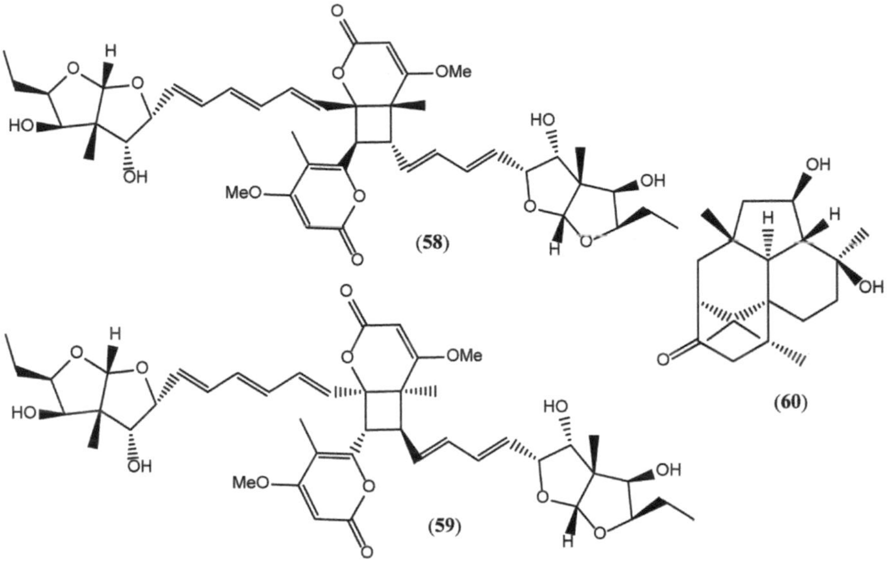

FIGURE 8.4 Structures of secondary metabolites obtained from marine sponges derived fungus.

TABLE 8.2 Inhibitory Effects of Chartarenes-A to -D (Images 50 to 53 in Figure 8.4) Against Tumor Cell-Lines

Compound	HepG2	HCT-116	BGC-823	A2780	NCI-H1650
	IC$_{50}$ (μM)				
Chartarenes A	3.95 ± 0.01	3.39 ± 0.01	2.87 ± 0.02	2.38 ± 0.02	>10
Chartarenes B	>10	5.58 ± 0.03	>10	>10	>10
Chartarenes C	2.09 ± 0.01	0.74 ± 0.02	>10	2.07 ± 0.02	2.58 ± 0.03
Chartarenes D	0.90 ± 0.02	1.48 ± 0.01	0.68 ± 0.04	0.69 ± 0.03	1.23 ± 0.02

Two isocumarins (Pestaloisocoumarins A (image 54 in Figure 8.4) and Pestaloisocoumarins B (image 55 in Figure 8.4)) were isolated from marine derived fungus (*Pestalotiopsis heterocornis*) that was obtained from marine sponge (*Phakelliafusca*) collected at Xiasha islands in China, for evaluation of antibacterial and antifungal activities. The both compounds (54 and 55) exhibited potent antibacterial activity against *B. subtilis* with MIC value of 50 μM and 25 μM and against *S. aureus* with MIC value of 25 μM and 25 μM, respectively; and exhibited moderate antifungal activity against *Cryptococcus neoformans* with MIC values of 100 μM and 100 μM [29].

TABLE 8.3 Inhibitory Effects of Compounds Chartarenes-A to -D (Images 50 to 53 in Figure 8.4) Against Tumor-Related Kinases

Compound	FGFR3	PDGFRb	IGF1R	WT	TRKB
	IC_{50} (µM)				
Chartarenes A	>25	>25	>25	>25	>25
Chartarenes B	2.4	10.4	6.9	12.9	7
Chartarenes C	1.1	5.3	3.0	6.3	2.7
Chartarenes D	0.1	0.8	0.1	0.7	0.7

Two meroterpenoids (Austalide S (image 56 in Figure 8.4) and Austalide U (image 57 in Figure 8.4)) were isolated from the endophytic fungus (*Aspergillus aureolatus*) that was obtained from unidentified sponge collected from Xiasha islands in China [46]. Two asteltoxin-containing dimers (diasteltoxins A (image 58 in Figure 8.4) and diasteltoxins B (image 59 in Figure 8.4)) were also isolated from the sponge-derived mutated strain of *Emericella variecolor* fungus by means of chemical elicitation. The Diethyl sulfate was used as a chemical elicitor to activate the sleeping genes of fungus strain. These diasteltoxins A and B were not observed in normal growth of the current strain but were activated as novel compounds by means of chemical elicitation. All four compounds were examined for anti-cancer potential. The diasteltoxins A and B exhibited weak inhibition against the growth of breast cancer (MCF) with IC_{50} values of 73 µM and 127 µM and cell lung cancer (H1299) cell lines with IC_{50} values of 188 µM, and 164 µM, respectively. Furthermore, all diasteltoxins A and B significantly inhibited thioredoxin reductase (TrxR) with IC_{50} values of 12.8 ±0.8 µM and 11.1 ± 0.2 µM, respectively.

The Trichodermanin C (image 60 in Figure 8.4) was examined for its cytotoxic activity. It exhibited potent cytotoxicity against leukemia Human HL-60, murine P388 and murine L1210 cell lines with an IC_{50} value of 6.8 µM, 7.9 µM and 7.6 µM, respectively [61].

8.2.5 CORALS

Coral reefs are invertebrates belonging to class *Anthozoa* and *phylum Cnidaria*. They are disseminated all over the oceans in tropics and subtropics. Their exterior and internal muscles are often inhabited by copious microbial species, especially symbiotic fungi. Corals have been recognized to be rich sources of structurally novel and biologically active secondary metabolites that have been explored by researchers for drug discovery [38].

Two flavones compounds (Aspergivones A (image 61 in Figure 8.5) and Aspergivones B (image 62 in Figure 8.5)) were isolated from the endophytic fungus (*Aspergillus candidus*) that obtained from the gorgonian coral (*Anthogorgiao chracea*) collected from South China Sea. Aspergivone A was reported to have 30% inhibitory rate against alpha-glucosidase while aspergivone B showed activity with an IC_{50} value of 244 µg/ml [38].

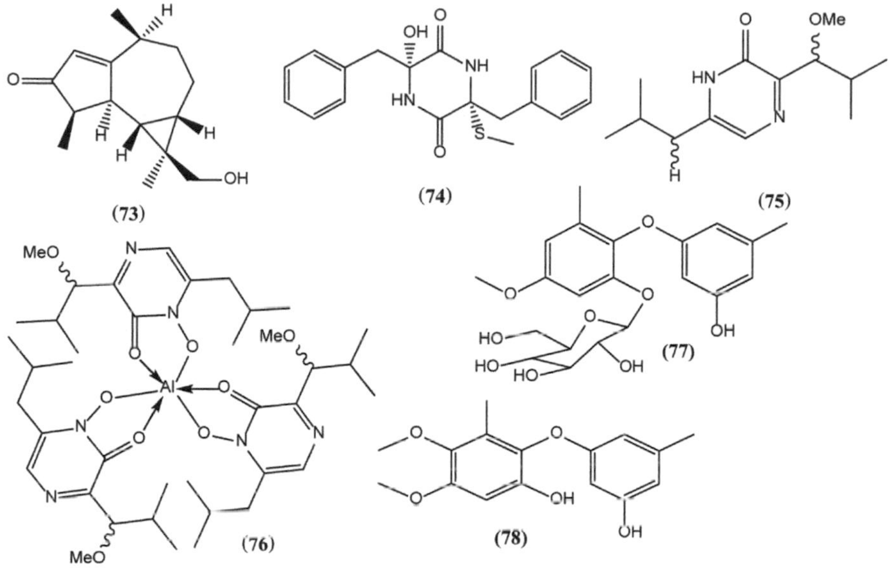

FIGURE 8.5 Structures of secondary metabolites obtained from fungus derived from marine corals.

Four fumiquinazoline-type alkaloids [(Versiquinazoline A (image 63 in Figure 8.5), Versiquinazoline B (image 64 in Figure 8.5), Versiquinazoline G (image 65 in Figure 8.6), Versiquinazoline K (image 66 in Figure 8.5)] were isolated from the marine derived fungus (*Aspergillus* versicolor) that were obtained from the gorgonian *Pseudopterogorgia sp.* in the South China Sea.

Versiquinazoline A and B containing the methanediamine unit and representing a unique subtype of fumiquinazolines were found from nature for the first time. All these four compounds exhibited weak inhibitory action against A2780 (human ovarian cancer) and A549 (lung adenocarcinoma) cell lines. Also, all four compounds displayed potent inhibitory effects against thioredoxin reductase with IC_{50} values of 20 ± 1 μM for Versiquinazoline A, 12 ± 2 μM for Versiquinazoline B, 13 μM for Versiquinazoline G and 13 μM for Versiquinazoline K [7].

The chondrosterin K (image 67 in Figure 8.5), chondrosterin L (image 68 in Figure 8.5), and chondrosterin M (image 69 in Figure 8.5) were obtained by the 1HNMR pre-screening method and by tracing the diagnostic proton signal of methyl groups. The chondrosterin K displayed potent activity with an IC_{50} values of 17.66 μM against CNE1, 12.03 μM against CNE2, 22.06 μM against HONE1, 16.44 μM against SUNE1, 23.51 μM against A549, 18.08 μM against GLC82 and 22.14 μM against HL7702. The chondrosterin

L also displayed a strong activity with an IC_{50} values of 33.5 µM against CNE1, 22.50 µM against CNE2, 34.60 µM against HONE1, 30.40 µM against SUNE1, 29.67 µM against A549, 37.47 µM against GLC82 and 34.26 µM against HL7702. The chondrosterin M also demonstrated strong cytotoxic efficacy with an IC_{50} values of 42.0 µM against CNE1, 44.08 µM against CNE2, 46.11 µM against HONE1, 58.83 µM against SUNE1, 49.58 µM against A549, 55.90 µM against GLC82, and 56.40 µM against HL7702, respectively [18].

The aspergillipeptides D (image 70 in Figure 8.5) aspergillipeptides (image 71 in Figure 8.5) were evaluated for their anti-viral efficacy. The aspergillipeptides D and E both exhibited an antiviral efficacy against herpes simplex virus type 1 (HSV-1) with IC_{50} values of 9.5 µM and 19.8 µM under their non-cytotoxic concentrations against a Vero cell line. In addition, the aspergillipeptides D also demonstrated its antiviral activity against acyclovir-resistant clinical isolates of HSV-1 [38, 39].

A hexahydrobenzopyran derivative (cytosporin L (image 72 in Figure 8.5)) was isolated from the marine derived fungus *Eutypellasp.* that was obtained from the gorgonian *Dichotella gemmacea* in the South China Sea. The cytosporin L exhibited antibacterial activity against *Micrococcus lysodeikticus* and *Enterobacter aerogenes* with the same MIC value of 3.12 µM. In addition, it strongly arrested the respiratory syncytial virus (RSV) with IC_{50} value of 72.01 µM [33].

The aromadendrane-type sesquiterpene diastereomer (pseuboydones A (image 73 in Figure 8.5) and diketopiperazine (pseuboydones C (image 74 in Figure 8.5) (74) were isolated from the culture broth of marine fungus (*Pseudallescheria boydii*) that was obtained from the soft coral (*Lobophytum crissum*). The pseuboydones A and pseuboydones C were examined for their cytotoxicity test against the Sf9 cells with MIC value of 2.2 ± 0.2 µM and 0.7 ±0.1 µM [27], respectively.

One pyrazin-2(1*H*)-one derivate (ochramide B (image 75 in Figure 8.5) and one ochralate (image 76 in Figure 8.5) were isolated from marine fungus *Aspergillus ochraceus* that was obtained from gorgonian coral *Dichotella gemmacea* in South sea of China. Both of these compounds exhibited antimicrobial activity against *Enterobacter aerogenes* with an MIC value of 40.0 µM and 18.9 µM, respectively [47].

Two diphenyl ethers (Phomaethers A (image 77 in Figure 8.5) and Phomaethers C (image 78 in Figure 8.5)) were isolated from marine fungus *Phoma* sp. that was obtained from the inner soft tissue of gorgonian *Dichotella gemmacea* in the South China Sea. It is the first report for discovery of the diphenyl glycoside derivatives from coral-derived fungus. Both compounds showed strong antibacterial activity against *S. aureus, E. coli,* and *S. albus* with MIC values of 0.312 µM, 1.25 µM, and 0.625 µM, respectively [49].

8.2.6 TUNICATES

Tunicates are marine invertebrates belonging to sub-phylum Tunicata, which is a part of the phylum Chordata. Secondary metabolites associated with tunicates are dynamic way to study the drug discovery and development [41].

Tunicates have long been studied for their bioactivity by different scientists. Usually inner tissues of tunicates have been examined to isolate bioactive microbes, especially fungi [40]; and different habitat associated tunicates possessed different varieties of microbes [9].

Two bioactive cerebrosides (Penicillosides A (image 79 in Figure 8.6) and Penicillosides B (image 80 in Figure 8.6) were isolated from the fungus *Penicillium* sp. that was obtained from the tunicate *Didemnum sp.* in Egyptian Red Sea coast at a depth of 1–2 m. The Penicillosides A exhibited anti-fungal efficacy against *Candida albicans* with inhibition zone of 23 mm, while Penicillosides B exhibited antibacterial activity against gram-positive *Staphylococcus aureus* with inhibition zone of 19 mm and gram-negative *E. coli* with inhibition zone of 20 mm. In addition, both compounds also showed weak anti-cancer activity against HeLa cells with IC_{50} value of ≥ 50 µg/ml [40].

A bioactive Terretrion D (image 81 in Figure 8.6) with 1,4-diazepane skeleton was isolated from the endophytic fungus *Penicillium* sp. that was obtained from inner tissue of tunicate *Didemnum sp.* in Suez Canal-Egypt at a depth of 1–2 m. Terretrion D possessed potent anti-migratory efficacy, much better than the positive control Z-4-ethylthio-phenylmethylene hydantoin (S-Ethyl) with IC_{50} value of 43.4 µM. The compound also showed weak anti-proliferation potential against HeLa cell lines with $IC_{50} > 50$ µM/ml. Furthermore, the compound also displayed antifungal activity [48] and it was evaluated for its anti-enzymatic activity [53]. The methanolic extracts of marine-derived fungi isolated from a tunicate *Didemnum* sp. that was obtained from Red sea were evaluated for anti-proliferative and cytotoxic activities. The extracts showed potent cytotoxic activity against MCF-7 cell lines with IC_{50} value of 5.93 µg/ml and 11.37 µg/ml, respectively. Therefore, marine tunicates-derived fungi have great potential as cytotoxic compounds [2].

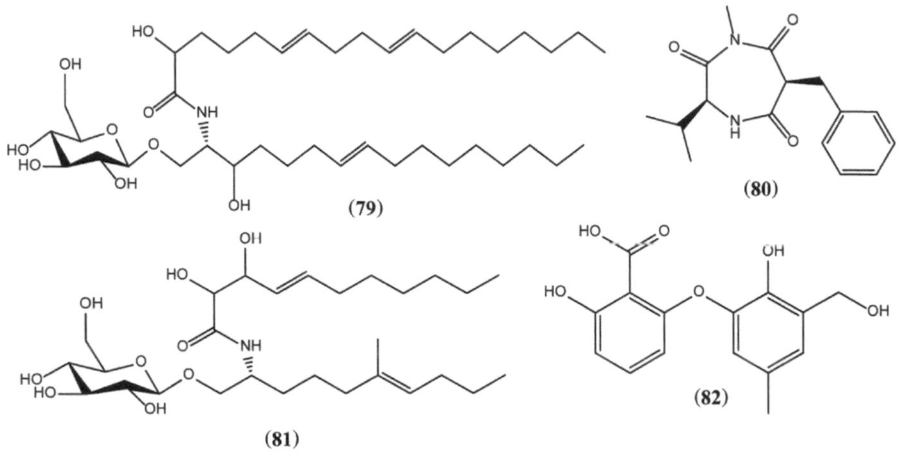

FIGURE 8.6 Structures of secondary metabolites obtained from marine tunicates derived fungus.

8.2.7 CRABS

Crabs are *Decapod crustaceans*. They are generally covered with a thick exoskeleton and have a single pair of claws. Since ancient times, crabs have been used as a food product and are source of antioxidant and antimicrobial metabolite compounds [51]. Crab possesses nutraceutical, pharmaceutical, and cosmo-ceutical potential [44].

The bioactive verrucosidin derivative (methyl isoverrucosidinol (image 83 in Figure 8.7)) was purified from marine endophytic fungus *Penicillium sp.* [50]. Three Cyclopenin derivatives (7-methoxycyclopeptin (image 84 in Figure 8.7), 7-methoxycyclopenin (image 85 in Figure 8.7), and 9-hydroxy-3-methoxyviridicatin (image 86 in Figure 8.7)) possessed inhibitory activities against *E. coli* with MIC value of 32 µg/ml [44].

Three isocoumarins (penicillisocoumarin A (image 87 in Figure 8.7), penicillisocoumarin B (image 88 in Figure 8.7) and penicillisocoumarin D (image 89 in Figure 8.7)) were examined for their cytotoxic and antibacterial activities. However, all these three compounds only exhibited potent antibacterial activity against *E. coli* with MIC value of 32 µg/ml [45].

8.2.8 SEA WORMS

Two meroterpenoids (Chrodrimanins K (image 90 in Figure 8.8), Chrodrimanins N (image 91 in Figure 8.8) and verruculide B2 (image 92 in Figure

8.8)) were isolated from the marine-derived fungus *Penicillium sp.* that was obtained from the marine worm *Sipunculus nudus* in Haikou Bay, China. All three compounds were evaluated for their antiviral and antibacterial activities. The compounds chrodrimanin K and chrodrimanin N showed anti-viral activity against H1N1 with IC_{50} values of 74 µM and 58 µM, respectively. The verruculide B2 exhibited weak inhibition action against *Staphylococcus aureus* with an MIC of 32 µg/mL [24].

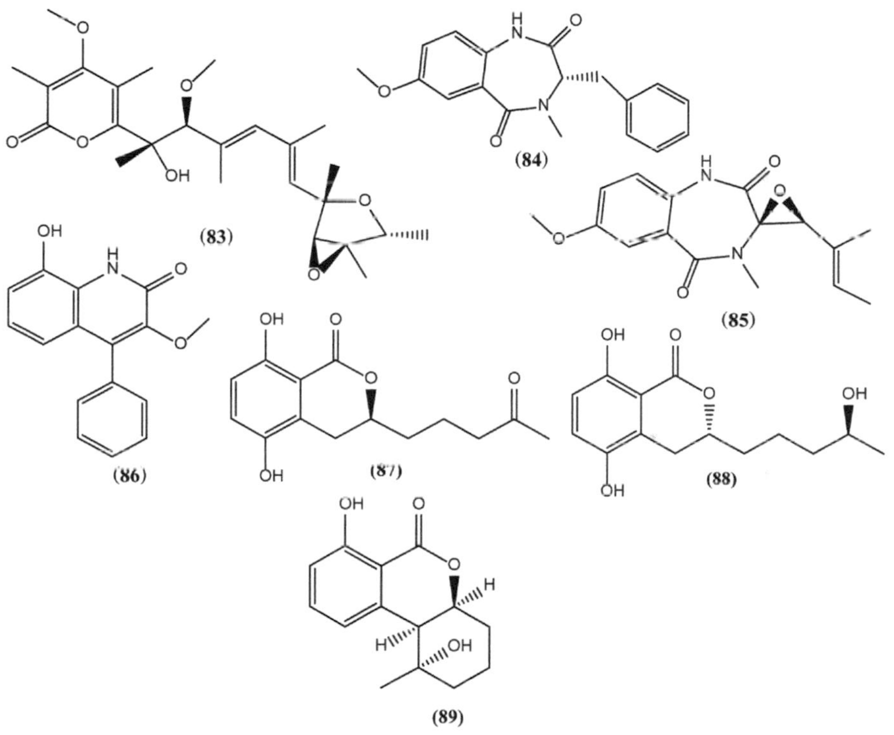

FIGURE 8.7 Structures of secondary metabolites obtained from marine crab derived fungus.

Three meroterpenoids (chrodrimanins O (image 93 in Figure 8.8), chrodrimanins R (image 94 in Figure 8.8) and chrodrimanins S (image 95 in Figure 8.8)) were isolated from the culture broth of marine fungus *Penicillium sp.* that was obtained from the marine worm *Sipunculus nudus* in Haikou Bay, China. All three compounds were evaluated for its anti-enzymatic and anti-tumor activities. But the compounds did not exhibit satisficed response towards these activities.

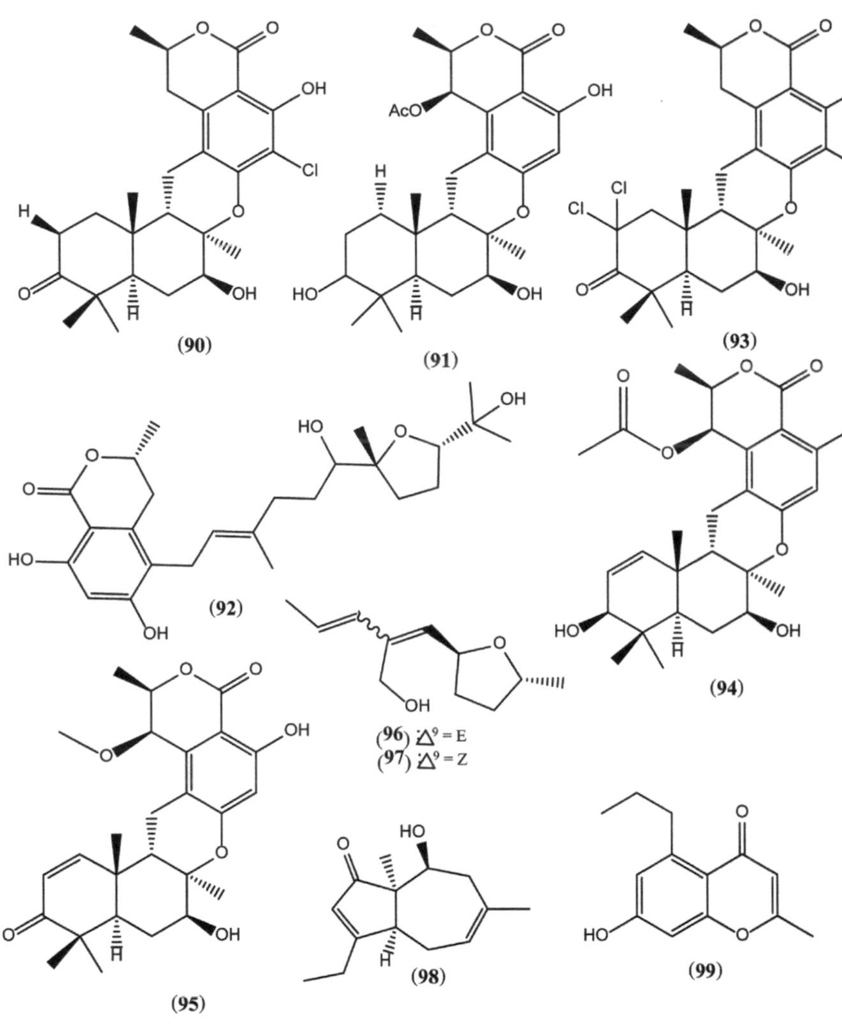

FIGURE 8.8 Structures of secondary metabolites obtained from sea worms derived fungus.

According to the previous results for novel compounds from same family towards target of protein tyrosine phosphatase 1B (PTP1B), a specific target for treatment of Type-2 diabetes, leads for an evaluation of novel compounds efficacy towards this target. The results revealed that these compounds have moderate efficacy towards this target with an IC_{50} value of 71.6 ± 12.3 µM for chrodrimanin O, 62.5 ± 10.32 µM for chrodrimanin R, and 63.1 ± 15.6 µM for chrodrimanin S, respectively. Therefore, PTP1B inhibitors can easily be achieved by appropriate structural modifications of chrodrimanins [23].

The Penicillar B (image 96 in Figure 8.8) and Penicillar C (image 97 in Figure 8.8) were isolated from the marine endophytic fungus *Penicillium* sp. that was obtained from the marine worm *Sipunculus nudus* in Haikou Bay, China. The Penicillar B and Penicillar C, at a concentration of 50 mg/mL, displayed weak inhibitory activity against acetylcholinesterase (AChE) with inhibition rate of 19.5% and 21.3%, respectively [26].

The Aculene E (image 98 in Figure 8.8) and Penicitor B (image 99 in Figure 8.8) were derived from the culture of the marine fungus *Penicillium sp.* that was obtained from the marine worm *Sipunculus nudus* in Haikou Bay, China. Both these compounds were examined for their quorum sensing inhibition activities [25].

8.2.9 SEA CUCUMBER

Sea cucumbers (also known as Holothuroids) are soft-bodied worms like marine invertebrates [52]. It is consumed as raw traditionally or is boiled in many countries [59]. Apart from this, sea cucumber is also used as traditional food tonic in Taiwan, Korea, and China.

Also, sea cucumber is of economic value in these regions. In the past, it has served as a therapeutic natural remedy for fatigue, sexual impotence, impotence because of aging, urinary incontinence, hypertension, anemia, arthritis, intestinal dryness, etc. Sea cucumber toxins also have anti-tumor, anti-viral, anti-pregnancy properties [1, 21].

The isopimarane diterpene (image 100 in Figure 8.9) was examined to check its anti-diabetic efficacy. The isopimarane diterpene displayed α-glucosidase inhibitory efficacy with IC_{50} value of 4.6 ± 0.1 µM. This is the first report of isopimarane diterpene with this activity [59].

Four diterpene glycosides (Virescenosides R1 (image 101 in Figure 8.9), Virescenosides R2 (image 102 in Figure 8.9), Virescenosides R3 (image 103 in Figure 8.9) and Virescenosides R4 (image 104 in Figure 8.9)) were isolated from the marine endophytic fungus *Acremonium striatisporum* that was obtained from sea cucumber *holothurian Eupentacta fraudatrix*. All four compounds were studied for enzymatic actions and inhibition of non-specific esterase activity in mouse lymphocytes. These compounds were applied at concentration of 100 mg/mL and these displayed inhibition of esterase activity of 56, 58, 36, and 40%, respectively [1].

The cladomarine (image 105 in Figure 8.9) was isolated from the marine fungus *Penicillium coralligerum* that was obtained from sea cucumber in Sao Pao Plateau, Brazil. It was evaluated for its anti-parasitic activity. It displayed potent activity against *Saprolegnia parasitica* with an MIC value of 64 µg/ml[-1]. The compound also exhibited moderate anti-oomycete activity against *Pythium* sp. [55].

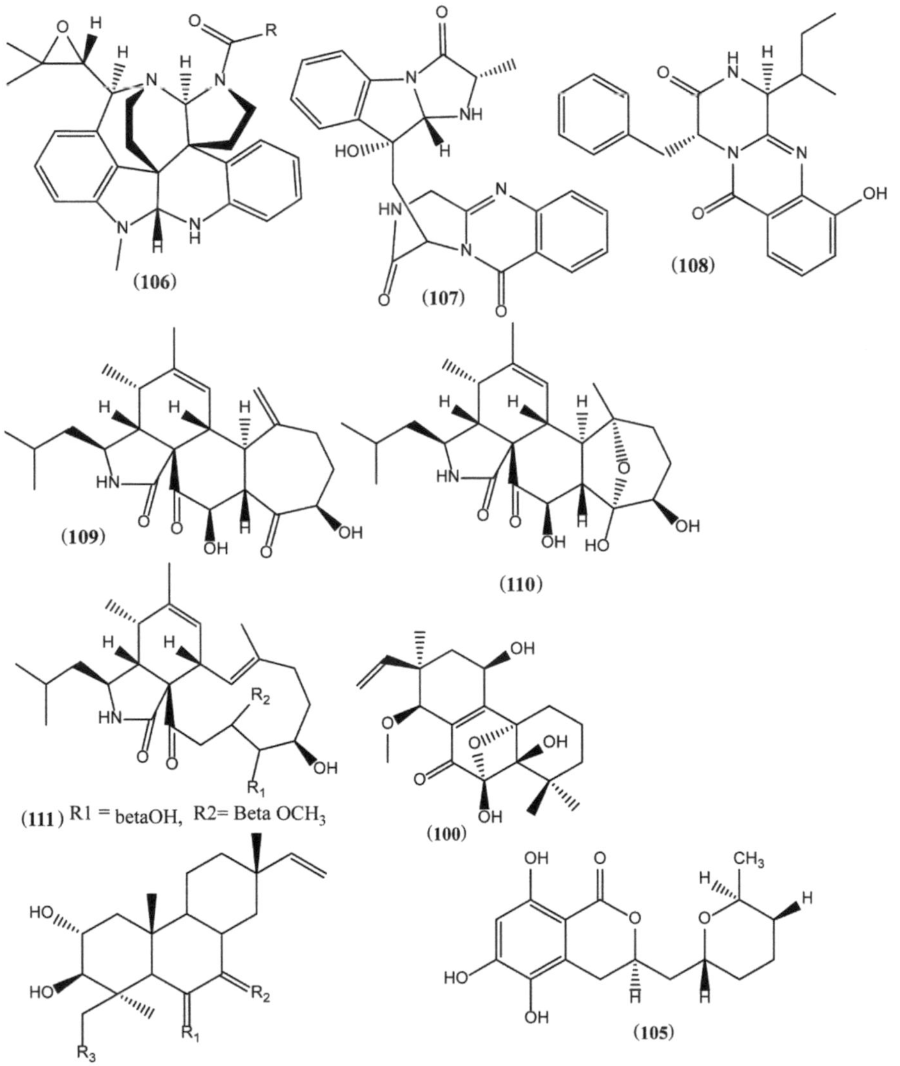

(106)
(107)
(108)
(109)
(110)
(111) R1 = betaOH, R2= Beta OCH$_3$
(100)
(105)

(101) R$_1$ = H , R$_2$ = H , R$_3$ = alpha-D-Glc$p$-(1-6)-Beta-D-Alt$p$[-]
(102) R$_1$ = H, H , R$_2$ = O , R$_3$ = alpha-D-Glc$p$-(1-6)-Beta-D-Alt$p$[-]
(103) R$_1$ = H, H , R$_2$ = H , R$_3$ = alpha-D-Glc$p$-(1-4)-Beta-D-Alt$p$[-]
(104) R$_1$ = O , R$_2$ = H , R$_3$ = Beta-D-Alt$p$-, ^{4}C$_1$

FIGURE 8.9 Structures of secondary metabolites from seawater and sea cucumber derived fungus.

8.2.10 SEAWATER

Three alkaloids (Communesin I (image 106 in Figure 8.9), Fumiquinazoline Q (image 107 in Figure 8.9) and Protuboxepin E (image 108 in Figure 8.9)) were tested for their cardiovascular effects in Zebra fish model. During the experiments on heart rate, the Fumiquinazoline Q and Protuboxepin E (at a concentration of 20 µg/mL, 50 µg/mL, and 100 µg/mL) exhibited very potent vasculo-genetic activity and the Communesin I demonstrated a moderate activity. This has been first study on these compounds, which exhibited cardiovascular effects in Zebrafish model [11].

Out of three compounds, such as tetracyclic cytochalasin (flavichalasines A (image 109 in Figure 8.9), pentacyclic cytochalasin (flavichalasines E (image 110 in Figure 8.9) and tricyclic cytochalasin (flavichalasines F (image 111 in Figure 8.9): only flavichalasines F exhibited moderate inhibitory activities against Jurkat with an IC_{50} value of 10.5 µM; HL60 with an IC_{50} value of 12.8 µM; NB4 with an IC_{50} value of 12.4 µM; HEP-3B with an IC_{50} value of 13.6 µM; HCT-116 with an IC_{50} value of 26.6 µM; and RKO with an IC_{50} value of 11.6 µM, respectively [58].

8.3 SUMMARY

Current articles were reviewed for marine sources of secondary metabolites for health care. Currently, thousands of compounds with their unique chemical structures and with their bioactivities have been reported from marine fungi. The development of these biomolecules can overcome the problems owing to the early entrance of first marine natural products in drug market [3]. Can a resolute channel of marine drugs be provided by an ocean? The chemical structure can be analyzed by various emerging techniques, such as sampling strategies [45], modern culturing technique [13], genetic engineering, and nanoscale NMR [14].

KEYWORDS

- **anti-cancer**
- **anti-diabetic**
- **hepatitis C virus**

- herpes simplex virus type 1
- marine fungi
- therapeutically active

REFERENCES

1. Afiyatullov, S. S., Kalinovsky, A., Antonov, A. S., & Zhuravleva, O. I., (2016). Isolation and structures of virescenosides from the marine-derived fungus *Acremonium striatisporum*. *Phytochemistry Letters*, *15*, 66–71.
2. Alahdal, A. M., Shaala, L. A., Noor, A. O., & Elfaky, M. A., (2017). Evaluation of the antiproliferative and cytotoxic activities of marine invertebrates-derived fungi. *Pak. J. Pharm. Sci.*, *30*(3), 1001–1006.
3. Anjum, K., Abbas, S. Q., Akhter, N., Shagufta, B. I., Shah, S. A., & Hassan, S. S. U., (2017). Emerging biopharmaceuticals from bioactive peptides derived from marine organisms. *Chem. Biol. Drug Des.*, *90*, 12–30.
4. Anjum, K., Abbas, S. Q., Shah, S. A. A., Akhter, N., Batool, S., & Hassan, S. S. U., (2016). Marine sponges as a drug treasure. *Biomol. Ther.*, *24*(4), 347–362.
5. Anneboina, L. R., & Kavi, K. S., (2017). Economic analysis of mangrove and marine fishery linkages in India. *Ecosystem Services*, *24*, 114–123.
6. Chen, S., Wang, J., Wang, Z., Lin, S., & Zhao, B., (2017). Structurally diverse secondary metabolites from a deep-sea-derived fungus *Penicillium chrysogenum* SCSIO 41001 and their biological evaluation. *Fitoterapia*, *117*, 71–78.
7. Cheng, Z., Lou, L., Liu, D., Li, X., Proksch, P., Yin, S., & Lin, W., (2016). Versiquinazolines A to K: Fumiquinazoline-type alkaloids from the gorgonian-derived fungus *Aspergillus versicolor* LZD-14-1. *J. Nat. Prod.*, *79*, 2941–2952.
8. Cui, H., Liu, Y., Nie, Y., & Liu, Z., (2016). Polyketides from the mangrove-derived endophytic Fungus *Nectria* sp. HN001 and their α-glucosidase inhibitory activity. *Mar. Drugs*, *14*, 86–89.
9. Dewapriya, P., Prasad, P., Damodar, R., Salim, A. A., & Capon, R. J., (2017). Talarolide A: Cyclic heptapeptide hydroxamate from an Australian marine tunicate-associated fungus, *Talaromyces* sp. (CMB-TU011). *Org. Lett.*, *19*, 2046–2049.
10. Du, F. Y., Li, X., Li, X. M., Zhu, L. W., & Wang, B. G., (2017). Indole diketopiperazinealkaloids from *Eurotium cristatum* EN-220, an endophytic fungus isolated from the marine alga *Sargassum thunbergii*. *Mar. Drugs*, *15*, 24–28.
11. Fan, Y. Q., Li, P. H., Chao, Y. X., Chen, H., Du, N., & He, Q. X., (2016). Alkaloids with cardiovascular effects from the marine-derived fungus *Penicillium expansum* Y32. *Mar. Drugs*, *13*, 6489–6504.
12. Hassan, S. S. U., Anjum, K., Abbas, S. Q., Akhter, N., & Shagufta, B. I., (2017). Emerging biopharmaceuticals from marine actinobacteria. *Environ. Toxicol. Pharmacol.*, *49*, 34–47.
13. Hassan, S. S. U., Jin, H. Z., Abu-Izneid, T., Rauf, A., Ishaq, M., & Suleria, H. A. R., (2019). Stress-driven discovery in the natural products: A gateway towards new drugs. *Biomed. Pharmacother*, *109*, 459–467.

14. Hassan, S. S. U., Shah, S. A. A., Pan, C., & Fu, L., (2017). Production of an antibiotic enterocin from a marine actinobacteria strain H1003 by metal-stress technique with enhanced enrichment using response surface methodology. *Pak. J. Pharm. Sci., 30*, 313–324.

15. Hassan, S. S. U., & Shaikh, A. L., (2017). Marine actinobacteria as a drug treasure house. *Biomed. Pharmacother., 87*, 46–57.

16. Hawas, U. W., Radwan, A. F., El-Kaseem, L. T. A., & Turki, A. J., (2016). Different culture metabolites of the Red Sea fungus *Fusarium equiseti* optimize the inhibition of hepatitis C virus NS3/4A protease (HCV PR). *Mar. Drugs, 14*, 190–199.

17. Henkel, T., Brunne, R. M., Müller, H., & Reichel, F., (1999). Statistical investigation into the structural complementarity of natural products and synthetic compounds. *Angew. Chem. Int. Ed., 38*, 643–647.

18. Huang, L., Lan, W. J., Deng, R., Feng, G. K., & Xu, Q. I., (2016). Additional new cytotoxic triquinane-type sesquiterpenoids chondrosterins-K to -M from the marine fungus *Chondrostereum* sp. *Mar. Drugs, 14*, 157–163.

19. Huang, Z., Nong, X., Ren, Z., Wang, J., Zhang, X., & Qi, S., (2017). Anti-HSV-1, antioxidant and antifouling phenolic compounds from the deep-sea-derived fungus *Aspergillus versicolor* SCSIO 41502. *Bioorganic and Medicinal Chemistry Letters, 27*, 787–791.

20. Huarong, H., Ting, L., Xiaoen, W., Junxi, G., Xiong, L., Qin, Z., & Zhang, K., (2017). A new antibacterial chromone derivative from mangrove-derived fungus *Penicillium aculeatum* (No. 9EB). *Natural Product Research, 31*(22), 2593–2598.

21. Janakiram, N. B., Mohammed, A., & Rao, C. V., (2015). Sea cucumbers metabolites as potent anti-cancer agents. *Mar. Drugs, 13*, 2909–2923.

22. Kong, D. X., Jiang, Y. Y., & Zhang, H. Y., (2010). Marine natural products as sources of novel scaffolds: Achievement and concern. *Drug Discov. Today, 15*, 884–886.

23. Kong, F. D., Zhang, R. S., Ma, Q. Y., Xie, Q. I., & Wang, P., (2017). Chrodrimanins -O to -S from the fungus *Penicillium* sp. SCS-KFD09 isolated from a marine worm, *Sipunculus nudus. Fitoterapia, 122*, 1–6.

24. Kong, F. D., Zhou, L. M., Ma, Q. Y., Huang, S. Z., & Wang, P., (2017). Chrodrimanins-K to-N and related meroterpenoids from the fungus *Penicillium* sp. SCS-KFD09 isolated from a marine worm, *Sipunculus nudus. J. Nat. Prod., 80*, 1039–1047.

25. Kong, F. D., Zhou, L. M., Ma, Q. Y., Huang, S. Z., & Wang, P., (2017). Metabolites with Gram-negative bacteria quorum sensing inhibitory activity from the marine animal endogenic fungus *Penicillium* sp. SCS-KFD08. *Arch. Pharm. Res., 40*, 25–31.

26. Kong, F. D., Zhou, L. M., Ma, Q. Y., Huang, S. Z., & Wang, P., (2016). Penicillars-A to -E from the marine animal endogenic fungus *Penicillium* sp. SCS-KFD08. *Phytochemistry Letters, 17*, 59–63.

27. Lan, W. J., Wang, K. T., Xu, M. Y., Zhang, J. J., Lam, C. K., & Zhong, G. H., (2016). Secondary metabolites with chemical diversity from the marine-derived fungus *Pseudallescheria boydii* F19–1 and their cytotoxic activity. *RSC. Adv., 6*, 76206–76213.

28. Lee, M. S., Wang, S. W., Wang, G. J., Pang, K. L., & Lee, C. K., (2016). Angiogenesis inhibitors and anti-inflammatory agents from *Phoma* sp. NTOU4195. *J. Nat. Prod., 79*, 2983–2990.

29. Lei, H., Lin, X., Han, L., Ma, J., Ma, Q., Zhong, J., & Liu, Y., (2017). New metabolites and bioactive chlorinated benzophenone derivatives produced by a marine-derived fungus *Pestalotiopsis heterocornis*. *Mar. Drugs, 15*, 69–73.

30. Li, H. L., Li, X. M., Li, X., & Wang, C. Y., (2017). Antioxidant hydroanthraquinones from the marine algal-derived endophytic fungus *Talaromyces islandicus* EN-501. *J. Nat. Prod., 80*, 162−168.

31. Li, H. L., Li, X. M., Liu, H., Meng, L. H., & Wang, B. G., (2016). Two new diphenylketones and a new xanthone from *Talaromyces islandicus* EN-501, an endophytic fungus derived from the marine red *alga Laurencia okamurai*. *Mar. Drugs, 14*, 223–227.

32. Li, X. D., Li, X. M., Li, X., Xu, G. M., Liu, Y., & Wang, B. G., (2016). Aspewentins-D to -H, 20-nor-isopimarane derivatives from the deep-sea sediment-derived fungus *Aspergillus wentii* SD-310. *J. Nat. Prod., 79*(5), 1347−1353.

33. Liao, H. X., Sun, D. W., Zheng, C. J., & Wang, C. Y., (2017). A new hexahydrobenzopyran derivative from the gorgonian-derived Fungus *Eutypella* sp. *Natural Product Research, 31*(14), 1640–1646.

34. Liu, H., Chen, S., Liu, W., & Liu, Y., (2016). Polyketides with immunosuppressive activities from mangrove endophytic fungus *Penicillium* sp. ZJ-SY$_2$. *Mar. Drugs, 14*, 217–223.

35. Liu, H., Li, X. M., Liu, Y., Zhang, P., Wang, J. N., & Wang, B. G., (2016). Chermesins-A to -D: Meroterpenoids with a drimane-type spirosesquiterpene skeleton from the marine algal-derived endophytic fungus *Penicillium chermesinum* EN-480. *J. Nat. Prod., 79*, 806−811.

36. Long, H., Cheng, Z., Huang, W., Wu, Q., Li, X., Cui, J., et al., (2016). Diasteltoxins-A to -C: Asteltoxin-based dimers from a mutant of the sponge-associated *Emericella variecolor* fungus. *Org. Lett., 18*, 4678–4681.

37. Luo, X., Zhou, X., Lin, X., Qin, X., Zhang, T., & Wang, J., (2017). Anti-tuberculosis compounds from a deep-sea-derived fungus *Aspergillus* sp. SCSIO Ind09F01. *Natural Product Research, 31*(16), 1958–1962.

38. Ma, J., Zhang, X. L., Wang, Y., Zheng, J. Y., & Wang, C. Y., (2017). Aspergivones A and B, two new flavones isolated from a gorgonian-derived *Aspergillus candidus* fungus. *Nat. P. Res, 31*(1), 32–36.

39. Ma, X., Nong, X. H., Ren, Z., Wang, J., Liang, X., Wang, L., & Qi, S. H., (2017). Antiviral peptides from marine gorgonian-derived fungus *Aspergillus* sp. SCSIO 41501. *Tetrahedron Letters, 58*, 1151–1155.

40. Murshid, S. S. A., Badr, J. M., & Youssef, D. T. A., (2016). Penicillosides A and B: New cerebrosides from the marine-derived fungus *Penicillium* species. *Revista Brasileira de Farmacognosia (Brazilian Journal of Pharmacognosy), 26*, 29–33.

41. Newman, D. J., & Cragg, G. M., (2004). Marine natural products and related compounds in clinical and advanced preclinical trials. *J. Nat. Prod, 67*, 1216–1238.

42. Ngan, N. T. T., Quang, T. H., Kim, K. W., Kim, H. J., Sohn, J. H., Kang, D. G., Lee, H. S., et al., (2017). Anti-inflammatory effects of secondary metabolites isolated from the marine-derived fungal strain *Penicillium* sp. SF-5629. *Archives of Pharmacal Research, 40*(3), 328–337.

43. Pan, C. Q., Shi, Y., Auckloo, B. N., Chen, X. G., & Wu, B., (2016). An unusual conformational isomer of verrucosidin backbone from a hydrothermal vent fungus, *Penicillium* sp. Y-50-10. *Mar. Drugs, 14*(8), 156.

44. Pan, C. Q., Shi, Y., Chen, X. G., & Wu, B., (2017). New compounds from a hydrothermal vent crab associated fungus *Aspergillus versicolor* XZ-4, *Org. Biomol. Chem., 15,* 1155–1163.

45. Pan, C. Q., Shi, Y., Hassan, S. S. U., Akhter, N., & Wu, B., (2017). Isolation and antibiotic screening of fungi from a hydrothermal vent site and characterization of secondary metabolites from a *Penicillium* isolate. *Mar. Biotechnol., 19*(5), 469–479.

46. Peng, J., Zhang, X., Wang, W., Zhu, T., Gu, Q., & Li, D., (2016). Austalides-S to -U: New meroterpenoids from the sponge-derived fungus *Aspergillus aureolatus* HDN14-107. *Mar. Drugs, 14,* 131–135.

47. Peng, X., Wang, Y., Zhu, T., & Zhu, W., (2018). Pyrazinone derivatives from the coral-derived *Aspergillus ochraceus* LCJ11-102 under high iodide salt. *Archives of Pharmacal. Research, 41*(2), 184–191.

48. Shaala, L., & Youseef, D., (2016). Identification and bioactivity of compounds from the fungus *Penicillium* sp. CYE-87 isolated from a marine tunicate. *Mar. Drugs, 13,* 1698–1709.

49. Shi, T., Qi, J., Shao, C. L., Zhao, D. L., Hou, X. M., & Wang, C. Y., (2017). Bioactive diphenyl ethers and isocoumarin derivatives from a gorgonian-derived fungus *Phoma* sp. (TA07-1). *Mar. Drugs, 15,* 146–150.

50. Shin, H. J., Pil, G. B., Heo, S. J., & Lee, S. J., (2016). Anti-inflammatory activity of Tanzawaic acid derivatives from a marine-derived fungus *Penicillium steckii* 108YD142. *Mar. Drugs, 14,* 14–20.

51. Sivaperumal, P., Kamala, K., Natarajan, E., & Dilipan, E., (2013). Antimicrobial peptide from crab haemolypmh of *Ocypodamacrocera* (Milne Edwards 1852) with reference to antioxidant: A case study. *Int. J. Pharm. Pharm. Sci., 5*(2), 719–727.

52. Suleria, H., Osborne, S., Masci, P., & Gobe, G., (2015). Marine-based nutraceuticals: An innovative trend in the food and supplement industries, *Marine Drugs, 13*(10), 6336.

53. Sumilat, D. A., Yamazaki, H., & Endo, K., (2017). New biphenyl ether derivative produced by Indonesian ascidian derived *Penicillium albobiverticillium. J. Nat. Med., 71*(4), 776–779.

54. Sun, W., Chen, X., Tong, Q., & Zhu, H., (2016). Novel small molecule 11β-HSD1 inhibitor from the endophytic fungus *Penicillium commune. Sci. Reports, 6,* 26418–26423.

55. Takahashi, K., Sakai, K., & Nagano, Y., (2017). Cladomarine: New anti-saprolegniasis compound isolated from the deep-sea fungus, *Penicillium coralligerum* YK-247. *The J. of. Antibiot.,* 1–4.

56. Tian, Y. Q., Lin, X. P., Wang, Z., Zhou, X. F., & Qin, X. C., (2016). Asteltoxins with antiviral activities from the marine sponge-derived fungus *Aspergillus* sp. SCSIO XWS02F40. *Molecules, 21,* 34–39.

57. Uchoa, P. K., Pimenta, A. T., & Braz-Filho, R., (2017). New cytotoxic furan from the marine sediment-derived fungi *Aspergillus Niger. Natural Product Research, 31*(22), 2599–2603.

58. Wei, G., Tan, D., Chen, C., Tong, Q., & Hunag, J., (2017). Flavichalasines-A to -M: Cytochalasan alkaloids from *Aspergillus flavipes. Sci. Repor.,* 42434–42438.

59. Xia, X., Qi, J., Liu, Y., Jia, A., & Zhang, Y. G., (2016). Bioactive isopimarane diterpenes from the fungus, *Epicoccum* sp. HS-1, associated with *Apostichopus japonicas. Mar. Drugs, 13,* 1124–1132.

60. Xiao, Z., Chen, S., Cai, R., Lin, S., Hong, K., & She, Z., (2016). New furoisocoumarins and isocoumarins from the mangrove endophytic fungus *Aspergillus* sp. 085242. *Beilstein J. Org. Chem. (Germany), 12*, 2077–2085.

61. Yamada, T., Suzue, M., Arai, T., Kikuchi, T., & Tanaka, R., (2017). Trichodermanins-C to -E: New diterpenes with a fused 6-5-6-6 ring system produced by a marine sponge-derived fungus. *Mar. Drugs, 15*, 169–175.

62. Zhang, P., Li, X. M., Mao, X. X., Mandi, A., Kurtan, T., & Wang, B. G., (2016). Varioloid A, a new indolyl-6, 10b-dihydro-5a*H*-[1]benzofuro[2,3-*b*]indole derivative from the marine alga-derived endophytic fungus *Paecilomyces variotii* EN-291. *Beilstein J. Org. Chem., 12*, 2012–2018.

63. Zhou, X. F., Fang, W., Tan, S., Lin, X., Xun, T., Yang, B., Liu, S., & Liu, Y., (2016). Aspernigrins with Anti-HIV-1 activities from the marine-derived fungus *Aspergillus Niger* SCSIO Jcsw6F30. *Bioorganic and Medicinal Chemistry Letters, 26*(2), 361–365.

64. Zhu, M., Zhang, X., Feng, H., Dai, J., & Li, J., (2017). Penicisulfuranols-A to -F: Alkaloids from the mangrove endophytic fungus *Penicillium janthinellum* HDN13-309. *J. Nat. Prod., 80*, 71–75.

CHAPTER 9

MARINE BIOTOXINS: SYMPTOMS AND MONITORING PROGRAMS

HUMA BADER UL AIN, FARHAN SAEED, HAFIZA SIDRA YASEEN, TABUSSAM TUFAIL, and HAFIZ ANSAR RASUL SULERIA

ABSTRACT

Biotoxins are toxic chemical agents produced by harmful marine micro- and macroalgae. All types of alga contain high amounts of poisonous compounds, which represent a risk to the biological community and human wellbeing. Many types of marine biotoxins are being identified recently with different laboratory diagnostic tests that vary according to the composition of toxins. Future work should also include developing of the molecular assays for the rapid, real-time detection of harmful species and toxin genes. Literature review demonstrated that marine biotoxin are of two types, such as hydrophilic toxins and lipophilic toxins. The most common and harmful of these poisons are domoic acid (DA), saxitoxin, azaspiracid, ciguatoxin, okadaic acid toxins, yesso toxins, and Pectenotoxins. This review chapter focuses on types of shellfish poisoning, their related biotoxins, symptoms, prevention, and treatments.

9.1 INTRODUCTION

Toxins are poisonous substances produced by microorganisms and others. Toxins act as primary factors for pathogenicity. There are about 220 known bacterial toxins 40% of which are harmful to humans by damaging the Eukaryotic cell membrane. Based on toxins, there are various classes of toxins, such as [11, 47]:

- Cytotoxins (cells);
- Dermatotoxins (skin cells);
- Enterotoxins (enteric system);
- Hemotoxins (blood cells);
- Hepatotoxins (liver tissue);
- Neurotoxins (nerve tissue).

These toxins are produced through different organisms, such as [11, 47]:

- Algae (microcystins, amnesic shellfish poisoning (ASP), paralytic shellfish poisoning);
- Bacteria (endotoxins, exotoxins);
- Fungi (tricothecenes);
- Higher animals (fish, insects, snakes, frogs);
- Plants (alkaloids, tannins, cyanogenic glycosides);
- Protozoa (endotoxins, phospholipase, protease);
- Viruses (stx phage, cytotoxins, lysins).

Biotoxin is a toxic substance produced by a living organism. Biotoxins become dangerous when they contaminate the food matrix, potentially causing injury or death to those, who feed either directly or indirectly on them. In the marine environment, biotoxins are produced by several types of algae, such as diatoms, and dinoflagellates, which can produce "harmful algal blooms" or "red tides." Marine toxins are naturally occurring chemicals that can contaminate certain seafoods. There is Europe-wide legislation to limit their levels in seafood that is placed on the market [47]. These toxins are tasteless, odorless, and heat- and acid stable.

Reference methods described in the European Union (EU) legislation to test for the presence of these toxins are based on chemical testing. To increase our knowledge of marine biotoxins, the EU established a network of European reference laboratories (EURL) and National Reference Laboratories (NRL).

This review chapter focuses on types of shellfish poisoning along with their associated symptoms, main toxins, causative organisms, diagnosis, prevention, and treatment methods.

9.2 TYPES OF SHELLFISH POISONING

9.2.1 HYDROPHILIC TOXINS

9.2.1.1 AMNESIC SHELLFISH POISONING (ASP)

Amnesic shellfish poisoning (ASP) is one of the main hydrophilic shellfish poisonings caused by domoic acid (DA). DA is a natural toxicant, which is obtained from microscopic marine diatom algae (*Pseudo-nitzschia*). All filter-feeding shellfish eat algae, which accumulate in their body and if the algae contain toxic chemical agents (such as DA), then it will assemble in shellfish's body. It is very harmful toxicant and poison due to DA; and it is called domoic acid poisoning (DAP) or ASP. In this poisoning, short-term memory loss has occurred. This toxin was responsible for many of human diseases in Prince Edward Island (PEI) in 1987. ASP has been linked to the diatom *Pseudo-nitzschia spp.* [11].

In late 1987 in eastern Canada, 153 DA human poisoning cases and four deaths of elderly patients have been reported [11, 47, 61]. Recent report shows that Pseudo-nitzschia species has more capability to induce ASP compared to the Nitzschia species such as *Nitzschia navisvaringica* and *Nitzschia bizertensis* [36, 55]. It is also described that macroalgae (such as *Chondria armata*, *Chondria baileyana*, and *Alsidium corallinum*) are also responsible for the ASP [30, 60].

9.2.1.1.1 Main Toxin and Structure

DA is the primary poison found in an assortment of shellfish. It is a normally happening compound belonging to kainoid class that has been separated from macroalgae and microalgae. About 10 isomers of DA (iso DAs A to H and DA 5′ diasteriomer) have been determined in marine samples, and these are collectively called kainates [68, 71]. DA is a crystalline amino acid, which is water-soluble and acidic in nature. The main mechanism of action of this toxin is to obstruct some glutamate receptors in the central nervous system (CNS), which are responsible for depolarization of neurons [12, 28].

The kainoids are a class of excitatory and excitotoxic pyrrolidine dicar-
boxylates that act at ionotropic glutamate receptors. Types of kainoids and
their derivatives are: Acromelic acid, Allokainic acid, Arylkainoid, CPKA,
Dihydrokainic acid, DA, HFPA, Homokainoid, Kainic acid and MFPA.
Examination of kainoids in *Chondriaarmata* brought the disclosure of minor
quantity of the geometrical isomers (such as isodomoic acid A, isodomoic
acid B and isodomoic acid C) and domoilactones [68]. By exposing the dilute
solution of DA to ultraviolet light, the geometrical isomers can be generated.
It has nearly a similar efficacy to the kainate receptor as DA itself [68].

9.2.1.1.2 *Symptoms*

Due to DA poisoning, many of the signs and symptoms appear in the patient
within 24 hours (Figure 9.1). These manifestations cover almost all body
systems, such as gastrointestinal tract, CNS. Therefore, the symptoms associated with gastrointestinal tract and CNSs are: nausea, vomiting, diarrhea,
headache, dizziness, confusion, amnesia, seizures, coma, and sudden death
respectively [48].

9.2.1.1.3 *Analysis*

High-performance liquid chromatography (HPLC)-ultraviolet light is the
basic laboratory method for identification of DA in shellfish. According to
the European Commission (EC) No 853/2004 [16], live bivalve mollusks set
available for human utilization in the market must not contain more than 20
mg/Kg of DA shellfish meat. Every single diatom cell can be separated with
the help of a microscope and thin glass tube; and then put into a seawater
culture medium. The main advantage of this test is that it has enabled to
determine, which environment is suitable for the development and generation of DA. Wide extensive range of temperature and saline conditions is
required to produce *Pseudo-nitzschia* species except for *P. seriata*, for which
just 15°C is needed. Adequate light is required for development, and enough
energy is needed for poison biosynthesis.

9.2.1.1.4 *Monitoring Methods*

The presence of toxins in molluscan shellfish is screened out by methods
described by CFIA (Canadian Food Inspection Agency). The combined

effort of CFIA and Prince Edward Island (PEI) Department of Agriculture, Fisheries, Aquaculture, and Forestry collects water samples each week from 15 locations. The purpose of this water collection is to monitor quantity of lethal phytoplankton, as a component of the PEI Mussel Monitoring Program. To evaluate the potential for the blossom to render the shellfish poisonous, water samples in season of early September to early December are sent to the industrial plant belonging to CFIA and DFO (National Research Council Canada, Fisheries, and Oceans Canada-University of PEI). Under contract to DFO, the *Pseudonitzschia* species are determined by utilizing the SEM at the digital microscopy facility (DMF) of Mount Allison University (Sackville, New Brunswick).

9.2.1.1.5 Diagnosis

Diagnostic techniques (which are required to determine the source of ASP) are complete blood count (CBC), electrocardiogram ECG (for arrhythmias of heart), electroencephalogram EEG (for CNS problems), comprehensive metabolic panel (CMP), enzyme-linked immunosorbent assay (ELISA). DA poisoning patients can also be diagnosed by different pathological and biochemical tests, which include: magnesium (Mg^{2+}), Calcium (Ca^{2+}), and Potassium (K^+) and feces testing (C and S, Gram recolor, ova, and parasites).

All other signs and symptoms (such as unsteadiness, generalized weakness, symmetric transient hyper/hyporeflexia, Babinski, fasciculation, distal atrophy, loss of distal sensitivity to pain, and hypo/hyperthermia) can be monitored by different tests [61]. All CNS related issues (such as symptoms of confusion, disorientation, and memory loss) can be determined by neuropsychological testing. Imaging technologies (such as computed tomography (CT), magnetic resonance imaging (MRI), and positron emission tomography (PET) scans) are also used to evaluate the auxiliary changes of hippocampus and amygdala.

9.2.1.1.6 Prevention

Patients (with different co-morbidities, hepatic, and renal) ought to maintain a strategic distance from seafood especially in late summer. The person should not try eating shellfish sold as decoy, because these items do not meet similar nourishment security controls as seafood for human utilization. One can check with nearby health authorities before gathering shellfish (particularly mussels, shellfish, mollusks, scallops, and certain verities of crab, especially Dungeness crabs) [11].

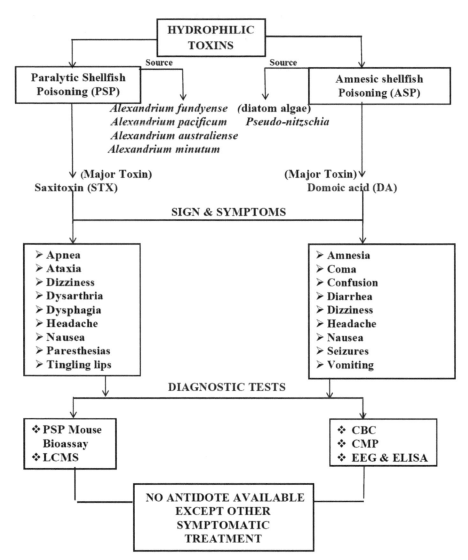

FIGURE 9.1   Flowsheet of hydrophilic shellfish poison.

9.2.1.1.7   Treatment

The basic treatment for a person with DA poisoning includes anticonvulsant drugs in spite of the fact that various anti-glutaminergic drugs are in the preclinical stage. Some of the symptomatic treatments are also available for DA poisoning, such as I.V. hydration with Lactated Ringer solution, Antiemetic's for emesis, regurgitating proper diastalsis to expand gastric emptying time, e.g., metoclopramide, Analgesics as painkiller, administration of IV antiepileptic drugs (such as Diazepam and Phenobarbital) to treat seizures. Furthermore, maintenance of cardiac rhythms are regularly exacerbated due to imbalance of electrolytes including Mg^{2+} and Ca^{3+}. The proper rectification of basic electrolyte anomalies in the body is performed by using a mixture of I.V. magnesium sulfate and additionally cardioversion if the patient is hemodynamically unsteady and shows organ hypoperfusion.

It is concluded that heat and cooling have no effect on this toxic chemical compound and no antidote can be used in DA poisoning and only symptomatic treatment is given. A few neurons experience excitotoxic cell apoptosis rapidly because of DA poisoning, which may lead to the initiation of different events of neurons cell death [64]. Many of research studies indicated that if troxerutin and ursolic acid are given to the DA poisoned mice, then amnesia can be cured.

9.2.1.2 PARALYTIC SHELLFISH POISONING (PSP)

The major source of paralytic shellfish poisoning (PSP) is a marine dinoflagellate species (*Alexandrium fundyense*, *Alexandrium pacificum*, *Alexandrium australiense*, *Alexandrium minutum*) and cyanobacteria [65]. Paralytic shellfish poisoning is a foodborne illness that is due to quickly acting nerve-damaging toxin having a structure of 3,4,6-trialkyltetrahydropurine. This poison can act by inhibiting the voltage-gated sodium channels in neurons and can impair the transmission of impulses in skeletal and cardiac muscle due to which sudden death may occur [64]. Paralytic shellfish poisoning poisons represent the most genuine risk to general health and cause limitless monetary loss [15].

Paralytic shellfish poisoning poisons were identified in European and North American waters. Chile, South-Africa, and Australia saxitoxins have been accounted for a major source of poisoning. Saxitoxin (STX) is the primary poison for paralytic shellfish poisoning and is the most lethal. Shellfish has the ability to gather poisons yet the shellfish itself is somewhat impervious to the destructive impacts of these poisons.

The safe and allowed level of saxitoxin 2-HCl by EU is 75 µg and 800 µg counterparts/kg shellfish, respectively [40]. Marine foods to cause PSP are:

mussels, cockles, clams, scallops, oysters, crabs, and lobsters found in colder waters, such as those of Pacific and New England Coasts.

9.2.1.2.1   Main Toxin and Structure

The basic structure of saxitoxin (STX) was illustrated by using x-ray crystallographic and magnetic resonance spectroscopy (MRS). The fundamental toxic effect of this poison is due to presence of dihydroxy or hydrated ketone group in the 5-ring, but this effect can be eliminated by formation of monohydroxy compound through reduction of dihydroxy group with simple hydrogen. The 60% activity of this toxin can be reduced by release of carbamoyl assemble side-chain on the 6-membered ring [40]. The paralytic shellfish poisoning toxins are oxidized under alkaline conditions and they are not heated sensitive at acidic pH except N-sulfo-carbamoyl [40].

9.2.1.2.2   Symptoms

The main mechanism of action of poison for PSP is to specifically block the Na+ channel in CNS so that the action potential will not be generated further [65]. The main indications of paralytic shellfish poisoning are: tingling sensation of lips and tongue, numbness of extremities, paresthesia, ataxia, apnea, nausea, headache, dizziness, vomiting, dysphagia, and dysarthria (Figure 9.1). Due to fast action of this chemical agent, respiratory distress and even death may result [15, 40]. Symptoms usually begin from 15 minutes to 2 hours after eating contaminated shellfish but can take as long as 10 hours to appear. Symptoms are usually mild in healthy people, but the illness can be severe or even fatalin those with compromised health.

9.2.1.2.3   Analysis

In view of the potential risk to the people, a technique may decide the existence of PSP poisons in shellfish. Initially, PSP toxins has been identified by mouse bioassay strategy, but the improvement of option is also increased due to the questionable issue of utilizing pharmacological assays, immunoassays, chemical, or separation assays for determination of toxins [40]. In 2005, the Lawrence method was adopted for the detection of STX, neoSTX, GTX2,3, GTX1,4 compounds in mollusks [6, 20].

First of all, the pre-screening of PSP poison is done with hydrogen peroxide and sodium periodate by fluorometric detection. As per Commission Regulation 2004, live bivalve mollusks must not contain crippled shellfish poison (PSP) surpassing 800 micrograms for each kilogram of shellfish meat for human use. Presently, the Association of Official Analytical Chemists (AOAC) created the mouse bioassay to deliver a fast and sensible exact estimation of aggregate paralytic shellfish poisoning poisons [28].

9.2.1.2.4 *Prevention*

The paralytic shellfish poisoning poison saxitoxin is not damaged by cooking, solidifying, or smoking and is imperceptible by sight or smell, so that identification of paralytic shellfish poisoning toxin is compulsory before the shellfish arrive the consumers [62]. During the period of April to October, blossoming of Alexandrium happens each year, particularly along the New England States, Alaska, California, and Washington. When the poison levels surpass 80 mcg/g in the United States, the closing of those territories is ordered due to possibility of high risk [41]. The best strategy for counteractive action of paralytic shellfish poisoning is to abstain from eating shellfish amid red tide poisonous alert, a significant number of which are declared by federal and state governments, particularly in the Americas, Scandinavia, and Western Europe (Figure 9.1).

9.2.2 *LIPOPHILIC TOXINS*

9.2.2.1 *CIGUATERA SHELLFISH POISONING (CFP)*

The most common Algal Bloom aliment named as ciguatera fish poisoning (CFP) is due to the utilization of coral reef fish, such as barracuda, grouper, and snapper. The expected worldwide rate of this disease is not less than 50,000–500,000 individuals for each year [50].

9.2.2.1.1 *Etiology*

Most of the Coral reef fish are defiled with marine poisons called ciguatoxin that are obtained from dinoflagellate alga *Gambierdiscus toxicus*. When fish uses the poisonous algae developing on coral reefs then exchange of ciguatoxins may happen. These contaminated fish are later consumed by larger predatory reef fish that can collect higher levels of ciguatoxins in their muscle tissue, organs, and fat.

9.2.2.1.2 Symptoms

CFP symptoms fluctuate impressively among people. The toxicity of CFP may last from days to months and even years. This poisoning mainly involves: CNS, CVS (Bradycardia, Hypotension), and gastrointestinal tract [7, 9, 10, 13]. Genomic highlights are paranesthesia and dysesthesia [10, 27]. Discomfort or weakness related with CFP can be exceptionally incapacitating and unending, which can cause fatigue disorder. Common nonspecific symptoms include nausea, vomiting, diarrhea, cramps, excessive sweating, headache, and muscle aches [26, 27, 44, 50], as shown in Figure 9.2.

9.2.2.1.3 Laboratory Studies

All patients with ciguatera shellfish toxicity are identified only through history and clinical presentation because no laboratory test for this poisoning has been discovered yet, thus diagnosis of ciguatera poisoning is not easy. In Australia, flare up out of two scuba divers, one initially faced paranesthesia, arthralgia, shortcoming, myalgia, and gastrointestinal side effects symptoms, but the second patient showed neuromuscular (NM) symptoms without gastrointestinal tract disturbance. Subsequently, ciguatera shellfish poisoning (CFP) with insignificant or no gastrointestinal side effects could have effectively been misdiagnosed as decompression sickness in the Australian flare-ups [43, 50].

9.2.2.1.4 Treatment

There is no neutralizing agent accessible for ciguatera poison. Regurgitating must be initiated if the patient is wake and has eaten ciguatera poison containing fish in the last 3 to 4 hours. Ipecac drug is used because regurgitating, but numerous specialists think ipecac may cause electrolyte imbalance in the body, so that IV liquids are must in worst sickness. Initially, the administration of activated charcoal within 3 to 4 hours after the ingestion of toxicant may adsorb the poison. Amitriptyline and gabapentin may diminish neural side effects. Intravenous Mannitol is the main CFP treatment assessed by randomized single-or double-blind studies to counteract the symptoms during the initial 2–3 days of poisoning [13, 43, 54]. Diphenhydramine, hydroxyzine, and NSAID's may help to treat irritation and eliminate painful sensation. Stay away from liquor, fish, nuts, and nut oils are strongly recommended after the use of ciguatera poison to reduce the chances of symptoms reoccurrence.

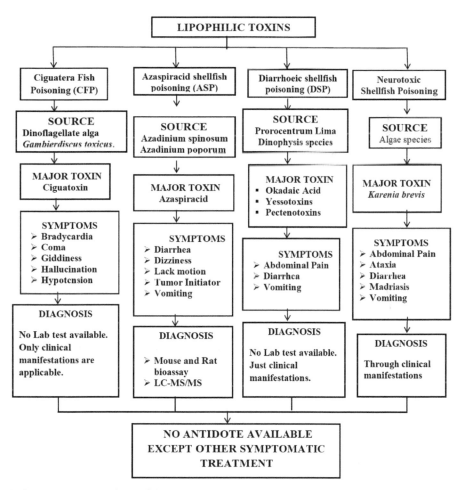

FIGURE 9.2 Flow sheet of lipophilic toxins.

9.2.2.1.5 Prevention

For determination of ciguatoxin in fish, no cost-effective method has been available yet. The Cigua-Check is a kit, which is accessible on the Internet.

Likewise, Ciguatera poison can also be recognized by mouse bioassay or compound immunoassay, nonetheless, the intoxication of the fish is sporadic [35, 38]. It is concluded that activation of Ciguatera poison is not influenced by cooking, solidifying, or stomach corrosive and cannot be recognized by sight, smell, or taste [46]. *G. toxicus* abundance and CFP flare-ups might be the relevant cause of multifactorial impacts of global warming, bleaching, and death of coral, farming compost and pesticide overflow, dock development, and rainstorm flood with untreated sewage spillover [38].

9.2.2.2   AZASPIRACID SHELLFISH POISONING (ASP)

Azaspiracid is a marine toxin that is responsible for an outbreak of diarrheic food poisoning associated with the consumption of contaminated shellfish. The azaspiracid toxin group can cause severe poisoning in consumers of mussels after being enriched in the shellfish tissues.

9.2.2.2.1   Main Toxin and Its Structure

Azaspiracid has been called killer Toxin-3 or KT3. Azaspiracid is a toxin comprising of a six-membered cyclic imine ring having a straight carbon chain, which is cyclized by ether bonds at numerous points in the structure [42, 52]. The major sources of this toxin are organisms, such as *Azadinium spinosum*, *Azadinium poporum*, and *Amphidoma languida* [37]. Different forms of Azaspiracid toxins are present in plankton like AZA-1, AZA-2, and AZA-3; while every other was recognized in shellfish and are viewed as shellfish metabolites [32–34]. Different investigations recommend that AZA4, dissimilar to any of the other Azaspiracid analogs, is a novel inhibitor of plasma film Ca2þ channels; and it represses Ca2þ section and works directly in human T-lymphocytes [4].

9.2.2.2.2   Symptoms

AZA1 is cytotoxic, and its toxicity is related to the concentration and time of exposure with toxin [63]. The fundamental manifestations of ASP are: stomach disturbances, nausea, vomiting, and diarrhea. Also, Azaspiracid is a tumor initiator. The Azaspiracid poison in mice prompted the improvement of lung tumors and an advancing loss of motion was seen in the mouse test utilizing the extracted solution of mussel [52].

9.2.2.2.3 Laboratory Studies

The "EU regulation on scientific philosophy for Azaspiracid" endorses the utilization of the entire body as the test sample because the initial hypothesis is that Azaspiracid can travel to the mussels from the digestive system [33]. Recent data has demonstrated that Azaspiracid ordinarily accumulates in glands of digestive system of mussels [29]. Basically, two techniques are used for the recognition of Azaspiracid. First biological method includes the mouse and rat bioassay. Despite developing the resistance against these mammalian assays, they are yet utilized. Many of the lipophilic poisons other than Azaspiracid are present in shellfish, which can give the positive result due to bioassays so that further corroborative testing to assess real danger is required [23].

Secondly, the chemical methods include: The fluid chromatography hyphenated with tandem mass spectrometry (LC-MS/MS). It may be the best method for the determination of AZA-1, AZA-2 and AZA-3. The extremely muddled lattice of shellfish contains many of the poisonous compounds, such as Azaspiracid, DSP, and spirolides [5, 32]. The maximum allowable level of Azaspiracid is 160 μg/kg-SM. Major difficulty is to get in adequate amounts to synthesize suitable antibodies. Initially, small amount of AZA-2 was identified in the Mediterranean Seafood. The Azaspiracid profile of mussels from the North-focal Adriatic Sea demonstrated a prevalence of AZA-2. However, it has the relevance with shellfish toxicity from the Atlantic shores of Morocco and Portugal. A few European ventures have been authorized to get a sufficient supply of unadulterated material.

9.2.2.3 DIARRHEIC SHELLFISH POISONING (DSP)

There are three types of poisons groups (such as okadaic acid toxins, yesso toxins and pectenotoxins) that are responsible for the diarrheic shellfish poisoning (DSP). All these groups belong to the polyether toxin group. *Prorocentrum lima* and *Dinophysis* species of the plant kingdom are also major source of DSP [51, 69]. It has been evidenced that these three poisons have distinctive natural impacts; and okadaic acid and its analogs are diarrhoeagenic [8, 24, 39].

9.2.2.3.1 Main Toxin and Structure

Major source of DSP is okadaic acid (OA) toxin and chemically related agents, such as dinophysistoxins (DTX1, DTX2) and okadates [70]. The difference between okadaic acid and DTX2 is just the position of CH_3 group; however, DTX1 has one extra CH_3 molecule in its structure. After metabolism in shellfish dinophysistoxins and okadaic acid are produced, which are involved in tumor development [56–59, 70]. The main mechanism of okadates is the blockage of serine/threonine phosphoprotein phosphatases in the body [25].

Most of the shellfish have the lipophilic sulfur-bearing polyether toxins called yesso toxins (YTXs). There are different types of Yesso toxins, such as 1a-homo-YTX, 45-hydroxy-YTX and 45-hydroxy-1ahomo-YTX. The origin of YTXs are dinoflagellate *Proceratium reticulatum* and *Lingulodinium polyedrum*. Major metabolites of yesso toxins are 45-hydroxy-YTX, carboxy-YTX, and their relevant compounds 1a-homologues [1, 18]. The intraperitoneally injection of YTX developed strong toxic effect in mice while administration by oral route only caused swelling of some heart muscle cells of mice [8]. It is concluded that Yesso toxins have no toxic effects on human [1, 2, 18]. According to the EFSA, if shellfish have the concentration of 3.75 mg of YTX per kg of shellfish, then it is safe for use.

Another type of toxin present in shellfish is called pectenotoxins (PTXs) [64]. Most common agents of this toxin are pectenotoxin-2 (PTX2), pectenotoxin-2-seco acid (PTX2sa) and 7-epi pectenotoxin-2seco acid [66, 67]. When this toxin was given as injection in mice, it caused hepatocytes death due to liver damage [22]. This toxin is not responsible to develop diarrhea but after oral use it caused liver and stomach pathological changes in mice. Studies showed that this toxin is not dangerous for the human beings, therefore the allowable dose of PTXs is 120 µg of PTX2 per kg [3]. Chemically PTXs contain cyclic polyether lactones, which differ in their chemical structure, such as Oxidation at C43 from methyl to carboxylic acid, in both rings epimerization of the spiroketal ring, in C1 to C33 lactone ring-opening [14, 49]. According to the EC announcement in March 2002, the highest amount of okadaic acid, DTXs, PTXs together, and YTXs in edible tissues of mollusks, echinoderms, tunicates, and marine gastropods shall be 160 mg okadaic acid equivalents per kg and 1 mg of YTX equivalents per kg, respectively.

9.2.2.3.2 Symptoms

DSP mainly affects gastrointestinal tract system of the body and causes nausea, vomiting, diarrhea, and abdominal pain.

9.2.2.3.3 Diagnosis

DSP is just detected by patient history and complaints as no diagnostic test is available today for its identification. However, in live bivalve mollusks, general marine biotoxins compound detection test (known as EU-RL LC-MS/MS method) can be used. This diagnostic method can detect okadaic acid, DTX1, DTX2, DTX3 including their esters, pectenotoxins group toxins (PTX1 and PTX2), and yeso toxins group toxins (YTX, 45 OH YTX, homo YTX, and 45 OH homo YTX) to some extent only.

9.2.2.3.4 Treatment

For the DSP, no antitoxin has been developed. Therefore, only symptomatic treatment is used and patients recover after 3 days either after medical attention or without such treatment.

9.2.2.3.5 Prevention

To avoid DSP toxicity, primary prevention is to separate the shellfish with such toxins; and it is necessary to stop the use of shellfish in red tide alert season.

9.2.2.4 NEUROTOXIC SHELLFISH POISONING (NSP)

Neurotoxic shellfish poisoning is associated with mussels, oysters, and clams found in the warmers waters of the western coast of Florida, the Gulf of Mexico, and the Caribbean Sea. This toxin is present in the United States algae areas, alongside Gulf Coasts of Florida and Texas.

9.2.2.4.1 Main Toxin and Structure

The main source of neurotoxic shellfish poisoning or brevetoxin is *Karenia brevis* (a dinoflagellate). Its mechanism of action is at the site 5 of voltage-gated sodium channels. The toxicity appears from fifteen minutes to three hours after the use of neurotoxic shellfish. Neurotoxic shellfish poisoning can likewise be contracted by means of inhalation [17, 41, 62].

9.2.2.4.2 Symptoms

The main symptoms of neurotoxic shellfish poisoning are: low blood pressure, numbness of extremities, dilation of pupil, muscle spasm nausea, vomiting, diarrhea, inebriation feel, dizziness, respiratory problems including apnea and ataxia. The severity of symptoms is low.

9.2.2.4.3 Treatment

In neurotoxic shellfish poisoning toxicity, only supportive therapy is medical attention; and patient is recovered within 2 days.

9.2.2.4.4 Diagnosis

Patients of neurotoxic shellfish poisoning can only be diagnosed by present illness complaints and complete history [17, 41, 62].

9.2.2.4.5 Prevention

The primary prevention is to avoid use of shellfish with neurotoxin. The detection and separation of such shellfish is necessary as early as possible. All toxicity caused due to inhalation route can be prevented if one stays away from tidal waterways and surfing during blooming [17, 41, 62].

9.3 SUMMARY

Most of the poisons in seafoods have harmful impacts on marine life. Major phytotoxins (such as saxitoxin DA, azaspiracid, ciguatoxin, okadaic acid, yesso toxins, and pectenotoxins) are produced from the algal blooms. Recently, illnesses have occurred due to dangerous effects of marine toxins. To study the etiology of poisoning, its related symptoms and medical therapy are major concerns in this review chapter. In addition, it is also necessary to understand the chemical structure and isomers of shellfish toxins. There is a need for improved data collection so that poisoning trends can be assessed worldwide.

KEYWORDS

- **azaspiracid shellfish**
- **ciguatoxin**
- **domoic acid**
- **marine biotoxins**
- **okadaic acid**
- **pectenotoxins**
- **saxitoxin**
- **yesso toxins**

REFERENCES

1. Aasen, J. A. B., Mackinnon, S. I., & IeBlanc, P., (2005). Detection and identification of spirolides in Norwegian shellfish and plankton. *Chemical Research in Toxicology, 18*, 509–515.
2. Aasen, J. A. B., Samdal, I. A., Miles, C. O., Dahl, E., Briggs, L. R., & Aune, T., (2005). Yessotoxins in Norwegian blue mussels (*Mytilus edulis*): Uptake from *Protoceratium reticulatum*, metabolism and depuration. *Toxicon., 45*, 265–272.
3. Alexander, J., Benford, D., Cockburn, A., & Cravedi, J., (2009). Marine biotoxins in shellfish-summary on regulated marine biotoxins scientific opinion of the "Panel on Contaminants in the Food Chain". *EFSA Journal, 1306*, 1–23.
4. Alfonso, C., Alfonso, P., Otero, P., & Rodriguez, P., (2008). Purification of five azaspiracids from mussel samples contaminated with DSP toxins. *Journal of Chromatography B, 865*, 133–140.
5. Alvarez, G., Uribe, E., & Avalos, P., (2010). First identification of azaspiracid and spirolides in *Mesodesma donacium* and *Mulinia edulis* from Northern Chile. *Toxicon., 55*, 638–641.
6. Anonymous, (2005). *AOAC (Association of Official Analytical Chemists) Method 2005.06* (pp. 99–130). Gaithersburg, MD, USA: Association of Official Analytical Chemists (AOAC).
7. Arena, P., Levin, B., Fleming, L. E., Friedman, M. A., & Blythe, D., (2004). Pilot study of the cognitive and psychological correlates of chronic ciguatera fish poisoning. *Harmful Algae, 3*, 51–60.
8. Aune, T., Sorby, R., Yasumoto, T., Ramstad, H., & Landsverk, T., (2002). Comparison of oral and intraperitoneal toxicity of yessotoxin towards mice. *Toxicon., 40*, 77–82.
9. Baden, D. G., Fleming, L. E., & Bean, J. A., (1995). Marine toxins. In: De Wolff, F. A., (ed.), *Handbook of Clinical Neurology: Intoxications of the Nervous System, Part II* (Vol. 21, No. 65, pp. 141–165). New York: Elsevier Science.

10. Bagnis, R., Kuberski, T., & Laugier, S., (1979). Clinical observation on 3009 cases of ciguatera (fish poisoning) in the South Pacific. *American Tropical Medical Hygiene, 28*, 1067–1073.

11. Bates, S. S., Bird, C. J., De Freitas, A. S. W., Foxall, R., & Gilgan, M., (1989). Pennate diatom *Nitzschia pungens* as the primary source of domoic acid: A toxin in shellfish from eastern Prince Edward Island, Canada. *Canadian Journal of Fisheries and Aquatic Sciences, 46*, 1203–1215.

12. Berman, F. W., & Murray, T. F., (1997). Domoic acid neurotoxicity in cultured cerebellar granule neurons is mediated predominantly by NMDA receptors that are activated because of excitatory amino acid release. *Journal of Neurochemistry, 69*, 69–703.

13. Blythe, G., De Sylva, D. P., & Fleming, L. E., (1992b). Clinical experience with IV mannitol in the treatment of ciguatera. *Bulletin de la Soci´ et´ e de Pathologie Exotique (Bulletin of the Society of Exotic Pathology), 85*, 425–426.

14. Burgess, V., & Shaw, G., (2001). Pectenotoxins: An issue for public health: Review of their comparative toxicology and metabolism. *Environment International, 27*, 275–283.

15. Campas, M., & Marty, J. I., (2007). Enzyme sensor for the electrochemical detection of the marine toxin okadaic acid. *Analytica Chimica Acta, 605,* 87–93.

16. Commission Regulation. Regulation (EC) No 853/2004 of the European Parliament and of the Council of 29 April 2004: Laying down specific hygiene rules for on the hygiene of foodstuffs, (2004). Report L139/55, The European Parliament and the Council of the European Union. *Official Journal of the European Union*, 151.

17. Daranas, A. H., Norte, M., & Fernandez, J. J., (2001). Toxic marine microalga. *Toxicon.*, *39*, 1101–1132.

18. Draisci, R., Ferretti, E., & Palleschi, L., (1999). High levels of yesso toxin in mussels and presence of yesso toxin and homo-yesso toxin in dinoflagellates of the Adriatic Sea. *Toxicon., 37,* 1187–1193.

19. Draisci, R., Lucentini, L., Giannetti, L., Boria, P., & Poletti, R., (1996). First report of pectenotoxin-2 (PTX-2) in algae (*Dinophysis fortii*) related to seafood poisoning in Europe. *Toxicon., 34*, 923–935.

20. EC, (2006). *Official Journal of the European Union* (Vol. 320, pp. 13–18). EUR-Lex.

21. Escalona, D. M. G., Feliu, J. F., & Izquierdo, A., (1986). Identification and epidemiological analysis of ciguatera cases in Puerto Rico. *Marine Fisheries Review, 48*, 14–18.

22. Espina, B., & Rubiolo, J. A., (2008). Marine toxins and the cytoskeleton: Pectenotoxins, unusual macrolides that disrupt actin. *FEBS Journal, 275*, 6082–6088.

23. FAO (Food and Agriculture Organization), (2004). *Marine Biotoxins* (p. 174). FAO Food and Nutrition Paper 80, Food and Agriculture Organization, Rome, Italy.

24. FAO/IOC/WHO, (2004). *Report of the Joint FAO/IOC/WHO Ad Hoc Expert Consultation on Biotoxins in Bivalve Mollusks* (p. 8). Report IOC/INF-1215, FAO Food and Nutrition Paper 80, Food and Agriculture Organization, Rome, Italy.

25. Fujiki, H., & Suganuma, M., (1999). Unique features of the okadaic acid activity on class of tumor promoters. *Journal of Cancer Research and Clinical Oncology, 125*, 150–155.

26. Gillespie, N. C., & Lewis, R. J., (1986). Ciguatera in Australia: Occurrence, clinical features, pathophysiology and management. *The Medical Journal of Australia, 145*, 584–589.

27. Glaziou, P., & Legrand, A. M., (1994). The epidemiology of ciguatera fish poisoning (Review article). *Toxicon., 32*, 863–873.

28. Hampson, D. R., & Manalo, J. L., (1998). The activation of glutamate receptors by kainic acid and domoic acid. *Natural Toxins, 6*, 153–158.

29. Hess, P., Nguyen, L., Aasen, J., & Keogh, M., (2005). Tissue distribution, effects of cooking and parameters affecting the extraction of azaspiracids from mussels, *Mytilus edulis*, prior to analysis by liquid chromatography coupled to mass spectrometry. *Toxicon., 46*, 62–71.

30. Impellizzeri, G., Mangialfico, S., & Oriente, G., (1975). Constituents of red algae, I: Amino acids and low-molecular weight carbohydrates of some marine red algae. *Photochemistry, 14*, 1549–1557.

31. Ito, E., Satake, E., Ofuji, M., & Higashi, K., (2002). Chronic effects in mice caused by oral administration of sublethal doses of azaspiracid, a new marine toxin isolated from mussels. *Toxicon., 40*, 193–203.

32. James, K. J., Fidalgo, M. J., Furey, A., & Lehane, M., (2004). Azaspiracid poisoning, the foodborne illness associated with shellfish. *Food Additives and Contaminants, 21*, 879–892.

33. James, K. J., Furey, A., & Lehane, M., (2002). Azaspiracid shellfish poisoning: Unusual toxin dynamics in shellfish and the increased risk of acute human intoxications *Food Additives and Contaminants, 19*, 555–561.

34. James, K. J., Sierra, M. D., Lehane, M., Brana, M. A., & Furey, A., (2003). Detection of five new hydroxyl analogs of azaspiracids in shellfish using multiple tandem mass spectrometry. *Toxicon., 41*, 277–283.

35. Karalis, T., Gupta, L., Chu, M., Campbell, B. A., Capra, M. F., & Maywood, P., (2000). Three clusters of ciguatera poisoning: Clinical manifestations and public health implications. *The Medical Journal of Australia, 172*, 160–162.

36. Kotaki, Y., Lundholm, N., Onodera, H., Kobayashi, K., & Bajarias, F. F. A., (2004). Wide distribution of *Nitzschianavis-varingica*, a new domoic acid-producing benthic diatom in Vietnam. *Fish Science, 70*, 28–32.

37. Krock, B., (2008). *Personal Communication*. Alfred Wegener Institute, Bremerhaven, Germany.

38. Lehane, L., & Lewis, R. J., (2000). Ciguatera: Recent advances but the risk remains. *Journal of Food Microbiology, 61*, 91–125.

39. Miles, C. O., Wilkins, A. I., & Jensen, D. J., (2004). Isolation of 41a-homo-yesso toxin and the identification of 9-methyl-41a-homo-yesso toxin and nor-ring A-yesso toxin from *Protoceratium reticulatum*. *Chemical Research in Toxicology, 17*, 1414–1422.

40. Morris, P., Campbell, D. S., Taylor, T. J., & Freeman, J. I., (1991). Clinical and epidemiological features of neurotoxic shellfish poisoning in North Carolina. *American Journal of Public Health, 81*, 471–473.

41. Morse, E. V., (1977). Paralytic shellfish poisoning: A review. *J. Am. Vet. Med. Assoc., 171*(11), 1178–1180.

42. Nicolaou, K. C., Frederick, M. O., Petrovic, G., Cole, K. P., & Loizidou, E. Z., (2006). Total synthesis and confirmation of the revised structures of azaspiracid-2 and azaspiracid-3. *Angewandte Chemie International Edition in English (Wiley), 45*, 2609–2615.

43. Palafox, N. A., Jain, L. G., & Pinano, A. Z., (1988). Successful treatment of ciguatera fish poisoning with intravenous mannitol. *Journal of the American Medical Association, 259*, 2740–2742.

44. Pearn, J., (1996). Chronic ciguatera: One organic cause of the chronic fatigue syndrome. *Journal of Chronic Fatigue Syndrome, 2*, 29–34.
45. Pearn, J., (2001). Neurology of ciguatera. *Journal of Neurology, Neurosurgery and Psychiatry, 70*, 4–8.
46. Perez, C. M., Vasquez, P. A., & Perret, C. F., (2001). Treatment of ciguatera poisoning with gabapentin. *NEJM, 344*, 629–693.
47. Perl, T. M., Bedard, L., Kosatsky, T., Hockin, J. C., & Todd, E. C. D., (1990). An outbreak of toxic encephalopathy caused by eating mussels contaminated with domoic acid. *The New England Journal of Medicine, 322*, 1775–1780.
48. Pulido, O. M., (2008). Domoic acid toxicologic pathology: A review. *Marine Drugs, 6*, 180–219.
49. Quilliam, M. A., (2003). Chemical methods for lipophilic shellfish toxins. In: Hallegraeff, G. M., Anderson, D. M., & Cembella, A. D., (eds.), *Monographs of Oceanographic Methodology* (pp. 211–224). Paris, France: Intergovernmental Oceanographic Commission (IOC: UNESCO).
50. Quod, J. P., & Turquet, J., (1996). Ciguatera in Reunion Island (SW Indian Ocean): Epidemiology and clinical patterns. *Toxicon., 34*, 779–785.
51. Reguera, B., & Riobo, P., & Rodriguez, F., (2014). Dinophysis toxins: Causative organisms, distribution and fate in shellfish. *Marine Drugs, 12*, 394–461.
52. Samdal, I. A., (2019). Practical ELISA for azaspiracids in shellfish via development of a new plate-coating antigen. *J. Agric. and Food Chem., 67*(8), 2369–2376.
53. Satake, M., Ofuji, K., & Naoki, H., (1998). Azaspiracid, a new marine toxin having unique spiro ring assemblies, isolated from Irish mussels, *Mytilus edulis. Journal of the American Chemical Society, 120*, 9967–9968.
54. Schnorf, H., Taurarii, M., & Cundy, T., (2002). Ciguatera fish poisoning: A double-blind randomized trial of mannitol therapy. *Neurology, 58*, 873–880.
55. Smida, D. B., Lundholm, N., Nooistra, W. H. C. F., & Sahraoui, I., (2014). Morphology and molecular phylogeny of *Nitzschia bizertensis* sp.: New domoic acid producer. *Harmful Algae, 32*, 49–63.
56. Suzuki, T., Horie, Y., & Koike, K., (2007). Yesso toxin analogs in several strains of *Protoceratium reticulatum* in Japan determined by liquid chromatography-hybrid triple quadrupole/linear ion trap mass spectrometry. *Journal of Chromatography A, 1142*, 172–177.
57. Suzuki, T., Miyazono, A., & Baba, K., (2009). lC-MS/MS analysis of okadaic acid analogs and other lipophilic toxins in single-cell isolates of several *Dinophysis* species collected in Hokkaido, Japan. *Harmful Algae, 8*, 233–238.
58. Suzuki, T., & Quilliam, M. A., (2011). lC-MS/MS analysis of diarrheic shellfish poisoning (DSP) toxins, okadaic acid and dinophysistoxin analogs, and other lipophilic toxins. *Analytical Sciences, 27*, 571–584.
59. Suzuki, T., & Watanabe, T., (2013). Matsushima, R. lCMS/MS analysis of palytoxin analogs in blue humphead parrotfish Scarusovifrons causing human poisoning in Japan. *Food Additives and Contaminants: Part A, 2013,* 1–7.
60. Takemoto, T., & Diago, K., (1958). Constituents of *Chondriaarmata. Chemical and Pharmaceutical Bulletin, 6*, 578–580.
61. Teitelbaum, J. S., Zatorre, R. J., & Carpenter, S., (1990). Neurologic sequelae of domoic acid intoxication due to the ingestion of contaminated mussels. *The New England Journal of Medicine, 322*, 1781–1787.

62. Trevino, S., (1998). Fish and shellfish poisoning. *Clinical Laboratory Science: Journal of the American Society for Medical Technology, 11*, 309–314.
63. Twiner, M. J., & Hess, P., (2005). Cytoskeletal effects of azaspiracid-1 on mammalian cell lines. *Toxicon., 45*, 891–900.
64. Vale, C., Alfonso, A., & Vieytes, M. R., (2008). *In vitro* and *in vivo* evaluation of paralytic shellfish poisoning toxin potency and the influence of the pH of extraction. *Analytical Chemistry, 80*, 1770–1776.
65. Vale, P., Botelho, M. J., & Rodrigues, S. M., (2008b). Two decades of marine biotoxin monitoring in bivalves from Portugal (1986–2006): A review of exposure assessment. *Harmful Algae, 7*, 11–25.
66. Vale, P., & Sampayo, M. A. D., (2002). Esterification of DSP toxins by Portuguese bivalves from the Northwest coast determined by lC-MS-a widespread phenomenon, *Toxicon., 40*, 33–42.
67. Vale, P., & Sampayo, M. A. D., (2002). Pectenotoxin-2 seco acid, 7-epi-pectenotoxin-2 seco acid and pectenotoxin-2 in shellfish and plankton from Portugal. *Toxicon., 40*, 979–987.
68. Wright, J. L. C., Falk, M., McInnes, A. G., & Walter, J. A., (1990). Identification of isodomoic acid D and two new geometrical isomers of domoic acid in toxic mussels. *Canadian Journal of Chemistry, 68*, 22–25.
69. Yasumoto, T., Oshima, Y., & Sugawara, W., (1980). Identification of *Dinophysis fortii* as the causative organism of diarrheic shellfish poisoning. *Bulletin of the Japanese Society for the Science of Fish, 46*, 1405–1411.
70. Yasumoto, T., Raj, U., & Bagnis, R., (1984). *Seafood Poisoning in Tropical Regions* (p. 9). Oral presentation at symposium on seafood toxins in tropical regions, Laboratory of Food Hygiene. Faculty of Agriculture, Tohoku University, Sendai, Japan.
71. Zaman, L., Arakawa, O., Shimosu, A., & Onoue, Y., (1997). Two new isomers of domoic acid from a red alga *Chondriaarmata. Toxicon., 3*, 205–212.

INDEX